Oliver Rivington Willis

A practical flora for schools and colleges

Oliver Rivington Willis

A practical flora for schools and colleges

ISBN/EAN: 9783337268893

Printed in Europe, USA, Canada, Australia, Japan

Cover: Foto ©berggeist007 / pixelio.de

More available books at **www.hansebooks.com**

A

PRACTICAL FLORA

FOR SCHOOLS AND COLLEGES

BY

OLIVER R. WILLIS, A.M., Ph.D.

INSTRUCTOR IN BOTANY, PHYSICS, AND CHEMISTRY IN THE
NEW YORK MILITARY ACADEMY

NEW YORK :·: CINCINNATI :·: CHICAGO
AMERICAN BOOK COMPANY

Printed by
Wm. Ivison
New York, U.S.A.

PREFACE.

BOTANIES without number have been published, giving scientific descriptions of plants of such character that the student is enabled by careful analysis of their structure and appearance to determine their names and physical characteristics. Such books are excellent in their way, and the information they contain is necessary for all students of the science; but all our pupils who take up this study as part of their curriculum have not the scientific mind which makes the acquirement of the science an end in itself. To engage the interest and enthusiasm of such students, it is necessary to show the practical aspects of the vegetable world, and its relations to the needs of every-day life; to reveal something of its history, which, in itself, becomes a fascinating study; and to show enough of its economic features to satisfy those who have neither the scientific mind nor the poetic temperament required for a love of the study for itself.

There has been a long-felt want for a work of such practical character, and this book has been prepared to meet the demand. It does not aim to be exhaustive, as such a treatment would make a book of many thousands of pages, which it would be impracticable to place in the hands of a pupil; but the author has made a careful selection of the most important food-producing trees, shrubs, and herbs, including ornamental plants, fruits, nuts, medicinal plants, and those which furnish oils, dyes, lumber, textile fabrics, etc.

So far as the scientific description and classification of these plants are concerned, the plan of this book does not differ from that adopted by the best botanists. The various genera are grouped together under their respective orders, and the species and varieties under their genera in the same way as in other books. Each order has a general statement which characterizes all the plants belonging to it, and each genus and species and variety a more specific description of such other characteristics as determine its classification. But in addition to this, and to supplement it, are introduced the features in which this book differs from those heretofore published. Thus, after the technical description of a plant will be found an account of its geographical range, the origin of its name, its history, including a statement of its birthplace and distribution over the globe, its uses, modes of cultivation, preparation, and propagation, and many statistics of economical and commercial interest.

The book is the outgrowth of a successful class-room experience, and the author recommends it to the notice of teachers and pupils, in the hope that they may find in it both interest and profit, and that it may tend to relieve the monotony of a strictly technical treatment of the subject, and enhance, if possible, the beauty and the usefulness of the study of Botany.

CONTENTS.

	PAGE
INTRODUCTION	ix
BRIEF STATEMENT OF THE SUBJECT AND ITS SUBDIVISIONS	xv
AUTHORS' NAMES AND ABBREVIATIONS	3
KEY TO THE ORDERS	5
SYNOPSIS OF ORDERS AND GENERA	12

	PAGE		PAGE
RANUNCULACEÆ	30–42	TERNSTRŒMIACEÆ	64–67
Anemone	30	Thea	64
Anemonella	33	MALVACEÆ	67–70
Ranunculus	33	Gossypium	67
Thalictrum	39		
Caltha	39	STERCULIACEÆ	70–72
Clematis	40	Theobroma	70
BERBERIDACEÆ	42–43	TILIACEÆ	72–74
Berberis	42	Corchorus	73
Podophyllum	42	LINACEÆ	74–77
PAPAVERACEÆ	43–46	Linum	74
Papaver	43	Erythroxylon	76
CRUCIFERÆ	46–54	ZYGOPHYLLACEÆ	77–78
Capsella	46	Guaiacum	77
Brassica	46	RUTACEÆ	78–83
Cochlearia	52	Citrus	78
Isatis	53	MELIACEÆ	83–85
Nasturtium	53	Swietenia	83
CAPPARIDACEÆ	54–56		
Capparis	54	ILICINEÆ	85–87
VIOLACEÆ	56–62	Ilex	85
Viola	56	RHAMNACEÆ	87–88
BIXINEÆ	63–64	Rhamnus	87
Bixa	63	Ceanothus	88

CONTENTS.

	Page
AMPELIDEÆ	88–91
Vitis	88
SAPINDACEÆ	91–93
Acer	91
ANACARDIACEÆ	93–95
Rhus	93
Anacardium	95
LEGUMINOSÆ	96–112
Indigofera	96
Astragalus	97
Arachis	99
Lens	100
Pisum	101
Phaseolus	103
Glycyrrhiza	105
Hæmatoxylon	105
Cassia	107
Ceratonia	108
Dalbergia	108
Cæsalpinia	109
Tamarindus	109
Acacia	111
ROSACEÆ	113–127
Rubus	113
Fragaria	116
Prunus	117
Pyrus	123
SAXIFRAGACEÆ	128–131
Ribes	128
COMBRETACEÆ	131–132
Terminalia	131
MYRTACEÆ	132–137
Myrtus	133
Eugenia	134
Bertholletia	137
LYTHRACEÆ	138–139
Punica	138
CUCURBITACEÆ	139–146
Cucumis	139
Citrullus	142
Cucurbita	143
UMBELLIFERÆ	146–156
Apium	146
Pimpinella	147
Fœniculum	148
Ferula	149
Peucedanum	151
Coriandrum	152
Cuminum	153
Daucus	154
Carum	155
RUBIACEÆ	157–164
Cinchona	157
Coffea	160
Cephaëlis	162
Rubia	163
COMPOSITÆ	165–170
Inula	165
Anthemis	166
Chrysanthemum	167
Tanacetum	168
Carthamus	169
CAMPANULACEÆ	170–171
Lobelia	170
VACCINIACEÆ	171–174
Gaylussacia	171
Oxycoccus	172
SAPOTACEÆ	174–176
Dichopsis	174
EBENACEÆ	176–177
Diospyros	176
OLEACEÆ	178–180
Olea	178
LOGANIACEÆ	180–182
Strychnos	180
BORRAGINACEÆ	182–183
Symphytum	182
CONVOLVULACEÆ	183–185
Ipomœa	183

SOLANACEÆ	185–195
Lycopersicum	185
Nicotiana	187
Atropa	191
Capsicum	192
Solanum	194
PEDALINEÆ	195–197
Sesamum	195
VERBENACEÆ	197–198
Tectona	197
LABIATÆ	198–207
Lavandula	198
Mentha	199
Origanum	201
Thymus	203
Salvia	204
Rosmarinus	205
Nepeta	206
Marrubium	207
CHENOPODIACEÆ	208–210
Beta	208
Spinacia	209
POLYGONACEÆ	210–213
Fagopyrum	211
Rheum	212
PIPERACEÆ	213–215
Piper	214
MYRISTICACEÆ	215–217
Myristica	215
LAURACEÆ	217–220
Cinnamomum	217
SANTALACEÆ	220–222
Santalum	220
EUPHORBIACEÆ	222–229
Euphorbia	222
Hevea	223
Buxus	225
Croton	226
Manihot	227
Ricinus	228
URTICACEÆ	230–238
Morus	230
Ulmus	231
Humulus	233
Ficus	234
Cannabis	236
JUGLANDACEÆ	238–242
Juglans	238
Hicoria	238
CUPULIFERÆ	242–256
Castanea	243
Quercus	245
Corylus	254
Fagus	255
SALICACEÆ	256–258
Salix	256
ORCHIDACEÆ	258–260
Vanilla	259
ZINGIBERACEÆ	260–268
Curcuma	260
Maranta	261
Elettaria	263
Zingiber	264
Musa	265
BROMELIACEÆ	268–270
Ananassa	269
IRIDACEÆ	270–271
Crocus	270
DIOSCOREACEÆ	271–272
Dioscorea	271
LILIACEÆ	272–278
Smilax	273
Asparagus	274
Phormium	276
Aloe	277
PALMÆ	278–285
Areca	279
Phœnix	280
Cocos	282
Metroxylon	283

	PAGE		PAGE
GRAMINEÆ	285–300	CONIFERÆ	300–315
Zea	286	Pinus	300
Triticum	288	Picea	307
Oryza	290	Abies	308
Saccharum	292	Larix	309
Sorghum	294	Juniperus	310
Secale	295	Thuja	312
Hordeum	296	Tsuga	312
Avena	298	Chamæcyparis	314
Setaria	300		

GLOSSARY 317
INDEX 339

INTRODUCTION.

DURING the last century the means of acquiring knowledge in every department of science has been wonderfully increased and facilitated. In the several departments of Natural History the improvement has been striking. Naturalists no longer study any branch as a whole, but divide it into parts, each of which affords employment for a lifetime. Thus, in Zoology, one naturalist devotes himself to Quadrupeds, another to Birds, a third to Fishes, and a fourth to Insects.

It is less than a hundred years since Botany was allowed a place as a branch of Natural History; nevertheless it has not only become a favorite study, but is now a very important branch of science. The field has so enlarged itself that it has been found convenient, as in other sciences, to break it up into several departments.

The number of species of plants was supposed by Linnæus not to exceed ten thousand, while at the present day about one hundred and fifty thousand are known; and new species, and even new genera, are yearly increasing the names already in the vast catalogue.

The constant expansion of this broad field is a source of the highest gratification to the student of Natural History. He need entertain no fears that his insatiable curiosity will ever lack food; a long life given to the study of Botany in any one of its departments, with the most zealous and enthusiastic devotion, will leave it still unfinished. No branch of science affords such unalloyed pleasure in its pursuit.

He who sees no beauty in the expanding Rose or the Tulip's variegated petals is, indeed, an object of pity. Even a common observer is struck with such prominent examples of Nature's beauties, and feels what the poet felt when he wrote: —

> "But who can paint
> Like Nature ? Can imagination boast,
> Amid its gay creation, hues like hers ?
> Or can it mix them with that matchless skill,
> And lose them in each other, as appears
> In every bud that blows ?"

But Botany is no longer regarded as a mere accomplishment; nor is it only to be studied for its beauty. It is a most important part of a useful education. Indeed, the value of this study is so great that it can scarcely be estimated.

Most of the early writers upon plants took into consideration only the subject of their usefulness in the art of healing. The study of structure and mode of growth was left to modern times — till after the microscope came to reveal the marvels of plant structure, and bring to light the tissues and vessels that carry on the life and activity of plant growth; till Chemistry discovered the materials that have built up, drop by drop and particle by particle, the plants of the world, from the gray Lichen of the hillside rock to the gigantic forest tree towering four hundred feet into the air.

A knowledge of the geographical locality and habits of plants enables us to select with certainty and transplant the trees and shrubs that beautify our lawns and gardens. To an intelligent agriculturist, an understanding of some of the departments of Botany is indispensable. He must know the habits of vegetables and their mode of growth in order to prepare his ground; he must know their composition to be able to apply the requisite food; and he must know their structure and organs to determine where and how to apply it. In many departments of industry a knowledge of Botany is found invaluable and it lends assistance to the various arts and elegancies of life.

One of the most interesting departments of this subject deals with the history of the cultivation and use of plants. Their history, in a measure, runs parallel with the world's history. When communities became populous they could no longer depend upon the chase, because the demand outgrew the supply; and the use of plants for food was a natural result. The needs of man undoubtedly led to husbandry, and as men became more numerous, they found it necessary to provide, not only for more certain supplies of food, but for more permanent abodes; hence the cultivation of food plants and the establishment of permanent homes must have been simultaneous.

Of the hundred and fifty thousand plants which make up the world's vast Flora, about three hundred constitute the food-bearing plants for man; of this number, by far the greater part is herbaceous. Among the *cereals*, *Wheat* is the most important; it is one of the most ancient of human food-plants, having been cultivated four thousand years before Christ. Reliable records point to the banks of the River Euphrates as its home. *Barley* has a history which makes it contemporaneous with Wheat, its early home being middle Asia. *Rice* had its origin in southern China, and is also believed to have been contemporaneous with Wheat. *Rye* and *Oats*, the other two great bread-plants, had their origin in southeastern Europe, and were under cultivation two thousand years before Christ.

These five cereals furnish bread for the teeming millions of the earth, and there are no records to show that they have ever been found in a condition that would place them among those plants known as weeds.

Among plants known as *table vegetables*, the most prominent, beginning with those that have been longest in use, are the *Turnip*, the *Onion*, and the *Cabbage*, which have been used for more than four thousand years. The home of the Turnip and Cabbage is middle Europe, and the northern shores of the Mediterranean.

The *Carrot*, *Beet*, *Parsnip*, and *Asparagus* have been in use two thousand years. The Carrot and Beet, so far as known, are natives to western Asia and eastern Europe. The home of the Parsnip is Europe, and the Asparagus is native to the countries of the Levant.

Among the fruits that minister to man's support as food, are the *Grape*, found in western Asia; the *Apricot*, a native of China; the *Apple*, *Pear*, and *Quince*, natives of central Europe, which have accompanied man in all his changes of locality where the climate permitted; the *Raspberry* and *Strawberry*, choice and favorite berries, whose cultivation near the great cities in the Temperate Zone has become an important industry in western Asia and Europe; the *Cherry*, which originated near the Caspian Sea; the *Plum*, a native of northern Persia; and the *Peach*, one of the most delicious, if not the most popular, of the fruits of the Temperate Zones, a native of China.

Sugar Cane is native to Cochin China, and southwestern China; it is known to have been in use more than two thousand years.

There are a few textile plants that deserve mention, the most important of which are *Flax*, *Hemp*, and *Jute*. Flax was cultivated near the eastern coast of the Mediterranean four thousand years ago. Hemp is of Siberian origin, and is known to have been under cultivation over four thousand years. Jute has been known to commerce less than two thousand years.

Of the three hundred species of cultivated plants that were found in the New World, the most important food-yielding plants are: the Potato, whose home is believed to be Chile and Peru; the Sweet Potato, whose home is in the warm regions of North America; Tobacco, originally found in Ecuador and neighboring regions; Indian corn, one of the most important cereals of the present age; and the Pineapple, found in Mexico, Central America, Panama, and Colombia. These constitute about one sixth of the important food-plants of the whole world.

This introduction will give the pupil an idea of the character and magnitude of the study of the Flora of Useful Plants. As a specimen of the plan of description pursued in this book, we give here a brief outline of a few illustrations, beginning with Wheat, which stands at the head of the cereals. After classifying it under the order Gramineæ and the genus Triticum, the description proceeds as follows: —

Triticum vulgare, L. (Wheat.) Stem or culm 2 to 5 feet high, tapering from the root to the base of the head or the ear, divided by nodes into several internodes, or lengths, from 4 to 7 inches long. At each node is a single, clasping, lance-shaped leaf, strongly veined and rough on the upper side. Flowers appear at the top of the culm in a close panicle.

The grains, or seeds, are oval in shape, a quarter of an inch in length, flat, and marked on the side next the rachis by a groove the whole length, outside convex. It is an annual, and when planted in early spring, it flowers and fruits the same season; when thus cultivated it is known as "summer wheat" or "spring wheat." The best wheat is biennial: it is planted in early autumn, in time to take root and form root or radical leaves before winter sets in; it ripens in July of the following year, and is called "winter wheat," because it remains in the ground during the winter.

Geography. — Wheat, the most important of all the cereal family, has not been found in what is considered the wild state, and though it has assumed several forms or varieties, it has not so departed from the typical form as to lead to the conclusion that it was ever what might be regarded as a weed.

Wheat when planted north of the fiftieth parallel does not fruit; neither will it bear well south of 30° in America. In Europe, it fruits as high as 51° in southern Russia, and as low as 37° in southern Italy. It is cultivated successfully in Turkey, Syria, northern and southern Africa, Brazil, Chile, Argentine Republic, and Australia.

Etymology. — *Triticum* is from the Latin *tero*, rub, referring to the mode of reducing the grain to flour. *Vulgare*, common, is from the Latin. The common name, *wheat*, is derived from the Sanscrit *seveta*, white, and is due to the white flour produced from the grain.

History. — No form of wheat has ever been found in a wild state. Its home is believed to have been western Asia. It was brought to America by a negro slave belonging to Cortes, and was first planted in Mexico.

Cultivation. — Wheat will not grow upon poor soil nor thrive upon scanty fare. It demands a deep, heavy soil, well tilled, and highly fertilized. In the United States, especially in the rich soils of the Central States, fifty bushels to the acre is not an unusual yield.

Use. — Wheat is the bread grain of man in all regions of the earth where it will grow. The straw is utilized for the manufacture of hats for both men women. It furnishes the material for the fine leghorn hats and bonnets.

Marts. — The great wheat markets are Odessa on the Black Sea, Riga on the Baltic, all the north German ports, Constantinople, London, Liverpool, Chicago, San Francisco, New York, and Toronto.

The building and structural material of the world is obtained from about eighty genera, from twenty-five orders. Of these eighty genera, thirty are cone-bearing plants; hence the greatest amount of building material is furnished by the Coniferæ.

We therefore choose for illustration one of its genera, the Tsuga, which yields the Hemlock.

Tsuga Canadensis, Carr. (Hemlock Spruce.) Trunk 50 to 100 feet in height, branching freely. Bark gray, smooth on young trees, but very rough and furrowed on old trees. Leaves solitary, flat, slightly toothed, blunt at the apex, in two ranks, half an inch long, and less than an eighth of an inch wide. Cones three fourths of an inch in length, and less than half an inch in diameter; scales suborbicular, half an inch long; wing less than half an inch broad. Flowers in June; seed matures in the following year in June.

Geography. — The geographical range of the hemlock is confined to a belt on both sides of the forty-fifth parallel, in the Northern Hemisphere, reaching down to Pennsylvania in mountainous regions, and even to North Carolina, and as far north as Oregon and Hudson Bay.

Etymology. — *Tsuga* is Japanese for "yew-leaved" or "evergreen." The name *Canadensis* comes from Canada, the home of the tree. The origin of *hemlock*, the common name, is not so easily determined; it is suggested that it comes from *hem*, the edge or border, and *loc*, fasten, inclose, alluding to the use of the tree in hedging. Again, *hem* means "injure" or "cripple," and may allude to the poisonous properties of the cicuta, called *hemlock*.

History. — The hemlock is an American tree; it was taken to Europe soon after the settlement of northeastern America, and grows well in the northern parts of England.

Use. — As an ornamental tree the hemlock is a favorite in large grounds. When not crowded it rises to the height of 40 to 80 feet, a perfect pyramid, its lower branches resting on the ground. Its foliage is the most delicate of all the Coniferæ. It bears the knife well, and makes a compact and beautiful hedge. The wood is soft, easily split, and has a very coarse grain; yet it is strong, holds a nail well, requires a great force to produce a cross fracture, and is very durable. It is largely used in the frames of edifices, for joists and for sheathing, being the cheapest of all the soft-wood lumber. The lumber is obtained from the stripped trunks, which are sawed into 13-feet lengths, which during the following winter are drawn to the frozen streams and left till the spring thaw, when they are floated down to the saw-mills, where they are sawed into boards, scantling, and ceiling laths, and thence sent to market.

The bark is highly charged with tannin, and is used in immense quantities for manufacturing leather. It is obtained by felling the tree in the early summer, when the sap is in its greatest activity; girdles are cut around the trunk and large branches by means of an axe, and with a wedge-shaped bar the bark is stripped; it is then piled to dry. Its value is estimated by the cord.

BRIEF STATEMENT OF THE SUBJECT AND ITS SUBDIVISIONS.

BOTANY may be subdivided into the following departments: Structural, Physiological, and Systematic Botany.

Structural Botany has for its object the investigation of the structure, mode of growth, and functions of the cells and vessels that make up the plant. *Organography* is a division of this department that has special reference to the organs. *Morphology* is properly a division of Structural Botany and notes the changes that take place in the cells and tissues of plants.

Physiological Botany takes into consideration the vital action in the reception, preparation, and disposition of the nourishment necessary to keep up the growth of the plant and to enable it to perform the offices of flowering and fruiting.

Systematic Botany embraces the classification of plants and their arrangement under classes, with accurate descriptions in scientific terms of those characters that determine the classification of a plant. The Descriptive Botany in this book comes under the head of Systematic Botany. *Glossology*, which is a division of Systematic Botany, has for its object the application of appropriate names to the organs and parts of a vegetable, by means of which a plant may be so described as to distinguish it from every other individual except one of its own species.

To the above may be added the *Art* or *Practice* of Botany, which consists in applying the principles investigated under the above heads to determining the class, order, etc., of an individual plant. It also includes the collecting, drying, labeling and arranging of botanical specimens.

The classification and nomenclature of the Vegetable Kingdom has taxed the ingenuity and talent of the most learned and skillful botanists. The primary object of classification is to arrange the individual plants or species that resemble each other in the greatest number of characters around a common center called a genus; those genera which possess the greatest number of constant common characters are clustered round a common center called an order; the orders in like manner are grouped into cohorts and classes; the classes in their turn have a center called a series, or a subkingdom, — the common center of the whole being the Vegetable Kingdom.

DESCRIPTIVE BOTANY.

NAMES OF AUTHORS

CITED IN THIS VOLUME, WITH ABBREVIATIONS.

Adanson	*Adans.*
Aiton	*Ait.*
Allemo.	
Alpinus.	
Andrews.	
Aublet	*Aub.*
Austin . . .	*Aust.*
Bauhin	*Bauh.*
Beauvois . .	*Beauv.*
Bentham .	*Benth.*
Bentley.	
Berg.	
Blackstone . .	*Bl.*
Boissier	*Boiss.*
Bonpland	*B.* or *Bonpl.*
Breyn.	
Britton . .	*Britt.*
Brown, R. .	*R. Br.*
Carrière .	*Carr.*
Chaix.	
Chamisso .	*Cham.*
Clinton.	
Coulter . . .	*Coult.*
Decaisne	*Decsne*
De Candolle (A.)	*DC. f.* or *A.DC.*
De Candolle (A. P.)	*DC.*
Delile.	
Don.	
Duchesne.	
Dunal.	
Du Roi.	
Elliott . .	*Ell.*

Endlicher	*Endl.*
Engelmann . . .	*Engl.*
Fischer	*Fisch.*
Forster	*Forst.*
Fremont.	
Gaertner	*Gaert.*
Geyer . . .	*Gey.*
Goldie.	
Gray	*Gr.*
Hayne.	
Heretier . .	*Her.*
Hoffman . . .	*Hoffm.*
Hooker	*Hook.*
Houttuyn	*Houtt.*
Humboldt	*H.* or *Humb.*
Humboldt, Bonpland, and Kunth . . .	*H. B. K.*
Koch.	
Kunth . .	*K.*
Labilladière . .	*Lab.* or *Labill.*
Lamarck . . .	*Lam.*
Lambert . .	*Lamb.*
Lawson.	
Lechenault . .	*Lech.*
Lehmann . . .	*Lehm.*
Lind.	
Lindley	*Lindl.*
Link	*Lk.*
Linnaeus	*L.* or *Linn.*
Linné, C. von (son of Linnaeus) . .	*L. f.*
Loddiges	*Lodd.*

Loudon . . .	Loud.	Risso	Riss.
Marshall . . .	Marsh.	Robbins.	
Marshall and Bieberstein	M. B.	Roscoe .	Rosc.
		Rottboell .	Rottb.
Martius . .	Mart.	Roxburgh . .	Rox. or Roxb.
Maton.		Savi.	
Medicus . . .	Medic.	Schlechtendal . .	Schlech.
Michaux . . .	Michx. or Mx.	Schrader . .	Schrad.
Michaux (the younger)	Mx. f.	Schultes . .	Sch. or Schult.
Miller . . .	Mill.	Sims.	
Mœnchausen . .	Mœnch.	Spach.	
Moquin . . .	Moq.	Swartz	Sw. or Swz.
Muhlenberg . . .	Muhl.	Thomas.	
Necker	Neck.	Thompson.	
Nees von Esenbeck	Nees.	Thuillier . .	Thu. or Thuill.
Nuttall . . .	N. or Nutt.	Thunberg . . .	Thunb.
Olivier	Oliv.	Torrey	T. or Torr.
Pavon	Pav.	Tournefort . .	Tourn.
Persoon . .	Pers.	Wahlenberg . . .	Wahl.
Plumier	Plum.	Waldstein . . .	W.
Pohl.		Wallich.	
Poiret . .	Poir.	Walter . . .	Walt.
Presl.		Waugenheim .	Wang.
Pursh	Ph.	Watson, S.	S. Wats.
Rafinesque-Schmaltz	Raf.	Wedderburn . . .	Wedd.
Reinwaldt . . .	Reinw.	Wight.	
Retzius	Retz.	Willdenow .	Willd.
Richard . . .	Rich.	Willis	Wil.

KEY

To trace a plant to its natural Order based upon those characters which are most obvious.

Subkingdom I. FLOWERING PLANTS (Phaenogams).

CLASS I. DICOTYLEDONOUS ANGIOSPERMS.

Leaves netted-veined. Flowers *rarely* 3-parted, mostly 4-5-parted. Wood, if any, in concentric rings.
Seed in a vessel. Stigmas present. Embryos with 2 cotyledons.

Cohort I.
Calyx and corolla present; petals separate, sometimes wanting, in which case the sepals are bright colored. **APOPETALÆ.**

Cohort II.
Calyx and corolla present; petals more or less united. **SYMPETALÆ.**

Cohort III.
Calyx present, but no corolla, or both wanting. **APETALÆ.**

CLASS II. MONOCOTYLEDONOUS ANGIOSPERMS.

Leaves parallel-veined (rarely netted). Flowers 3-parted. Bark and wood commingled with pith. Root not axial. Embryo with one cotyledon.

Cohort IV.
Flowers in a spadix. **SPADICIFLORÆ.**

Cohort V.
Floral envelope in two 3-parted whorls; outer one green or colored (Lilies, etc.). **PETALOIDEÆ.**

Cohort VI.
Floral envelope chaff-like (Grasses and Grains). **GLUMIFERÆ.**

CLASS III. GYMNOSPERMS.

Stigma wanting. Seed naked. Bark and wood fibers in concentric rings.

Cohort VII.
Cone-bearing plants (Pines, etc.). **CONIFERÆ.**

6   KEY TO ORDERS.

Subkingdom I.

Class I.

COHORT I. APOPETALOUS DICOTYLEDONOUS ANGIOSPERMS.

Herbaceous Plants.

Calyx and corolla present.

Stems sometimes climbing. Leaves alternate, or opposite, 2-3-ternately compound, or merely lobed. Flowers regular.
Ranunculaceæ.

Stems bifid. Leaves peltate. Sepals falling. Valves of anthers not opening upwards. Flowers regular.
Podophyllum in Berberidaceæ.

With milky juice. Sepals falling. Ovary compound, 1-celled, several placentæ on wall. Flowers regular.
Papaveraceæ.

Stems simple or branched above. Leaves alternate. Flowers regular, 5-merous. Fruit a loculicidal 10-seeded capsule.
Linaceæ.

With stems. Leaves alternate, stipulate. Sepals 5. Petals 5, differing in size and shape, one terminating in a spur. Flowers irregular.
Violaceæ.

Without stems. Leaves all radical. Petals and sepals 5, one petal ending in a spur. Flowers irregular.
Violaceæ.

Leaves alternate. Flowers cruciform, regular. Sepals 4, falling. Petals 4, hypogynous. Stamens 6, 2 shorter. Fruit a 2-celled pod.
Cruciferæ.

Stems prostrate, creeping, or clambering. Flowers sympetalous, monœcious or diœcious. Limb of calyx and corolla 5-lobed. Fruit a Pepo.
Cucurbitaceæ.

Leaves alternate. Flowers in umbels. Stamens 5, styles 2. Fruit dry, dividing into halves when ripe.
Umbelliferæ.

Shrubs.

Calyx and corolla present.

Leaves alternate. Sepals and petals 5-merous; sepals imbricate. Stamens many. Fruit a berry.
Bixineæ.

Leaves alternate. Calyx free, involucelled, 5-lobed. Corolla 5-merous. Stamens many, monadelphous, attached to the petals. Pistils several, ovaries in a ring around an axis.
Malvaceæ.

In regions of little frost. Leaves alternate, palmate. Sepals 4-8, free or coherent. Petals 4-8. Flowers regular, cruciform.
Capparidaceæ.

With watery juice, climbing by tendrils. Stamens same in number as petals and opposite to them. Calyx lobes very small or wanting. Petals valvate.
Ampelideæ.

Leaves alternate, stipules small. Flowers perfect or without petals. Stamens alternate with sepals. Fruit a 1-seeded capsule.
Rhamnaceæ.

Leaves alternate, feather-veined. Calyx persistent. Stamens on base of petals. Flowers regular.
Ternstrœmiaceæ.

Leaves alternate, simple, 3–5-lobed. Stipules adnate or 0. Stamens 5–10. Styles 2–3, distinct. Flowers regular.
Saxifragaceæ.

Leaves alternate, simple. Calyx minute. Petals ovate 4–6, separate, or united at base. Fruit a drupaceous berry. Flowers regular.
Ilicineæ.

Leaves alternate, simple. Calyx tubular, limb 5-toothed. Corolla 0. Seed a drupe. Flowers regular.
Combretaceæ.

Leaves alternate, compound. Flowers small, inconspicuous, regular, 5-merous. Stigmas 3. Fruit drupaceous.
Anacardiaceæ.

TREES.
Calyx and corolla present.

Leaves alternate. Sepals 5, persistent. Petals clawed, and tongue-shaped. Stamens in a lobed tube. Flowers regular. Fruit a fleshy septicidal capsule.
Sterculiaceæ.

Leaves opposite, pinnate, stipulate. Calyx 5-parted. Sepals 5. Stamens 10. Flowers regular. Fruit a loculicidal capsule.
Zygophyllaceæ.

Leaves alternate, pinnate. Calyx obscure. Petals 5. Stamens united in a tube. Ovary many-celled. Seeds numerous. Flowers regular.
Meliaceæ.

Leaves alternate. Calyx falling. Petals 4–5. Stamens double or multiple of the parts of corolla. Flowers regular.
Tiliaceæ.

Leaves simple. Petals sometimes 0. Calyx 4–5-cleft. Stamens equal in number to petals. Ovary superior. Flowers regular.
Rutaceæ.

Leaves opposite, 3-nerved. Calyx superior. Limbs 5-parted. Petals 5. Stamens many. Ovary 2- or more-celled. Flowers regular.
Myrtaceæ.

Leaves opposite. Calyx colored, 5-many-parted. Petals 5–7, on throat of calyx. Stamens many. Flowers regular. Berry globular. Seeds many.
Lythraceæ.

Leaves opposite, simple, lobed. Stamens 5–8–10. Ovary 2-celled. Flowers regular. Fruit winged.
Sapindaceæ.

HERBS, SHRUBS, AND TREES.

Leaves alternate, stipulate, compound. Flowers papilionaceous. Stamens united in 1–2 groups. Seed vessel a legume or bean.
Leguminosæ.

Leaves alternate, stipulate. Stamens on the calyx, distinct. Pistils 1 to many, usually distinct, sometimes united, each with a stigma. Ovary simple, 1-celled, 1-ovuled.
Rosaceæ.

KEY TO ORDERS.

Leaves trifoliate, radical, petioles long. Sepals 5 with 5 alternate bractlets. Petals white. Stamens and pistils many. Fruit a heart-shaped edible receptacle.
Fragaria in **Rosaceæ**.

Leaves compound in 2–3 pairs and a terminal one. Fruit an aggregation of little drupes on a dry receptacle. In the blackberry the receptacle is succulent and edible.
Rubus (raspberry and blackberry) in **Rosaceæ**.

Leaves alternate. Flowers in racemes. Fruit a smooth drupe stone or pit flattened.
Prunus (plum) in **Rosaceæ**.

Leaves alternate. Flowers in twos and threes or solitary. Fruit an ovoid drupe. Stone smooth.
Prunus (plum and apricot) in **Rosaceæ**.

Flowers in racemes. Fruit a smooth globular or heart-shaped drupe. Stone smooth, globular.
Prunus (cherry) in **Rosaceæ**.

Leaves appearing after the flowers. Fruit a tomentose drupe, ovoid. Stone furrowed or wrinkled.
Prunus (peach and almond) in **Rosaceæ**.

Leaves ovate, crenate, serrate, woolly beneath, smooth above. Flowers in corymbs. Fruit a fleshy pome. Carpels 2–5, inclosed in a fleshy calyx.
Pyrus (apple and pear) in **Rosaceæ**.

Leaves oblong, or broad ovate, blunt at base and sharp at apex. Carpels 5 or more inclosed in the fleshy calyx. Fruit a fleshy, fragrant pome.
Pyrus (quince) in **Rosaceæ**.

COHORT II. SYMPETALOUS DICOTYLEDONOUS ANGIOSPERMS.

Herbaceous Plants.

Calyx and corolla present.

Leaves alternate. Flowers crowded in an involucrate head composed usually of a disc of yellow flowers, encircled by a ring of white, pink, or purple ones.
Compositæ.

Leaves alternate. Juice milky. Flowers perfect and regular, bell-shaped. Limb 5-lobed. Stamens 5, free.
Campanulaceæ.

Stem twining or clambering. Leaves alternate. Calyx 5-lobed. Corolla bell-shaped. Limb 5-lobed. Flowers regular, axillary.
Convolvulaceæ.

Leaves alternate. Calyx and corolla 5-lobed. Stamens 5 on the corolla. Fruit usually 2-celled, many seeded.
Solanaceæ.

Leaves alternate. Calyx 5-lobed. Corolla 2-lipped, 5-lobed. Stamens 5, on corolla tube. Flowers irregular. Fruit a drupe.
Pedalineæ.

Hairy and rough. Leaves alternate. Calyx 5-parted. Corolla 5-lobed. Stamens on the tube. Ovary 4-lobed.
Borraginaceæ.

Stems square. Leaves opposite and aromatic. Flowers irregular. Corolla 2-lipped. Ovary 4-lobed.
Labiatæ.

Leaves alternate. Flowers solitary. Ovary inferior, many-celled. Stamens double the number of corolla lobes.
Vacciniaceæ.

HERBS AND TREES.

Leaves opposite, stipulate. **Calyx tubular or 0.** Corolla limb 4-6-lobed. Stamens 4-6 on corolla tube.
Rubiaceæ.

Leaves alternate. Flowers regular. Stamens inserted at the bottom of corolla tube, and twice as **many** as lobes.
Ebenaceæ.

Juice milky. Leaves alternate. Corolla furnished with scales on the inner surface. Ovary 5-many-celled.
Sapotaceæ.

Leaves opposite, pinnate. Calyx 4-cleft. Corolla 4-cleft or 0. Stamens 2. Fruit a drupe.
Oleaceæ.

Leaves opposite, entire, substipulate. Calyx and corolla 5-lobed. Stamens 5. Flowers regular.
Loganiaceæ.

Leaves opposite. Calyx toothed. Corolla 4-5-lobed. Stamens 4. On the corolla sometimes only 2. Flowers irregular.
Verbenaceæ.

COHORT III. APETALOUS DICOTYLEDONOUS ANGIOSPERMS.

Calyx present, but no Corolla, or both wanting.

HERBS.

Leaves alternate. Flowers small, inconspicuous, regular. Calyx free. Stamens equal to number of calyx lobes.
Chenopodiaceæ.

Leaves alternate, sheathing. Calyx and corolla wanting. Stamens 1-9, usually 6-8.
Fagopyrum in **Polygonaceæ.**

SHRUBS.

Leaves opposite or whorled, 5-nerved. Calyx wanting. Ovary 1-celled. Flowers crowded on a long spadix. Fruit a berry.
Piperaceæ.

TREES.

Leaves alternate. **Calyx oval, 3-lobed. Stamens 3-15.**
Myristicaceæ.

Leaves alternate. Sepals 4-6, **colored.** Ovary 1-celled, 1-ovuled, anther lid opening upwards.
Lauraceæ.

Leaves alternate. **Sepals 5. Stamens 5.** Ovary inferior. Fruit a nut.
Santalaceæ.

Leaves alternate, pinnate. Calyx 2-4-lobed or entire. **Stamens 3** to many. Ovary 1-celled. Ovule 1. Fruit a nut.
Juglandaceæ.

SHRUBS AND TREES.

Leaves alternate and stipulate. Flowers in **catkins, naked.** Fruit 1-celled. Pod 2-4-valved. Seeds clothed with a silky down.
Salicaceæ.

Leaves alternate. Stipules falling. **Fertile** flowers solitary. Sterile ones in catkins. Fruit **a 1-celled,** 1-seeded nut.
Cupuliferæ.

KEY TO ORDERS.

HERBS, SHRUBS, AND TREES.

Leaves alternate or opposite. Juice milky. Flowers sometimes with petals. Many staminate flowers each with a naked stamen.

Euphorbia in **Euphorbiaceæ**.

Stamens inserted round a central column

Hevea in **Euphorbiaceæ**.

Ovarian cells 1-ovuled. Staminate flowers without rudimentary ovary.

Croton in **Euphorbiaceæ**.

Stamens polyadelphous. No rudiment of an ovary.

Ricinus in **Euphorbiaceæ**.

Leaves alternate, stipulate, penninerved. Flowers with a calyciform envelope, or 0. Ovary 1-celled.

Urticaceæ.

Floral envelope single or 0. Styles 1-2. Fruit a cylindrical mass of little drupes.

Morus in **Urticaceæ**.

Flowers fascicled. Perianth slightly colored, or leaf-like, 4-5-8-lobed.

Ulmus in **Urticaceæ**.

Perianth herbaceous. Sepals 5. Stamens 5. Flowers in strobiloid spikes.

Humulus in **Urticaceæ**.

Fruit a smooth, dry, 1-seeded pericarp.

Cannabis in **Urticaceæ**.

CLASS II.

COHORT IV. SPADICIFLOROUS MONOCOTYLEDONOUS ANGIOSPERMS.

Trees and shrubs with unbranched trunks. Flowers on a branching spadix bursting from a spathe.

Palmæ.

COHORT V. PETALOIDEOUS MONOCOTYLEDONOUS ANGIOSPERMS.

HERBS.

Floral envelope in two 3-parted whorls, outer one green, frequently colored. Rhizome tuberous and creeping. Flowers with double perianth. Outer foliaceous, inner petaloid, irregular, made up of petals and staminodes. Ovary inferior, 3-celled. Flowers irregular.

Zingiberaceæ.

Leaves at base of stem or scape, sheathing rigid, outer floral envelope calyx-like. Flowers spiked, regular.

Ananassa in **Bromeliaceæ**.

Leaves equitant in 2 ranks. Floral envelope petaloid. Stamens 3. Ovary inferior, 3-celled, many seeded. Spathe 2-valved. Filaments equal.

Crocus in **Iridaceæ**.

Stem simple, or branched near the summit. Leaves simple, sheathing, or amplexicaul. Perianth tubular. Limb 6-lobed. Stamens on the throat.

Phormium in **Liliaceæ**.

Perianth tubular. Stamens on the receptacle or tube of perianth. Leaves fleshy.

Aloe in **Liliaceæ**.

Herbs becoming woody. Leaves alternate. Stem a woody vine. Parasitic upon large trees. Flowers irregular. Fruit a lengthened pod, many-seeded. Seeds fragrant.

Vanilla in **Orchidaceæ**.

Herbs becoming woody, twining or clambering vines. Leaves opposite, nettedveined. Flowers small, in racemose spikes, axillary. Perianth 6-parted. Stamens 6. Fruit 3-angled, winged.

Dioscorea in **Dioscoreaceæ**.

COHORT VI. GLUMIFEROUS MONOCOTYLEDONOUS ANGIOSPERMS.

Floral envelope chaff-like or wanting, mostly herbaceous.

Stem cylindrical, tapering, and hollow, with closed joints. Leaves alternate, sheathing, sheath split opposite the blade. Flowers glumaceous, in spikelets. Perianth imperfect or wanting. Stamens usually 3, sometimes 6. Grasses and grains.

Gramineæ.

Spikelets in twos or threes. Stamens 3. Stigmas on long filiform styles.

Saccharum and Sorghum in Gramineæ.

Spikelets all fertile in a panicle. Stamens 6, sometimes 3.

Oryza in Gramineæ.

Spikelets monœcious or polygamous, in a spicate panicle. Stamens 2–3. Styles elongated filamentous.

Zea in Gramineæ.

Spikelets all fertile in a panicle. Stamens 3. Stigmas sessile.

Avena in Gramineæ.

Spikelets all fertile. Glumes 2, seldom 1, varying in length. Stamens 3, rarely 1. Stigmas sessile, or nearly so.

Triticum, Hordeum, and Secale in Gramineæ.

CLASS III.

COHORT VII. CONIFEROUS GYMNOSPERMS.

Cone-bearing plants, mostly trees. Juice resinous. Leaves awl- or needle-shaped. No calyx or corolla. Pistil on an open scale. Ovules naked.

Coniferæ.

SYNOPSIS OF ORDERS AND GENERA.

Subkingdom I. FLOWERING PLANTS (Phaenogams).

Plants that produce proper flowers and bear seeds.

CLASS I. DICOTYLEDONOUS ANGIOSPERMS.

Embryo dicotyledonous. Leaves netted-veined. Flowers mostly 5- or 4-merous.

Apopetalæ.

(Sometimes without petals.)

ORDER I. RANUNCULACEÆ.

Herbs or small shrubs. Stamens numerous. Carpels not united. Sometimes without petals, and sepals colored like petals.

Involucre, consisting of 2-3 divided leaves an inch or more below the flower. Radical leaves, 3-7-parted. Carpels many without grooves, ending in a short beak. Receptacle hemispherical, conical, or cylindrical.
         Windflower, **Anemone.**

Involucre compound. Sepals 5-10, white. Petals 0. Stigma, broad, flat, terminal. Leaves radical compound. Flowers umbellate.
        Rue Anemone, **Anemonella.**

Sepals 3-5. Petals, 5 or more, with a scale or pit at base. Stamens numerous, seldom few. Akenes numerous, flattened, ovate, pointed, in globular or cylindrical heads. Leaves mostly radical. Flowers terminal, solitary, or corymbose.
      Buttercup, Crowfoot, **Ranunculus.**

Flowers in compound panicles, greenish or white. Sepals 4-5-7, concave, falling early. No petals. Filaments larger above, longer than sepals. Numerous akenes, ribbed or swollen. Leaves usually ternately compound, leaflets stalked.
        Meadow Rue, **Thalictrum.**

Growing in wet places or in the water. Sepals golden yellow, 5-9. Petals wanting. Leaves round, kidney-shaped, usually crenate, glabrous. Follicles 5-10.
      Cowslip, Marsh Marigold, **Caltha.**

Sepals 4, petal-like. Petals none. Akenes numerous. Styles feathery or hairy. Leaves opposite, mostly climbers.
       Virgin's Bower, **Clematis.**

Order II. BERBERIDACEÆ.

Herbs and shrubs. Parts of the perianth 2-3-seriate. Stamens **opposite the petals.** Anthers mostly valvular. Carpel solitary, **1-celled.**

Calyx of 6 sepals. Corolla of 6 petals. Anthers 6. Style 0. Berry 1-celled.
Barberry, Berberis.

Flowers white. Petals 6-9. Stamens 12-18. Anthers opening by slits. **Leaves,** two in number, alternate, 5-9-lobed. Lobes toothed.
May Apple, Mandrake, Podophyllum.

Order III. PAPAVERACEÆ.

Herbs with milky or colored juice. Flowers regular. Sepals 2, rarely 3, falling off early. Petals 4-12. Stamens many. Syncarpous ovary, 1-chambered. Placentæ on **the wall.**

Leaves large, pinnate, and glaucous. **Leaflets deep** cut or toothed. Flowers showy. Pods globose. Juice milky.
Poppy, Papaver.

Order IV. CRUCIFERÆ.

Herbs. Sepals and petals growing in the form of a cross. Usually 6 stamens, 4 long and 2 shorter. Ovary 2-celled, with 2 united carpels.

Pods short, triangular; septum narrow; many-seeded. Flowers minute, white.
Shepherd's Purse, Capsella.

Pods long, tapering, four-sided, beaked. **Seeds globose, arranged in** a row. Flowers yellow.
Turnip, Mustard, Brassica.

Pods globose **or** egg-shaped. **Seeds** obovate, flattened, arranged in two rows in each cell. Leaves **large, lower** ones pinnate, upper ones entire. Flowers white.
Horse Radish, Cochlearia.

Pods oval or elliptical, flat, 1-seeded. Valves boat-shaped. Leaves thick, large. Flowers small, yellow, and in terminal panicles.
Woad, Isatis.

Pods round or oval. Sepals spreading. Valves, convex, nerveless. **Seeds in two rows in each cell.**
Water Cress, Nasturtium.

Order V. CAPPARIDACEÆ.

Herbs, shrubs, and trees. Flowers cruciate. Stamens 6 or more, **never** tetradynamous. Ovary 1-celled; **two** placentæ on the wall. Leaves simple, or digitate, spiny.

Sepals 4. Petals 4, alternating with the sepals. Stamens numerous. Pistil one, longer than the stamens.
Caper, Capparis.

Order VI. VIOLACEÆ.

Herbs. Stamens 5, 2 of them with spurs. Flowers irregular. One petal dissimilar, mostly prolonged into a hollow spur. Ovary free. Fruit, a capsule. Placentæ 3, on the wall. Leaves alternate.

Sepals with or without ears. Herbs acaulescent, or with stems. Flowers white, blue, yellow, or violet. Petals 5, one of which is broader than the others and prolonged into a spur.

Violet, Viola.

Order VII. BIXINEÆ.

Shrubs. Flowers usually regular, with many stamens, perfect, axillary or terminal, solitary or fascicled, racemose or panicled. Leaves alternate.

Sepals 5, fleshy, spatulate, eared, and some multiple of 5. Style thread-like. Fruit 1-celled, in a bristled pod.

Annatto, Bixa.

Order VIII. TERNSTRŒMIACEÆ.

Shrubs and small trees. Leaves alternate, simple, entire, or toothed, sometimes with pellucid dots. Flowers axillary or terminal.

Calyx 5-parted. Petals, 5 or more, united at base. Stamens numerous, distinct, or united at their base into groups.

Tea, Thea.

Order IX. MALVACEÆ.

Herbs and shrubs. Leaves alternate, monadelphous. Petals 5, large, twisted in the bud. Calyx cup-like, 5-toothed. Involucre 3-leaved. Styles united. Stigmas 3–5.

Capsules 3-5-celled, many-seeded. Seeds immersed in a wool-like substance, which is the cotton of commerce.

Cotton, Gossypium.

Order X. STERCULIACEÆ.

Trees and shrubs. Like Malvaceæ, except that the anthers are extrorse and 2-celled. Capsules united into a 2–5-celled ovary.

Leaves large, evergreen, oblanceolate, alternate. Calyx 5-parted. Petals 5, cordate. Strap-shaped stamens, united at base, extending upwards in ten divisions. Pistil thread-like. Fruit, in form and size like an ordinary cucumber, 5-angled, warty, with 20–40 seeds imbedded in pulp.

Cocoa, Theobroma.

Order XI. TILIACEÆ.

Trees and shrubs. Leaves alternate, occasionally opposite; simple or palmately lobed, coriaceous, stipulate. Fruit 2–10-celled.

Leaves alternate, serrate, stipulate. Calyx 5-parted. Petals alternating with sepals. Stamens two or three times as many as the petals. Flowers solitary or in small terminal and bracteate cymes.

Yellow Jute, Corchorus.

Order XII. LINACEÆ.

Herbs. Flowers perfect, regular in terminal racemes or corymbs. Ovary 5-1-celled, or spuriously 10-8-celled. Cells 2-ovuled. Styles 3-5, free. Fruit a globular capsule. Seeds flat.

Leaves sessile, entire, simple, alternate, occasionally opposite. Calyx 5-parted. Corolla, with 5 petals. Stamens 5. Styles 5, alternating.

Flax, Linum.

Sepals 5. Petals 5, with a scale on the inner side of base. Stamens 10-12. Leaves alternate. Flowers axillary. Fruit a drupe. Shrub.

Coca, Erythroxylon.

Order XIII. ZYGOPHYLLACEÆ.

Trees, shrubs, and herbs. Flowers perfect, regular, or irregular, axillary, solitary, or in twos. Fruit a loculicidal capsule, pentagonal, 5-celled. Cells 1-seeded.

Leaves, opposite, pinnate, stipulate, sometimes with spines. Calyx 5-parted. Petals 5. Stamens 5-10.

Lignum-vitæ, Guaiacum.

Order XIV. RUTACEÆ.

Small trees and shrubs. Flowers inferior or perigynous and fragrant. Sepals and petals imbricate, 4-5 in number. Fruit a berry. Seed imbedded in juicy pulp.

Leaves ovate, alternate, frequently dotted, tapering to a point on a winged petiole. Calyx 5-sepaled. Petals 5-10, white. Stamens numerous, and some multiple of 5. Filaments flat at base, grouped in sets.

Orange, Lemon, Citrus.

Order XV. MELIACEÆ.

Trees. Flowers in axillary panicles or thyrses. Inferior. Sepals and petals imbricate. **Stamens united in a tube. Fruit** pear-shaped, woody, 3 or 4 inches in diameter, 5-celled, 5-valved. Seeds numerous and winged.

Leaves alternate, compound, with 4 pairs of leaflets, dark-green, shining; leaflets opposite, entire, ovate-lanceolate, unequal at base. Calyx 5-cleft. Petals 5. Stamens 10, united into a tube with 10 teeth inclosing the anthers. Style short. Stigma 5-rayed.

Mahogany, Swietenia.

Order XVI. ILICINEÆ.

Trees and shrubs. Flowers perfect, small. Solitary or grouped in the axils of the leaves. 4-8-parted. Fruit a drupaceous berry, bright red, small, smooth.

Leaves alternate, oval, crenate, glossy, leathery, evergreen, darker above, veined below. Calyx with 4 or 5 teeth. Corolla 4- or 5-cleft, wheel-shaped. Stamens 4 or 5, alternating with segments of corolla.

Paraguay Tea, Ilex.

Order XVII. RHAMNACEÆ.

Trees and shrubs. Flowers small, in axillary clusters, perfect, regular, parts 4–5, frequently without petals. Fruit a 1-seeded capsule.

Leaves opposite or alternate. Calyx pitcher-shaped, or 4–5-cleft. Petals notched, sometimes wanting. Yellow berries.

Buckthorn, Rhamnus.

Leaves ovate, acuminate, serrate, pubescent beneath, alternate and stipulate. Calyx a semi-globular tube, with 5 segments. Petals clawed, rolled in at the edges. Stamens with ovate 2-celled anthers. Styles 3. Stigmas diverging. Fruit 3-berried.

New Jersey Tea, Ceanothus.

Order XVIII. AMPELIDEÆ.

A woody vine. Flowers, in compound panicles, green, and opposite the leaves. Stem climbing by tendrils. Fruit globular or elliptical, a pulpy berry, with 4–5 seeds.

Leaves simple, alternate, stipuled, palmately veined. Tendrils opposite leaves. Calyx small, 5-toothed. Petals 5. Stamens 5. Stigma sessile.

Grape, Vitis.

Order XIX. SAPINDACEÆ.

Trees and shrubs. Flowers polygamous or diœcious. Petals sometimes wanting. Ovary 2-lobed and 2-celled, 2 ovules in a cell, maturing one seed in a cell. Fruit with 2 diverging wings.

Leaves opposite, lobed. Lobes toothed or cut. Calyx 5-parted. Petals 5 or 4–12. Stamens 4–12. Anthers 2-lobed.

Maple, Acer.

Order XX. ANACARDIACEÆ.

Trees and shrubs. Flowers perfect, diœcious or monœcious, regular, small, in spikes or panicles. Ovary 1-celled. Fruit a little globose or kidney-shaped drupe.

Leaves alternate, simple or compound, with 8–15 pairs of sessile leaflets and a terminal one which is petioled, all dark above, light below. Calyx with 5 small persistent sepals. Petals 5, ovate, spreading. Stamens 5–10 or none. Styles 3, sometimes united.

Sumach, Rhus.

Leaves elliptical, green, leathery, alternate, obcordate, or deeply emarginate. Calyx 5-toothed. Corolla 5-parted. Stamens 5. Styles 3.

Cashew Nut, Anacardium.

Order XXI. LEGUMINOSÆ.

Herbs, shrubs, and trees. Flowers regular or irregular, perfect (usually axillary). Ovary superior, single. Fruit a legume. Seed flat, kidney-shaped, or globular, with large embryo and no endosperm.

Leaves compound, bluish green, with about 6 leaflets, and a terminal one. Calyx 5 acute segments. Petals 5; standard roundish, and emarginate; keel spurred on each side, reflexed. Stamens 10, generally united, or 1 free. Style simple. Pod many-seeded.

Indigo, Indigofera.

Leaves many, spreading. Leaflets in 7-12 pairs. Calyx tubular, swollen, with 5 short, nearly equal teeth. Petals long-clawed; standard ovate or pandurate; wings unequal, keel shorter than wings. Stamens 10, 9 united, 1 free. Ovary sessile. Ovules many, in 2 series. Style straight. Stigma small.

Gum Tragacanth, Astragalus.

Leaflets 2 pairs. Calyx of staminate flower, a slender tube. Limb 2-lipped, upper lip 4-toothed. Corolla resupinate. Stamens 10, 9 united, 1 abortive. Pistillate flowers. Calyx and corolla none. Ovary on a peduncle, lengthening downwards, forcing the pollenized pistil under ground. Legume usually with 2 ovoid seeds.

Peanut, Arachis.

Stem weak. Leaves of several pairs of oblong leaflets, with a branched tendril. Flowers in a raceme. Pods short, broad. Seeds lens-shaped.

Lentil, Lens.

Leaves in 2-3 pairs, of elliptical, entire, obtuse, mucronate leaflets, stalk terminating in long branched tendrils. Stipules large. Calyx free, leafy segments, 2 shorter. Petals 5, upper one broad, and turned back. Stamens 10, 9 united, 1 free. Pods oblong. Seeds globular.

Pea, Pisum.

Leaves trifoliate. Flowers white in racemes. Calyx with two bracts at base, bell-shaped, 2-lipped, upper lip bifid. Corolla with a beaked keel, which with the stamens and style is spirally twisted. Pods linear, curved, flat or cylindric. Seed kidney-shaped.

Bean, Phaseolus.

Leaves imparipinnate. Anther cells confluent. Pods prickly, short, almost indehiscent.

Liquorice, Glycyrhiza.

Leaves in 4 or 5 pairs of irregular obcordate leaflets. Flowers in terminal spikes. Calyx cup-shaped, hemispherical. Sepals 5, imbricated. Corolla papilionaceous. Stamens 10, 5 shorter. Ovary free, 2-ovuled. Pods furnished with lance-shaped, flattened beans.

Logwood, Hæmatoxylon.

Leaves alternate, in 4-6 pairs of leaflets. Sepals barely united at base. Petals 5, unequal. Stamens 5-10, some imperfect. Pods many-seeded, with cross partitions. Flowers yellow.

Senna, Cassia.

Calyx 5 segments. Petals 0. Stamens 5. Anther pods opening lengthwise. Styles short. Stigma peltate. Leaves abruptly pinnate. Stipules minute or 0. Flowers in short racemes.

Carob Tree, Ceratonia.

Leaves unequally pinnate or solitary. Flowers papilionaceous, white. Calyx unequally 5-toothed; standard obovate or orbicular, wings oblong blunt. Stamens 10 or 9-bifid-didymous. Ovary stipitate. Ovules 2 or more. Pods oblong linear, flat, thin.

Rosewood, Dalbergia.

Leaves bipinnate. Calyx tube short; 5 segments. Petals 5, orbiculate. Stamens 10. Seeds transverse.

Brazil Wood, Cæsalpinia.

Leaves alternate, stipules falling; 8–16 pairs of leaflets, small, crowded oblong, blunt, unequal. Calyx funnel-shaped, segments ovate, lance-shaped, acute. Petals 3, 1 posterior, 2 lateral, yellowish-white, with red veins. Stamens 3, filaments long and free. Ovary stalked, 1-celled. Ovules many. Style long and hooked. Pods long, flat, broad, curved, three strong woody fibers extending from end to end, along the pulp with which the pod is filled. Seeds 2–8, large, flat.

Tamarind, Tamarindus.

Leaves alternate or fascicled, bipinnate, rhachis slender, tomentose, ending in a gland with one also at the base. Flowers perfect or polygamous, small, in heads or spikes. Calyx 4–5-toothed. Petals united below. Stamens free, or united below, many longer than petals. Style thread-like. Pods 2-valved, sometimes indehiscent, flat, or cylindric. Seeds many, flat.

Gum Arabic, Acacia.

Order XXII. **ROSACEÆ.**

Trees, shrubs, and herbs. Flowers perfect, regular, terminal, solitary, cymose, or in umbels. Sepals 5 or less, united. Petals 5 or 0. Stamens many, in series, free or cohering, inserted with the sepals on the disk. Seeds 1 or few in each carpel. Leaves alternate and stipulate.

Leaves compound, of 1–2 pairs of leaflets and a terminal one, blunt at base, sharp at apex, white or glaucous below, darker above. Calyx and corolla 5-parted. Stamens many. Ovaries many. Akenes little drupes, pulpy, aggregated on a succulent receptacle.

Raspberry, Rubus.

Leaves on long radical petioles, trifoliate, pubescent, dentate, lateral leaflets oblique, nearly sessile. Flowers in cymes, stalk hairy. Calyx concave, deeply cleft. Sepals 5, with 5 alternate bractlets. Petals obcordate, white, large. Stamens many. Styles numerous, akenes naked on the surface of a subglobular, heart-shaped, pulpy, edible receptacle.

Strawberry, Fragaria.

Leaves 3–5-foliate. Stipules subulate. Leaflets ovate or oblong-lanceolate, villous beneath, petioles and midrib aculeate. Flowers in a raceme, white. Fruit ovoid, oblong, or cylindric, changing from green to red and black when ripe.

Blackberry, Rubus.

Leaves oblong, linear or lanceolate, tapering to the base, serrate and glabrous. Flowers solitary or in twos or threes, appearing before the leaves. Fruit a smooth drupe. Stone smooth, flattened.

Plum, Prunus.

Leaves conduplicate in the bud. Flowers with the leaves, in racemes or umbels. Fruit smooth, globular. Stone smooth, globular.

Cherry, Prunus.

Leaves convolute. Flowers solitary or in pairs. Fruit a drupe, soft, velvety. Stone smooth, and flattened.

Apricot, Prunus.

Leaves as above. Flowers solitary, rose-colored. Fruit a tomentose drupe. Stone flattened and corrugated, or wrinkled.

Peach, Prunus.

Leaves conduplicate, appearing after the flowers. Fruit a tomentose drupe. Stone furrowed and flattened.
Almond, Prunus.

Leaves ovate, serrate, acute, crenate, woolly underneath, glabrous above. Flowers in corymbs, roseate, appearing with the leaves. Fruit a fleshy pome. Carpels 5 or 2, inclosed in the fleshy calyx-tube.
Apple, Pear, Pyrus.

Leaves oblong or broad, ovate, blunt at base and sharp at apex. Seeds 5 or more.
Quince, Pyrus.

Order XXIII. SAXIFRAGACEÆ.

Shrubs and herbs. Flowers perfect, regular. Sepals, petals, and stamens 4–5. Stamens alternating with petals. Leaves alternate or opposite. Fruit capsular or berry-shaped.

Leaves 3–5-lobed, smooth above, pubescent below, unequally toothed. Flowers in pendent racemes. Calyx tube adherent to the ovary, 5-toothed. Petals 5. Stamens 5, alternating with petals. Fruit a many-seeded berry.
Currant, Ribes.

Leaves as above, villose. Flowers solitary or in twos. Fruit a globular or ellipsoidal many-seeded berry.
Gooseberry, Ribes.

Order XXIV. COMBRETACEÆ.

Shrubs and trees. Flowers perfect, or imperfect by arrest, in axillary or terminal spikes, or racemes. A bract to each flower, also two lateral opposite bractlets. Leaves alternate or opposite.

Leaves simple. Calyx tube cylindric, adhering to ovary limb, bell shaped, 4–5-toothed. Corolla 0. Stamens 10, on the calyx. Ovary inferior. Fruit a drupe, size of a prune.
Myrobalans, Terminalia.

Order XXV. MYRTACEÆ.

Trees. Flowers perfect, superior, regular, axillary or in spikes, cymes, corymbs, or panicles. Stamens numerous. Leaves opposite or whorled, entire, exstipulate. Fruit a berry or capsule, 2- or more-celled, 1-many-seeded. The Periwinkle (Vinca) of Apocynaceæ is often incorrectly called myrtle.

Leaves opposite, with punctured spots, ovate, lanceolate, evergreen. Calyx 4–6-parted, tube attached to ovary. Petals 4–6, together with the many stamens inserted in the neck of the calyx. Filaments free. Style solitary. Seeds on a central column.
Myrtle, Myrtus.

Leaves opposite, entire, dotted with pellucid spots. Calyx 4–5-parted. Petals 4–5, free or united. Stamens numerous, on the throat of the calyx. Flowers in cymes, or cyme-like panicles, 2-bracted, white or purple. Fruit olive-shaped, but smaller. Seed solitary.
Cloves, Eugenia.

Leaves opposite, evergreen, lanceolate, blunt, prominently veined. Calyx and corolla 5-parted. Stamens numerous. Style simple. Fruit a berry, with 2 or more cells.
Allspice, Eugenia.

Leaves 2-3 feet long, broad, leathery, and prominently veined beneath. Calyx 4-parted. Corolla composed of 4 fleshy petals. Stamens united at base in 5 concentric circles. Filaments short. Stigma sessile, and cruciform. Ovary inferior 4-5-celled. Flowers in terminal panicles. Fruit 3-5 inches in diameter, globular. Nuts numerous, obovoid, triangular.
Brazil Nut, Bertholletia.

Order XXVI. LYTHRACEÆ.

Tropical trees. Flowers perfect, symmetrical, calyx inclosing ovary, but free. Leaves mostly opposite, entire. Fruit hard. Seeds many.

Leaves opposite, or fascicled, on short stalks. Calyx large, broadly tubular, thick and leathery. Lobes 5-7, triangular, acute, smooth, valvate. Petals 5-7 on the calyx, spreading, imbricated, crumpled, crimson. Stamens many on the calyx tube beneath the petals. Style tapering. Stigma simple.
Pomegranate, Punica.

Order XXVII. CUCURBITACEÆ.

Herbs. Weak, long prostrate stems, creeping over ground. Flowers monoecious or dioecious, seldom perfect, solitary or fascicled, white or yellow, bell-shaped.

Leaves large, angularly lobed. Calyx tubular, bell-shaped, with 5 long teeth. Petals 5, attached to calyx. Stamens 5, in 3 groups. Stigmas 3. Fruit cylindrical, many-seeded. Seeds whitish, flat.
Cucumber, Cucumis.

Leaves heart-shaped or reniform, 3-5 inches long. Flowers as above. Fruit globular, sometimes a prolate spheroid, but usually flattened at the poles, and ribbed. Seeds many, flat.
Muskmelon, Cucumis.

Leaves 3-6 inches long, lobed. Lobes pinnately divided, glaucous beneath. Calyx, corolla, and stamens as above. Fruit globular, or shaped like a prolate spheroid, 6 inches to 2 feet in length, and 6-15 inches in diameter. Seeds many, flat.
Watermelon, Citrullus.

Leaves 5-angled, heart-shaped. Calyx, corolla, and stamens as above. Fruit wheel-shaped, and dished about the stem, convex on the opposite side.
Squash, Cucurbita.

Leaves broad, heart-shaped, or reniform. Calyx egg-shaped. Corolla bell-shaped. Petals united half way. Flowers monoecious, axillary. Fruit globular, flattened, or prolonged at the poles. Seeds many, flat.
Pumpkin, Cucurbita.

Order XXVIII. UMBELLIFERÆ.

Herbs. Flowers small, 5-parted, superior in simple or compound umbels. Calyx lobes minute. Ovary 2-celled. Fruit, 2 dry indehiscent akenes, each akene with 5 primary, and often 4 secondary ribs.

Leaves pinnate or serrate. Calyx teeth 0. Petals white, base of style flat. Carpels nearly straight; umbels opposite leaves.
Celery, Apium.

Leaves decompound. Calyx teeth 0. Petals white. Carpophore 2-cleft. Bracts of involucre few, small or 0.
Anise, Pimpinella.

Leaves triangular in outline, 3-4 times pinnate, divisions bristly; foot-stalks short and clasping. Flowers bright yellow; pedicels short; umbels large, 10-30 rayed. Involucre 0. Calyx, limb indistinct. Petals roundish, obovate, entire, truncate.
Fennel, Fœniculum.

Leaves decompound near the root, numerous on the stem, alternate; lower ones bipinnate, sheathing. Sheaths larger near the middle of stem. Calyx entire, or barely toothed. Petals broad, acuminate, short, and turned in. Fruit orbicular or egg-shaped. Flowers yellow, common. Involucre falling.
Asafœtida, Ferula.

Leaves pinnate, stem channelled. Flowers yellow. Fruit flattened. Base of style flat.
Parsnip, Peucedanum.

Secondary ribs most prominent. Flowers white. Fruit globose. No primary involucre, bracts of secondary involucre thread-like.
Coriander, Coriandrum.

Leaves long-stalked and clasping below, sessile above, ternately divided. Calyx teeth bristle-like, outer ones longer. Petals deeply 2-lobed. Style short, erect. Flowers white or rose-colored, few; umbels stalked, irregular, few-rayed. General involucre composed of a few long, spreading, and deflexed narrow, stiff, 3-parted, or entire bracts. Involucels 2-4; small bracts.
Cumin, Cuminum.

Fruit oblong ovate, bristly. Bracts of involucre dissected.
Carrot, Daucus.

Leaves pinnate. Petals white, notched. Base of style conical. Carpophore 2-cleft. Fruit oblong ovate.
Parsley, Carum.

Sympetalæ.

Order XXIX. **RUBIACEÆ.**

Trees, shrubs, and herbs. Flowers perfect, seldom unisexual, mostly regular. Calyx tubular, 4-5-toothed or 0. Petals united, limb 4-5-parted, valvate in the bud. Stamens 4-5 on tube of corolla. Ovary inferior. Style simple. Leaves opposite or whorled.

Leaves opposite, elliptical, entire, smooth above, hairy beneath. Calyx cup-shaped, 5-toothed. Corolla tubular, limb 5-parted. Stamens epipetalous Pistil divided at top. Capsule 2-celled, opening at base. Flowers panicled and pinkish. Fruit winged.
Peruvian Bark, Cinchona.

Leaves elliptical, lanceolate, crenate, or wavy, opposite and evergreen. Calyx tubular, 5-toothed. Corolla funnel-shaped, limb divided into 5 reflexed lanceolate divisions. Flowers white, in axillary, nearly sessile clusters. Fruit a dark-red berry, cherry-shaped. Seeds imbedded in a glutinous pulp, 2 in number, plano-convex, grooved on the longer axis of the plane.
Coffee, Coffea.

Leaves opposite, 6 in number, oblong, obovate, acute, entire, 4-6 inches long, 1-2 wide, rough above, downy and veined below. Stipules clasping, much divided. Calyx bell-shaped, toothed. Corolla tubular, inflated at throat, 5-parted. Stamens 5, stigma bifid. Flowers in a head, enveloped in 5 leaves. Berry 2-seeded.

Ipecacuanha, Cephaëlis.

Stems weak, 4-angled, trailing and clambering. Leaves in whorls of 6, lanceolate, midrib and margins aculeate. Calyx tube egg-shaped, 5-toothed. Corolla rotate, 5-parted. Stamens 5, short. Styles 2, united at base. Fruit berry-like, in twos, subglobular. Flowers brownish-yellow, terminal, in twos.

Madder, Rubia.

Order XXX. COMPOSITÆ.

Herbs. Flowers collected into dense heads surrounded by an involucre. Calyx tube attached to the ovary. Limb consisting of bristles (pappus), awns, or scales, or a cup. Corolla tubular or funnel-shaped, lobed or strap-like. Stamens equal to lobes of corolla, usually 5. Style bifid at top. Ovary 1-celled, 1-ovuled. Fruit an akene. Herb with stout stem.

Leaves alternate, clasping above, petioled near the root. Ovate, rough, downy underneath, very large, 2 feet long, serrate, midrib large. Heads large, involucre imbricated, outer scales leaf-like. Ray flowers, pistillate, yellow. Disk flowers perfect.

Elecampane, Inula.

Stems, 6 to 12 inches high, perennial. Leaves 1-2 inches long, sessile divisions linear. Flower-heads terminal, on long axillary pedicels. Rays white. Floral envelope hemispherical. Rays many and pistillate. Receptacle convex.

Camomile, Anthemis.

Stems 18 inches high. Leaves smooth, bipinnate. Segments of pinnæ acute. Ray florets 20-30 pale pink, ligulate nerved and 3-toothed. Disk florets numerous, 4-5-toothed. Receptacle flat or convex. Scales short.

Chrysanthemum, Chrysanthemum.

Stems 2 to 3 feet high, strong, angular, and branched. Leaves alternate, clasping, bipinnate. Segments oblong, cut, and serrate. Heads in corymbose cymes. Staminate flowers in the central part of the pistillate, with a tubular 3-5-toothed corolla. Seed-vessel ribbed, 3-5 ridges.

Tansy, Tanacetum.

Stems 3 to 4 feet high, smooth, much branched. Leaves ovate, lanceolate, sessile, and subclasping. Teeth armed with sharp spines. Flowers orange-colored. Heads discoid, involucre imbricated, outer bracts leaf-like. Florets tubular, perfect.

Safflower, Carthamus.

Order XXXI. CAMPANULACEÆ.

Herb varying from 8 inches to 2 feet high, simple or branched, pubescent. Leaves alternate, ovate, or lanceolate, irregularly toothed. Flowers blue, red, or white. Pods inflated.

Indian Tobacco, Lobelia.

SYMPETALOUS DICOTYLEDONOUS ANGIOSPERMS.

ORDER XXXII. VACCINIACEÆ.

Shrub, 1 to 8 feet high, branching. Leaves alternate. Flowers white or reddish, small, in lateral bracted racemes. Calyx adherent, 5-toothed. Fruit a black or dark-blue berry, globular, 10-celled, 10-seeded.

Huckleberry, Gaylussacia.

Herb, stem slender, 1 to 5 feet in length, prostrate, throwing up assurgent branches. Leaves one half an inch long, elliptical. Calyx 4-parted; anthers twice as long as filaments. Fruit a berry, varying from bell-shape to globular.

Cranberry, Oxycoccus.

ORDER XXXIII. SAPOTACEÆ.

Tree, 60 to 70 feet high, 2 to 3 feet in diameter. Leaves alternate, stipules falling; petioles long, stout, thickened at base; blade obovate oblong, leathery, clothed beneath with rusty yellow short woolly pubescence. Flowers axillary, stalked.

Gutta Percha, Dichopsis.

ORDER XXXIV. EBENACEÆ.

Tree, 30 to 50 feet high, 10 to 18 inches in diameter. Leaves elliptical, bluntly acuminate, entire, dark-green, paler underneath. Flowers diœcious and polygamous, 4-6-lobed. Corolla bell-shaped, 4-6-parted, rolled together in the bud. Stamens 4-8 or numerous. Calyx an inch or more in diameter, fleshy and persistent. Fruit globular, 4-8-celled, 8-12-seeded.

Persimmon, Ebony, Diospyros.

ORDER XXXV. OLEACEÆ.

Small tree, 20 to 30 feet in height, much branched. Leaves lanceolate, entire, deep-green above, light, hoary, beneath, and evergreen. Flowers axillary, in short compact racemes. Small and white calyx, short, 4-toothed, persistent. Tube of corolla short, limb 4-parted. Fruit a fleshy oily drupe.

Olive, Olea.

ORDER XXXVI. LOGANIACEÆ.

Small tree, 20 to 30 feet high. Bark smooth, gray, much branched; branches swollen or knotted at the nodes. Leaves 5-nerved, with 2 ribs each, side of midrib extending from base to apex; ovate pointed. Calyx somewhat bell-shaped, with 4 lobes which just meet. Stamens 4 or 5 on the corolla. Filaments short, attached to the back of the anthers. Fruit globular, size of a medium-sized orange, rind hard, smooth, yellow, inclosing a fleshy pulp in which are imbedded a number of flat circular seeds, concave on one side and convex on the other, an inch in diameter and a quarter of an inch thick. Wood hard and bitter.

Nux Vomica, Strychnos.

ORDER XXXVII. BORRAGINACEÆ.

Herb, 4 feet high, branching near the top. Leaves large, coarse, petioled, lower ones broad, lanceolate. Flowers in terminal racemes. Calyx 5-parted. Corolla tubular, bell-shaped.

Comfrey, Symphytum.

ORDER XXXVIII. CONVOLVULACEÆ.

Herb, stem trailing. Calyx 5-parted. Corolla bell-shaped; limb spreading. Stamens 5 within the tube. Style simple. Stigma capitate, 2-lobed. Capsule 4-celled, 4-valved, 4-seeded. Juice milky.

Sweet Potato, Ipomœa.

ORDER XXXIX. SOLANACEÆ.

Herb. Calyx 5-6-parted. Corolla rotate, tube short. Stamens 5-6, exserted. Anthers connate, dehiscing lengthwise. Berry many-seeded.

Tomato, Lycopersicum.

Herb. Calyx urn-shaped, 5-toothed. Corolla funnel-shaped, 5-lobed. Stamens 5. Capsules 2-celled, 2-4-valved. Flowers terminal.

Tobacco, Nicotiana.

Herb, 2 to 5 feet high, 3-forked. Leaves in pairs, unequal, entire, pointed. Petioles short, 8-12 inches long. Flowers large, axillary, pendent, brownish purple. Calyx leafy, 5-parted. Corolla bell-shaped, 5-cleft. Stamens shorter than the corolla. Fruit a berry, globular, 2-celled, black when ripe.

Deadly Nightshade, Atropa.

Herb, 18 to 30 inches high, smooth, branching. Leaves ovate, smooth, entire. Flowers solitary, axillary, white. Calyx tubular, with 5 small divisions. Corolla wheel-shaped, in 5 lapping pointed divisions. Fruit a berry, with an inflated pericarp, globose, conical, or oblong, solitary or in pairs.

Red Pepper, Cayenne Pepper, Capsicum.

Herb, erect, prostrate, or assurgent. Stem 2-4 feet long, angular, and branched towards the top. Leaves interruptedly pinnate, every alternate pair of leaflets very small. Flowers blue or white. Calyx persistent, 5-parted. Corolla rotate, bell-shaped, tube short; limb 5-cleft. Anthers connivent. Fruit a globular berry, 2-celled, many-seeded, roots swelling into tubers.

Potato, Solanum

ORDER XL. PEDALINEÆ.

Herb. Leaves opposite or alternate, simple exstipulate. Flowers perfect, irregular, axillary, solitary, racemed or spiked, usually with two bracts. Calyx 5-lobed, sometimes split on one side. Corolla sympetalous. Tube cylindrical. Stamens 5 on the corolla, 1 sterile, 4 fertile ones didymous. Ovary superior, 1-2-4-celled. Fruit a drupe.

Sesame, Sesamum.

ORDER XLI. VERBENACEÆ.

Tree, 80 to 150 feet high, 3 to 6 feet in diameter, branchlets 4-sided. Leaves opposite on ternate verticils, rough above, downy beneath, entire. 1-2 feet long, 6-18 inches wide. Flowers small, sessile, white, in terminal compound dichotomous panicles. Fruit lens-shaped, 4-celled drupe. Calyx bell-shaped, short, 5-6-cleft. Tube swollen below, contracted near the mouth. Corolla tube short, limb gaping, 5-6-cleft, lobes short. Stamens 5-6, attached to corolla near the base.

Teak, Tectona.

ORDER XLII. LABIATÆ.

Herb, 15 to 20 inches high or more, branching near the ground. Leaves crowded near base of branches, whitish, downy, oblanceolate, tapering to the base, sessile, revolute, upper ones narrow. Flowers in an interrupted spike. Calyx spindle-shaped, 13-15 striæ, 5-toothed, upper tooth largest. Corolla tube exserted, upper lip 2-lobed, lower one 3-lobed. Stamens shorter than corolla.

Lavender, Lavandula.

Herb, 12 to 20 inches high, from a creeping root. Leaves opposite, wrinkled, sub-sessile, lanceolate acute, cut serrate. Bracts narrow, lanceolate, bristly. Flowers in verticils, small, crowded, short-stalked. Calyx 5-toothed. Corolla 4-cleft, a little longer than calyx. Whole plant possesses a strong, agreeable odor.

Spearmint, Pennyroyal, Mentha.

Herb, 1 to 2 feet high, hairy, purple, leafy, and branched above. Leaves opposite, nearly entire, sprinkled with resinous dots. Flowers in a terminal 3-forked panicle, in globular compact heads. Calyx egg-shaped, obscurely 13-nerved, 5-toothed, throat hairy. Corolla 2-lipped, upper one notched, lower longer, 3-lobed. Stamens 4, ascending and spreading.

Marjoram, Origanum.

Herb, 12 inches high, slender, woody branches. Leaves sessile, linear-lanceolate or ovate, revolute, hoary beneath. Flowers small, purple, in terminal globose heads. Calyx bilabiate, 10-13 striæ, 5-toothed, 3 upper teeth short, lower pair linear. Corolla 2-lipped, upper lip notched, lower one 3-lobed, middle one sometimes larger. Stamens 4, exserted.

Sweet Thyme, Thymus.

Herb, 1 to 2 feet high, woody. Leaves elliptical, wrinkled, crenulate. Flowers in two opposite sets of 10-12 flowers. Calyx striate, 2-lipped, upper lip 3-toothed or entire, lower one bifid. Corolla 2-lipped, gaping, sometimes notched, lower one 3-lobed. The cross filament has a perfect half-anther on one end and a defective half-anther on the other.

Sage, Salvia.

Shrub, 4 feet high, much branched, hairy branchlets 4-sided and downy. Leaves opposite, an inch long, narrow, linear, obtuse, entire, revolute, dark-green above, smooth and shiny, woolly, veined, and silvery beneath. Flowers axillary and terminal, blue. Calyx bell-shaped, a little flattened, 2-lipped, upper lip minutely 3-toothed. Corolla gaping, downy, pale-blue, variegated with purple and white, tube longer than calyx; upper lip bifid, lower one slit into 3 segments, middle segment larger. Four nutlets at bottom of calyx.

Rosemary, Rosmarinus.

Herb, 3 feet high, square, branched, hoary. Leaves cordate, crenate, toothed, and petioled, hoary. Flowers in large hoary spikes, whorled, white or purplish. Calyx cylindrical, 5-toothed, marked with 15 striæ. Corolla slender below, swollen in the throat, upper lip emarginate, lower one spreading; 3-lobed, middle lobe largest, crenate, marked with crimson dots. Whole plant giving off a pleasant odor.

Catnip, Nepeta.

Herb, 12 to 18 inches in height, many stems from same root, whole plant hoary. Leaves ovate, rounded at base, crenate, toothed, wrinkled, and woolly. Flowers white, sessile, in dense globose verticils. Calyx tubular, woolly, 5-10-toothed, with a corresponding number of striæ. Corolla 2-lipped, upper lip erect, sometimes divided; lower one spreading, 3-lobed; middle lobe largest and notched.

Hoarhound, Marrubium.

Apetalæ.

Order XLIII. **CHENOPODIACEÆ.**

Herb, 2 to 5 feet high, angled and branched. Leaves alternate, 6-15 inches long, 4 to 8 inches wide, upper ones smaller; dingy copper-color to dark-purple; ovate lanceolate or spatulate. Flowers greenish-white in slender spikes, arranged in leafy panicles. Calyx hollow and contracted at the mouth, 5-cleft, persistent, becoming hardened at the base. Stamens 5. Stigmas 2. Seeds rugose or wrinkled. Root conical or napiform.

Beet, Beta.

Herb, 18 to 24 inches high, slightly branched or simple. Leaves alternate, petioled, 3-5 inches long, frequently hastate and lanceolate. Flowers diœcious, bractless, axillary, clustered, staminate flowers with a 4-5-parted calyx. Calyx of the fertile flower tubular, swelled in the middle, 3-toothed. Ovary egg-shaped, 1-celled, 1-ovuled, with 4 lengthened stigmas. Fruit 1-seeded, included within the hardened 2-4-horned calyx. Seeds flattened.

Spinach, Spinacia.

Order XLIV. **POLYGONACEÆ.**

Herb, 1 to 3 feet high, furrowed, stout, and hollow. Leaves cordate, triangular, or hastate. Flowers in terminal and axillary racemes, rose-colored or white. Calyx composed of 5 colored, equal sepals. Stamens 8. Styles 3. Fruit with three triangular faces.

Buckwheat, Fagopyrum.

Herb, 4 feet high, furrowed, stout, hollow. Leaves stipulate, large, sheathing, entire, cordate, smooth, upper leaves smaller. Flowers in racemose, paniculate fascicles. Calyx colored. Sepals 6, in double series, persistent. Stamens 9. Akenes 3-angled, edges winged. Root large, fleshy, yellow within.

Rhubarb, Rheum.

Order XLV. **PIPERACEÆ.**

Shrub, 5 to 8 feet long, climbing, nodes swollen. Leaves opposite, ovate, cordate, uneven at base, leathery, glossy above, 5-7-nerved, 5 inches long. Flowers without perianth. Stamens 2-4 or 5. Ovary 1-celled. Stigmas 2-5. Fruit a small berry, globular or egg-shaped. Tropical and subtropical.

Pepper, Piper.

Order XLVI. **MYRISTICACEÆ.**

Trees and shrubs. Juice astringent, turning red when exposed to the air. Leaves alternate, coriaceous, simple, entire, penninerved, clothed with hairs or scales. Flowers diœcious, axillary, inconspicuous, white or yellow, in racemes, panicles, or heads, with a solitary bract. Staminate flowers with 3-15 stamens, monadelphous, filaments united into a column. Pistillate flowers with a solitary carpel. Seed-vessel fleshy. Seed enveloped in a laciniate aromatic aril.

Nutmeg, Myristica.

Order XLVII. **LAURACEÆ.**

Tree, 30 to 80 feet high, 1 to 2 feet in diameter. Leaves alternate, bright-green above, glaucous beneath, evergreen, thick. Flowers small, in cymes, perfect or polygamous. Receptacle funnel-shaped, perianth in 6 parts. Stamens 12, 3 of which are sterile. Fruit a berry.

Camphor Tree, Cinnamon, Cinnamomum.

ORDER XLVIII. **SANTALACEÆ.**

Herbs, shrubs, and trees, frequently parasitic. The Santalum is a large tree, of Southern Asia. Leaves opposite, oblong, entire, penninerved in 5 pairs. Stamens equal to sepals, and opposite to them. Ovary inferior. Fruit dry, 1-seeded. Flowers terminal.

Sandal-wood, Santalum.

ORDER XLIX. **EUPHORBIACEÆ.**

Herbs, shrubs, and trees. Juice milky, acrid, sometimes watery and poisonous. Leaves alternate, rarely opposite or whorled, sometimes very small. Flowers with single, double, or no perianth.

Spurge,	Euphorbia.
Caoutchouc,	Hevea.
Box,	Buxus.
Croton-oil Plant.	Croton.
Tapioca,	Manihot.
Castor-oil Plant,	Ricinus.

ORDER L. **URTICACEÆ.**

Trees, shrubs, and herbs. Flowers diclinous. Stamens as many as the calyx lobes, and opposite to them. Ovary 1-celled. Style simple or 2-cleft. Sap of trees milky or watery. Leaves alternate and stipulate; stipules falling.

Mulberry,	Morus.
Elm,	Ulmus.
Hop,	Humulus.
Fig.	Ficus.
Hemp,	Cannabis.

ORDER LI. **JUGLANDACEÆ.**

Trees. Flowers diclinous, perianth of staminate flower a scale; that of pistillate flower 2-4-toothed. Ovary 1-celled, 1-ovuled. Fruit a drupe. Endocarp 2-valved. Leaves pinnate, stipulate.

Butternut,	Walnut,	Juglans.
Hickory nut,	Pecan Nut,	Hicoria.

ORDER LII. **CUPULIFERÆ.**

Trees. Staminate flowers with an unequally lobed calyx. Calyx of pistillate flowers 6-toothed. Ovary 2-6-celled. Fruit one to three 1-seeded nuts in an involucre. Leaves alternate, simple, stipulate.

Chestnut,	Castanea.
Oak,	Quercus.
Hazelnut,	Corylus.
Beech,	Fagus.

Order LIII. SALICACEÆ.

Shrubs and trees. Flowers in a catkin. Diœcious. Stigma 2-3-lobed. Leaves alternate, stipulate.

Willow, Osier, Salix.

Class II. MONCOTYLEDONOUS ANGIOSPERMS.

Leaves parallel-veined.

Order LIV. ORCHIDACEÆ.

A woody vine. Flowers perfect, but very irregular in form. Stem climbing, 16 to 20 feet long. Fruit 6 to 10 inches in length.

Vanilla, Vanilla.

Order LV. ZINGIBERACEÆ.

Herbs. Flower perfect, irregular. Perianth 6-parted. 1 stamen. Anther 1-celled.

Turmeric,	Curcuma.
Arrowroot,	Maranta.
Cardamom,	Elettaria.
Ginger,	Zingiber.
Banana, Manilla,	Musa.

Order LVI. BROMELIACEÆ.

Herbs. Flowers perfect, mostly regular. Perianth 6-parted, with 6 perfect stamens.

Pineapple, Ananassa.

Order LVII. IRIDACEÆ.

Herbs. Flowers perfect, bell-shaped.

Saffron, Crocus.

Order LVIII. DIOSCOREACEÆ.

Shrubs. Flowers diœcious, regular and axillary. Racemes inconspicuous.

Yam, Chinese Yam, Dioscorea.

Order LIX. LILIACEÆ.

Herbs. Flowers perfect, usually terminal and solitary. Perianth 6-parted. Fruit a capsule. Leaves sheathing.

Sarsaparilla,	Smilax.
Asparagus,	Asparagus.
New Zealand Flax,	Phormium.
Aloes,	Aloe.

CONIFEROUS GYMNOSPERMS.

Order LX. **PALMÆ.**

Shrubs, small and large trees. Flowers in a branched spadix. Perianth in 6 parts arranged in 2 series. Stamens usually 6. Fruit various in form and size.

Betel Nut, Areca.
Date Palm, Phœnix.
Cocoanut, Cocos.
Sago Palm, Metroxylon.

Order LXI. **GRAMINEÆ.**

Herbs. Flower envelope usually with 2 (rarely more) small scales. Fruit grooved on one side. Stem tapering and usually hollow. Leaves sheathing.

Indian Corn, Zea.
Wheat, Triticum.
Rice, Oryza.
Sugar Cane, Saccharum.
Broom Corn, Sorghum.
Rye, Secale.
Barley, Hordeum.
Oats, Avena.
Millet, Setaria.

Class III. GYMNOSPERMS.

Order LXII. **CONIFERÆ.**

Shrubs and trees, mostly trees. Flowers without perianth. Fruit naked in the scales of a cone or in a berry-like cup. Leaves opposite, whorled, or fascicled, simple, with simple nerves.

Pine, Pinus.
Spruce, Picea.
Fir, Abies.
Larch, Larix.
Cedar, Juniperus.
Arbor Vitæ, Thuja.
Hemlock, Tsuga.
Cypress, Chamæcyparis.

DESCRIPTIVE BOTANY.

Order I. **RANUNCULACEÆ.**

FLOWERS perfect, regular or irregular, rarely diœcious, mostly terminal, solitary, racemed or panicled, white or yellow. SEPALS 3-many, usually 5, free, sometimes petaloid, imbricate, seldom valvate. PETALS equal to and alternate with the sepals, hypogynous, free, clawed, imbricate, equal or unequal, varied in form, frequently minute or wanting. STAMENS numerous, many-rowed, hypogynous; *filaments* thread-like or clavate, free; *anthers* terminal, 2-celled. CARPELS few or many, seldom solitary; *style* simple; *stigma*, usually on the inner surface of the top of the style: fruit pointed or feathery akenes. SEEDS with coriaceous testa. LEAVES radical, or alternate on the stem and branches, seldom opposite, simple or compound, exstipulate; petiole broadened or clasping. Mostly herbs, or woody climbers, occasionally shrubs, with sharp, bitter, mostly poisonous juice. There are 30 genera in this Order varying greatly in form, and 540 species, growing in temperate and cold climates.

ANEMONE, L. (Windflower.) Sepals, 5 or many, petal-like. Petals wanting or rudimentary. Stamens numerous, short. Fruit in roundish or subcylindrical head. Akenes mucronate. Involucre open and below the flowers, which are terminal. Herbs, perennial, with radical leaves.

1. **A. cylindrica**, Gray. (Long-fruited Anemone.) Stem 1 to 2 feet high, silky, pubescent. Leaves 2-3 inches wide, 3-parted, parts wedge-shaped, deeply lobed, and toothed. Side lobes 2-parted, middle one 3-parted, lobes toothed and gashed at the apex; petioles 3 to 6 inches long; involucre long-petioled. Flowers on long, naked, 2-flowered peduncles, 3-6 in number, occasionally 1- or more-involucred. Sepals 5, silky, greenish white, blunt. Fruit in cylindrical heads an inch or more long. May.

Geography.— Dry copses. Mass. to New Jersey, and west to Colorado.

2. **A. decapetala**, L. (**A. Caroliniana**, Walt.) (Carolina Anemone.) Stem 3 to 10 inches high, pubescent above; tuber round, sending up several leaves and one stem or scape. Leaves long-stalked, 3-parted, much divided into wedge-shaped linear divisions. Involucre below middle of scape, 2- or 3-leaved, each 3-parted, segments 3-cleft. Flowers showy, sepals 10-20, nearly linear, outer ones dotted with purple. Fruit in oblong cylindrical head. April, May.

Geography. — Carolina to Arkansas and Nebraska.

3. **A. dichotoma**, L. (**A. Pennsylvanica**, L.) (Pennsylvanian Anemone.) Stem 12 to 20 inches high, frequently less than 12, — dichotomous and hairy. Leaves of the root 3-7-parted, segments cuneate, 3-lobed and acuminate, or pointed, parts large and veiny; those of the main involucre 3-parted, acuminate-lobed and toothed, those of involucres, sessile. First flower appears on a naked peduncle from the base of which rise two branches, each with a 2-leaved

RANUNCULACEÆ.

ANEMONE HEPATICA (Round-lobed Hepatica).

involucre, and one terminal flower. Flowers white, large; sepals 5, obovate. Fruit in a globular head. June to Aug.

Geography. — Canada to Penn. and west to Ind. and Wis. sparingly.

4. **A. hepatica**, L. **(Hepatica triloba**, Chaix.) (Round-lobed Hepatica.) Leaves 3-lobed; lobes ovate, obtuse, or rounded, entire, all radical, on long, hairy petioles, smooth and evergreen; purplish underneath. Flowers single, on scapes, 3 to 4 inches long, purplish blue or nearly white. Sepals 6-9. Akenes several, in a small loose head, pointed and hairy. March to May.

This plant is one of the earliest harbingers of spring, often putting forth its blossoms in the neighborhood of some lingering snowbank.

Geography. — Canada to Georgia, and west to the Mississippi valley. In damp woods, not rare, identical with the European plant.

5. **A. acutiloba**, Lawson. **(Hepatica acutiloba**, DC.) (Sharp-lobed Hepatica.) Lobes of the leaves acute; number of lobes sometimes 5; lobes of the involucre also sharp. Sepals 7-12, pale-purple, or nearly white.

Geography. — The A. acutiloba is found in the same geographical limits as the A. hepatica.

6. **A. multifida**, DC. (Many-cleft Anemone.) Stem 6 to 12 inches high, clothed with silky hairs. Radical leaves, 3 divided segments wedge-shaped, slit into 3 narrow sharp lobes, petioles 3-4 inches long. Leaves of the involucre 2-3 on short petioles, divided as the root leaves. Flowers purple, varying to yellowish-white. Sepals 5-8, blunt, small. Fruit in a globular head. June.

ANEMONE NEMOROSA (Windflower).

Geography. — Vermont, northern N. Y., and north and west to the Pacific, rare.

7. **A. nemorosa**, L. (Windflower. Wood Anemone.) Stem 5 to 10 inches high, smooth, from a filiform, frequently knotty, root stock. Radical leaf solitary, ternate, leaflets usually undivided, occasionally 3-parted or cleft; leaves of the involucre petioled, 3 in number, and near the summit of the stem, just above which is the solitary flower; sepals 4–7, oval or elliptical, white, pinkish or purplish outside. Fruit in a head, carpels oblong, tipped with a hooked beak. April, May.

Geography. — Northern United States and British America, in open woods near the base of old trees.

8. **A. parviflora**, Mx. (Small-flowered Anemone.) Stem 3 to 10 inches high, pubescent. Leaves 3-parted, parts 3-cleft and wedge-shaped, divisions crenate; involucre 2- or 3-leaved, nearly sessile, divided as the other leaves. Flowers white, sepals 5–6, oval. Fruit in a globular head. May to June.

Geography. — Canada, near Lake Superior, west to the Colorado Mountains, and north to the Arctic Ocean.

9. **A. patens**, L. (Var. **Nuttaliana**, Gray.) (Pasque-flower.) Stem 6 to 12 inches high, clothed with silky hairs. Leaves on long petioles, silky, ternately divided segments, cut into linear and wedge-shaped sections, the middle segment stalked and 3-parted, involucre below the middle of the stem, sessile and finely dissected, concave or cup-shaped. Flower solitary, appearing before the leaves, sepals 5–6 or 7, nearly an inch long, pale-purple and showy, silky outside.

Geography. — Dry hills. Illinois, Wisconsin, and west and north to the Rocky Mountain region.

10. **A. Virginiana**, L. (Virginian Anemone. Thimble weed.) Stem or scape 2 to 3 feet high, hairy, usually divided above into 2 or 3 long peduncles, with involucres of two bracts at the middle, or 1 naked, main involucre 3-leaved. Leaves on petioles 6 to 10 inches long, stalks of the bracts shorter, leaf 3-parted, parts ovate-lanceolate, toothed and lobed; those of the side 2-parted, middle one 3-cleft. Sepals 5, greenish-yellow or whitish. Fruit in oblong, woolly heads. June to August.

Geography — Canada, south to Carolina. Woods and damp copses, common.

Etymology and History. — Anemone is from the Greek word ἄνεμος, wind; the ancients believed the plant always appeared in places exposed to the wind. The specific names are from the Latin, and are explained by the common names, which are translations, as follows: *Parviflora*, small-flowered; *Multifida*, many-cleft; *Caroliniana*, Carolina Anemone, etc. *Hepatica* from the Greek ἡπατικός, the liver, due to the fancied resemblance of the 3-lobed leaves to the shape of the liver. Most of the species are natives of Europe.

Cultivation. — By cultivation the size of the flower may be increased; the colors are modified, and many of the stamens are often changed into small petals. The anemone prefers a light soil; the root is taken up after flowering, the plant being propagated by parting the roots as well as by seed. Seeding plants do not flower till the second or third year.

Use. — Several species of anemone are used for ornamental purposes. They are easily raised from the seed, and a bed of the single varieties is a valuable addition to a flower-garden, as it affords in a warm situation an abundance of handsome and brilliant spring flowers, appearing almost as early as the snow-drop and the crocus. In Europe it is used as borders in planted grounds, and

RANUNCULACEÆ.

some species are such favorites with florists and amateurs as to have an important commercial value, especially in England and Germany. The anemone acutiloba is used by empirics for the cure of pulmonary disorders.

ANEMONELLA, Spach. Involucre not close to the flower, composed of 2 ternate sessile leaves. Calyx regular, composed of 3 to many colored sepals. Corolla wanting. Ovaries numerous, free, forming a subglobular head. Akenes with short beak. Leaves radical.

A. **thalictroides**, Spach. (**Thalictrum anemonoides**, Mx.) (Rue Anemone.) Stem smooth, 5 to 10 inches high. Leaves glabrous, biternate, or triternate, common leaf-stalk 2 to 5 inches long. Leaflets roundish, 3-lobed at the end, cordate at base. Flowers subumbellate, involucre of two ternate leaves. Several white or pale-purple sepals, sometimes lobed like the leaves. Flowers in early spring.

This plant is one of the few that greet us in early spring in the Northern States, and upon which the novice in botany takes his first lessons.

Geography. — Canada to Georgia, and west through the Mississippi valley, in open woods, near the roots of trees, and especially in windy exposures.

RANUNCULUS, L.. (Buttercup. Crowfoot.) Sepals 5. Petals 5 or more, a scale or pit at the base. Stamens numerous, seldom few. Akenes many,

ANEMONELLA THALICTROIDES (Rue Anemone).

flattened, ovate, pointed, arranged in globular or cylindrical heads. Herbs, annual or perennial. Leaves usually radical. Flowers terminal, solitary, or in imperfect corymbs, yellow, sometimes white.

1. **R. abortivus**, L. (Small-flowered Crowfoot.) Stem branching, smooth, 6 to 30 inches high. Leaves at the base petiolate, cordate-orbicular, crenate, frequently 3-parted; stem-leaves in threes, 3-5 cleft, with linear, oblong, nearly entire segments; upper ones sessile, foliage varying greatly in form.

Flowers small, yellow. Sepals reflexed, longer than the petals. Carpels in a globular head tipped with a short recurved beak. May to June.

Geography. — Common throughout the northeastern States, and west to California. Damp and shaded places.

Var. **micranthus**, Gray. (**R. micranthus**, Nutt.) Has whole plant more or less clothed with soft hairs; root leaves seldom cordate, some of them 3-parted; divisions of those on the upper parts of the stem more linear and entire; peduncles more slender.

RANUNCULUS BULBOSUS (Bulbous Crowfoot).

Geography. — Mass., New York, Miss., and West. In dryer, more open grounds than the species.

2. R. acris, L. (Buttercup, Wayside Crowfoot. Garden Buttercup, Biting Crowfoot.) Stem erect, branched, 1 to 3 feet high, hairy, round, hollow. Leaves on long stalks at the base and on the lower parts of stem, upper ones on short sheathing petioles, divided into 3 parts or leaflets; leaflets lobed, segments acute, parts sometimes linear. Flowers large, bright-yellow, shining, becoming double by cultivation. Petals obovate, larger than the spreading sepals. Carpels roundish, smooth, compressed, terminated by a roundish recurved beak. June to August.

Geography. — This is a European plant. It was brought to northeastern N. America in seed-grain by European colonists, has spread over the Atlantic States and Canada, and is reaching towards the West. Common in fields, especially damp meadows, and roadsides.

3. R. ambigens, S. Wats. (**R. alismæfolius**, Gray, not Geyer.) (Water Plantain, Spearwort). Leaves entire. Stem hollow, 1 to 2 feet high, falling when young, rooting at lower joints afterwards assurgent. Leaves 3 to 6 inches long, narrow, lanceolate, entire or toothed, acute, subpetiolate, clasping, especially below, nearly sessile above. Petals 5 or 7, golden yellow, larger than sepals. Flowers solitary. Petioles 2 to 3 inches long. Carpels flattened, large, and armed with long, fine beak. June to August.

Geography. — Northeastern North America, South Carolina, west to Oregon. In damp places, edges of still water, coves of sluggish brooks.

RANUNCULACEÆ.

4. **R. aquatilis**, L. (Var. **trichophyllus**, Gray.) (White Water-crowfoot.) Stem 1 to 2 feet long, slender, weak, round, smooth, jointed, floating. Leaves stalked, dichotomously divided into many diverging hairlike segments, submerged, or some floating, rounded, 3–5-lobed. Petals white, narrow. June to August.

Geography. — Found sparingly from Arctic America to South Carolina, west to the Rocky Mountains. In ponds and sluggish streams.

5. **R. bulbosus**, L. (Bulbous Crowfoot. Buttercup.) Stem 8 to 13 inches high, hollow, erect, sparingly clothed with appressed pubescence, or densely covered with stiff, spreading hairs, somewhat branched, enlarging at the base into a bulb. Leaves mostly radical, on long stalks, ternate, middle leaflet stalked, lateral divisions sub-sessile, lobed, with crenate or acute divisions; stem leaves on short sheathing petioles, or nearly sessile; lobes much cut into linear divisions. Flowers bright-yellow, large, showy, becoming double by cultivation. Petals rounded, wedge-shaped at base, much longer than the reflexed sepals, frequently 6–7 in number. Carpels tipped with a short beak. May to August.

Geography. — This is eminently a British plant, and was no doubt introduced into northeastern North America by British colonists in their seed-grain, etc. Abundant in the damp meadows and pastures of the Atlantic States, especially in New England and eastern New York and New Jersey.

RANUNCULUS FASCICULARIS (Fascicle-rooted Crowfoot).

6. **R. Cymbalaria**, Pursh. (Seaside Crowfoot.) Stem slender, 3 to 8 inches long, creeping and rooting. Leaves clustered near the root, cordate, kidney-shaped, crenate-dentate. Flowers bright-yellow, scapes 3 to 6 inches long. 1–7 flowered, mostly without leaves. Petals 5–8, oval. Carpels striate, beak short. June to August.

Geography. — Coast of New Jersey, northward to Canada, along the borders of salt marshes, especially coasts of the Bay of Fundy; near salt springs; inland along the Great Lakes; west to California.

7. **R. fascicularis**, Muhl. (Fascicle-rooted Crowfoot. Early Crowfoot.) Stem erect, 6 to 10 inches high, clothed with silky hairs; root a bundle of fleshy fibers. Leaves of the upper part of the stem on short petioles, the radical and

lower ones on stalks from 3 to 8 inches long; blade of the radical leaves pinnate, or very much divided; the terminal division stalked, lateral ones sessile. Flowers large; petals yellow, spatulate, or oblong, with a scale at base, much longer than the sepals. April, May.

Geography. — Throughout the Atlantic States in southern exposures of rocky hillsides and open woods.

8. **R. Flammula, L.** (Var. **intermedius**, Hook.) (Smaller Spearwort.) Stem prostrate, upright, or assurgent, frequently rooting below, usually less than a foot high. Leaves lanceolate, entire, or slightly toothed, linear lanceolate, lower ones wider on short petioles, or sessile. Petals 5, 6, or 7, golden yellow, larger than sepals. Carpels flattish, each armed with a sharp point. Flowers from July to September.

Var. **reptans**, Gray. (Creeping Spearwort.) Diminutive form. Stem less than 6 inches long, prostrate, rooting at all the nodes. Leaves small, varying from linear to oblong or spatulate. Flowers from June to September.

Geography. — Northern part of New York, and northward on sandy shores. Northward rare. The following form is more common.

9. **R. multifidus**, Pursh. (Much-divided leaved Crowfoot. Yellow Water Crowfoot.) Stem long, slender, submerged, or floating. Leaves in 3-forked, thread-like, linear, or wedge shaped divisions, varying in outline; floating leaves lobed. Sepals reflexed, shorter than petals. Petals bright-yellow, 5-8. Carpels smooth, in a subglobular head, crowned with spine-like tips.

Geography. — Northeastern North America, in ponds, sluggish streams, and muddy places.

Var. **terrestris**, Gray. Does not grow in the water; has ascending stems, bearing each a small panicle of flowers at its summit. Leaves in the form of linear or oblong bracts.

Geography. — Michigan, near Ann Arbor, Minn., Alaska.

10. **R. muricatus, L.** (Prickly-seeded Crowfoot) Stem erect, branched, 12 inches high, glabrous. Leaves roundish, cordate, 3-lobed; lobes coarsely crenate-toothed; all similar and petioled; petioles 1 to 5 inches long; bracts near the flower simple. Flowers small, few, yellow; petals obovate; carpels large, aculeate, strongly margined, ending in a stout, ensiform, recurved beak. May to July.

Geography. — Seed brought from Europe in grain. Plant naturalized in southern United States, Virginia to Louisiana. Also seen by Dr. Wood in California. Loves damp places.

11. **R. oblongifolius**, Ell. (Oblong-leaved Crowfoot.) Stem usually erect, slender, sometimes hairy below, much branched above, about a foot high. Leaves lance-ovate, lanceolate, linear, or oblong, serrate or toothed, lower ones or all petioled. Flowers golden yellow. Petals 5, very much larger than the sepals. Stamens 20 or more. Carpels small, globular, crowned with a little spot. (R. pusillus var. Torr. & Gray.) Flowers in June.

Geography. — Southern United States. Wet prairies. Salem, Ill.

12. **R. parviflorus, L.** (Small-flowered Crowfoot.) Stem 6 to 12 inches high, slender, branched. Leaves all petiolate, small, roundish, cordate, 3-lobed, segments sharply toothed. Flowers very small. Yellow petals and sepals, about the same length. Carpels globular, small, tipped with a very short beak, arranged in a globose head. May to June.

Geography. — Naturalized from Europe. Found in gravelly places. From Virginia to Louisiana.

RANUNCULACEÆ.

13. R. Pennsylvanicus, L. f. (Bristly Crowfoot. Pennsylvanian Crowfoot.) Leaves all 3-parted. Stem stout, 1 to 3 feet high, erect, much branched. Leaves ternate, villous, segments sub-petiolate, acutely 3-lobed, somewhat ovate, incisely serrate; whole plant clothed with stiff, spreading hairs. Flowers small, pale yellow. Calyx reflexed. Sepals longer than the petals. Carpels crowned with a short, straight beak, massed into an oblong head. July and August.

Geography. — Found in wet places, in Canada, eastern United States, and west to Colorado.

14. R. pusillus, Poir. (Puny Crowfoot.) Stem slender, erect, sometimes prostrate, 6 to 12 inches high, branched. Leaves petioled, lower ones ovate, orbicular or cordate, entire or sparingly toothed, upper ones linear-lanceolate, obscurely toothed, nearly sessile. Flowers small, pale-yellow, on long peduncles, 1-flowered. Petals 1-5, sometimes 3, barely longer than the sepals. Stamens 5-10. Carpels crowned with small blunt point. June to August.

Geography. — Southern New York, and along the eastern and southern parts of the Southern and Gulf States to Louisiana, in wet places.

15. R. recurvatus, Poir. (Hooked Crowfoot. Wood Crowfoot.) Stem erect, 8 to 18 inches high, whole plant clothed with roughish hairs, sometimes forkedly branched. Leaves ternate, or deeply 3-parted, leaflets or segments broad, wedge-shaped, and acute; lateral ones 2-lobed; lower petioles long, sheathing at base; upper ones much shorter. Flowers small, pale-yellow, on short peduncles; petals shorter than the reflexed sepals. Carpels in a globular head, margined, and crowned with the sharp-hooked style. Whole plant pale-green. May to July.

Geography. — Labrador to Florida, throughout northeastern North America. Shady woods and damp places.

16. R. repens, L. (Creeping Crowfoot.) Stem 6 to 15 inches long, runners sometimes longer, hairy when in dry ground, glabrous when in wet places; sparingly branched. Root fascicled and large. Leaves ternate, on long stalks; leaflets wedge-shaped, 3-lobed, incisely toothed, middle one petioled, lateral ones on short petioles or nearly sessile; hairy on the veins and edges when in dry ground, veins conspicuous underneath. Flowers large, bright yellow; petals obovate, larger than the pilose spreading sepals. Carpels with a straight point, strongly margined. May to August.

Foliage and general appearance of the plant very variable. When found in a damp meadow and spreading up a dry hillside, it seems to run into No. 13, in form of leaf and stem.

Geography. — Atlantic States, and west to the Pacific. It loves damp ground, but is frequently found on the lower edges of hillsides.

17. R. rhomboideus, Goldie. (Rhomboid-leaved Crowfoot.) Stem 3 to 6 inches high, much branched, hairy. Leaves at the root, on long stalks, rhomboid-ovate, crenate-dentate; stem leaves below similar, 3-5-lobed, upper ones nearly sessile, lobes linear. Flowers deep-yellow. Petals larger than the sepals. Carpels globular, beak small, head spherical. April to May.

Geography. — Michigan, Ill., Wis., southwest to Colorado, and north to British America.

18. R. sceleratus, L. (Wicked Crowfoot. Celery-leaved Crowfoot. Cursed Crowfoot.) Stem thick, hollow, 10 to 15 inches high, glabrous, and branched. Leaves at the base 3-lobed, long-petioled; lobes divided, those on the lower stem 3 parted, on shorter petioles; upper ones nearly sessile, lobes oblong, linear, entire, or toothed. Flowers small, numerous, pale-yellow. Carpels small, numerous, in cylindrical heads. Juice very acrid. May to August.

Geography. — Canada to Georgia, and west to Colorado. Sparingly in damp places, edges of ditches, and near living water, brooksides, etc.

19. **R. septentrionalis**, Poir. (**R. palmatus**, Ell.) (Hand-shaped-leaf Crowfoot.) Stem 12 to 18 inches high, pubescent, slightly branched, branches slender. Leaves on long stalks, pentangular in outline, pubescent, 3-5-palmately cleft; segments all sessile, and cut-toothed or lobed, upper leaves composed of 3 linear segments. Flowers few, small, yellow. Carpels few, margined, and straight-beaked. April and May.

Geography. — Southern U. S., in pine woods, from Carolina to Florida.

Etymology. — Ranunculus, the generic name, is derived from the Latin rana, a frog, due to the circumstance that many of the species grow in wet places, the home of the frog. The specific names are derived from the following Latin words. *Abortivus*, not bringing forth properly, due to its small flowers. *Acris*, from *acer*, sharp, or biting, due to the sharp, acrid taste of the juice. *Aquatilis*, living in or near the water. *Bulbosus*, having small heads or bulbs, named from the bulb-shaped root. *Cymbalaria*, boat-like cup, said to have been applied on account of the fancied resemblance of the calyx to a boat. *Fascicularis*, from *fasciculus*, a little bundle, applied because the fibrous roots of this species appear in groups. *Flammula*, flame, due to the bright, flame-like color of the petals of this species. *Multifidus*, many-cleft as to the leaf. *Muricatus*, abounding in sharp points. *Oblongifolius*, from *oblongus*, oblong, and *folium*, a leaf: hence oblong-leaved. *Parviflorus*, from *parvus*, small, and *flos*, a flower;

RANUNCULUS SCELERATUS (Wicked Crowfoot)

hence small-flowered. *Pusillus*, small, referring to the size of the plant. *Recurvatus*, bent back, the carpels have a hooked beak. *Repens*, creeping. *Rhomboideus*, like a rhombus, due to the shape of the leaves. *Sceleratus*, wicked, or biting, due to the burning, acrid taste of the juice of this species. *Septentrionalis*, northern, *i. e.*, growing in the north.

Use. — Some of the species are very showy, and early attracted the attention of gardeners and cultivators of flowers, and are still favorites. The medicinal properties of the Ranunculus are little known. The acrid juice of the R. sceleratus and some other species blisters the skin very rapidly.

The leaves of the R. ficaria are used for salad in France. The root of this species was formerly used as a remedy in the cure of piles. The roots of the R. bulbosus are edible when cooked. R. aconitifolius, L., *Von-Mais* of France, is a favorite in cultivation for its white flowers.

THALICTRUM, L. (Meadow Rue.) No involucre. Calyx usually colored. Sepals 4–5 or 7, concave, falling early. No corolla. Filaments generally enlarged above and longer than the calyx, numerous. Flowers in panicles, diœcious, or polygamous. Ovaries many. Akenes usually sessile, occasionally stipulate, ribbed or swollen, pointed with the short style. Leaves usually ternately compound. Leaflets stalked. Perennial herbs.

1. **T. clavatum,** DC. Stem 2 to 3 feet high, smooth. Leaves biternate, on petioles an inch long; leaflets roundish, obtusely 3–5-lobed, glaucous beneath. Flowers in loose panicles. Fruit swollen, obovate, striate, acute, and as long as the stipe. June.
Geography. — Southern Virginia and North Carolina.

2. **T. dioicum,** L. (Meadow Rue.) Stem smooth, pale-green, or bluish, 1 to 2 feet high, slender. Leaves ternately decompound, on short general petioles. Leaflets roundish, obtusely 3–5-7 lobed, paler beneath. Flowers purplish or pale-green, filaments threadlike, longer than the calyx; anthers linear, yellowish, mucronate. Sepals 5, obtuse. Inflorescence a panicle. Fruit strongly ribbed and pointed. May.
Geography. — British America to Georgia and Alabama. In hilly, rocky woods.

3. **T. polygamum,** Muhl. **(T. Cornuti,** Gray). (Tall Meadow Rue.) Stem 3 to 5 feet high, branching, smooth, or slightly pubescent. Leaves variable in form, dark-green above and paler beneath, smooth or pubescent; stem leaves without general petioles, decompound; leaflets roundish-obovate or oblong, 3-lobed at the apex, lobes sharp, peduncles longer than the leaves. Flowers perfect, white, in large panicles, very compound diœcious or polygamous; filaments somewhat club-shaped. Carpels strongly ribbed, sharp at both ends, longer than the style. June and July.
Geography. — Atlantic States, and west to Colorado.

4. **T. purpurascens,** L. (Purplish Meadow Rue.) Stem 2 to 4 feet high, purplish, smooth, or finely pubescent. Leaves roundish, or longer than wide, wider towards the end, mostly 3-lobed, veins prominent, paler underneath, margins rolled over. Flowers in compound panicles, purplish or with a greenish tinge; anthers nearly linear; filaments broader at the ends; anthers drooping. May to July.
Geography. — New York, Southern New England, south to Georgia, and west to the Mississippi Valley.

Var. **ceriferum,** C. F. Austin. Differs from T. purpurascens in having the lower surface of the leaves and the fruit beset with waxy particles, and when bruised exhales a peculiar, strong odor.

Etymology. — *Thalictrum* is from the Greek word θάλλω, spring forth green, in allusion to the bright green foliage of the young shoots. The specific names are: *Dioicum,* thus named because the flowers are sometimes diœcious or dioicous. *Purpurascens,* purplish thalictrum. *Ceriferum,* wax-bearing. *Cornuti,* for a French physician, Cornutus. *Clavatum,* club-shaped, due to the form of the filaments of this species.

CALTHA, L. (Cowslip. Marsh Marigold.) Sepals 5 to 9, bright yellow, petal-like. Petals wanting. Pistils 5 to 10, styles very short.

DESCRIPTIVE BOTANY.

Pods flattened, spreading, many-seeded. Whole plant glabrous. Leaves undivided, but toothed and large. Perennial herb.

1. **C. leptosepala,** DC. Differs from the above in the leaf, which is oblong-ovate, lower ones serrate. Flowers usually solitary, rarely two. White or bluish.

Geography. — From New Mexico to Alaska. (Coulter.)

2. **C. palustris,** L. Stem hollow, from 8 to 12 inches high, cylindrical, grooved, and sometimes prostrate, forkedly branched. Root large and branched. Leaves dark-green, veiny, and smooth; lower ones 2 to 4 inches wide, on long stems; upper ones sessile, round-reniform. Flowers bright yellow, axillary, 3-5.

CALTHA PALUSTRIS (Cowslip).

Geography. — Found in wet meadows and swamps, from Canada to Carolina, and west to Oregon.

Number of species about a dozen, of which six are found in North America.

Etymology and History. — *Caltha,* the generic name, is from the Greek word κάλαθος, a goblet, due to the form of the calyx, which resembles a golden cup. *Palustris* is from the Latin word *paluster,* marshy, on account of the fondness of the plant for such localities. *Leptosepala* is from the Greek λεπτός, weak, and the Latin *sepio,* to inclose or surround, alluding to the size and thinness of the sepals.

Use. — The leaves of both these species are used for greens, and when very young for salad.

CLEMATIS, L. (Virgin's Bower.) Calyx 4, sometimes 5-8-sepaled usually colored, and pubescent. Petals wanting, or rudimentary. Filaments many. Anthers linear. Akenes numerous, in heads tipped with the long, persistent, feathery styles. Clinging to and climbing over shrubbery, by means of the leaf-stalks. Perennial. Leaves mostly opposite and compound.

1. **C. crispa,** L. **(cylindrica,** Sims.) Stem climbing, smooth. Leaves varying in form; leaflets 5 or 9, broad, ovate, or lanceolate, slightly cordate at base, entire, occasionally 3-5-lobed, prominently veined, thin. Flowers terminal, large, nodding, campanulate, bluish-purple. Calyx cylindrical, bell-shaped. Sepals dilated above and spreading, edges thin and wavy. Tails of the fruit silky, pubescent.

Geography. — Near Norfolk, Virginia, and south to Georgia.

2. **C. ochroleuca,** Ait. Stem 8 to 10 feet long, usually smaller, and silky. Leaves simple, ovate, silky, hairy underneath, sessile, entire, occasionally 3-lobed, 2 to 4 inches long; veins prominent, upper surface smooth. Flowers terminal, nodding, bell-shaped. Sepals silky outside, creamy white within. Plumes of the fruit long and straw-colored. May.

Geography. — Copses and river banks. New York to Georgia. Rare.

RANUNCULACEÆ. 41

3. **C. Pitcheri**, Torr. and Gray. Leaves pinnate, 3-9; leaflets rough, veins prominent, slightly cordate, ovate, entire, or 3-lobed, leathery, upper ones frequently simple. Flowers nodding. Calyx bell-shaped. Sepals ovato-lanceolate, dull purple; points narrow, and recurved, nearly an inch in length. Fruit tipped with thread-like plumes, naked or slightly pubescent. June.
Geography. — Mississippi Valley in Arkansas, Iowa, and Illinois.

4. **C. verticillaris**, DC. (Whorled-leaved Virgin's Bower.) Stem 10 to 20 feet long, climbing on small trees by means of its coiling petioles, woody, nearly smooth. Leaves in whorls or clusters of 4, ternate; leaflets acute, ovate, slightly notched or lobed. Flowers appearing in 2's, at the nodes, with the leaves; sepals lanceolate, acute, an inch long, bluish-purple. Filaments about 24, outer ones spatulate, or petaloid, tipped with rudimentary anthers. May to June.
Geography. — Atlantic States from Maine to North Carolina, and west to California, in upland woods. Not common.

CLEMATIS VIRGINIANA (Common Virgin's Bower).

5. **C. viorna**, L. (Leather Flower. Way - adorner.) Stem climbing, 10 to 15 feet long, round or striate, pubescent, purple, woody. Leaves opposite, pinnately decompound, with 9-12 leaflets, parts entire or 3-lobed, ovate and acute. Flowers bell-shaped, axillary, purple, nodding on long peduncles, with a pair of simple entire leaves near the middle; sepals very thick and leathery, acuminate, and connivent and reflexed at the apex. Plumes of the fruit from 1 to 2 inches long. June, July.
Geography. — Pennsylvania to Georgia, west to Ohio. Rich, open woods.

6. **C. Virginiana**, L. (Common Virgin's Bower.) Stem climbing, or clambering over shrubbery, 8 to 30 feet long, slender, woody, and channelled. Leaves opposite, ternate; leaflets ovate, acute, coarsely toothed; teeth mucronate; more or less 3-lobed. Flowers white, axillary, abundant, diœcious; sepals 4, oblong, ovate, blunt. Fruit tipped with long plumose tails, very showy in autumn. July, August.
Geography. — Canada to Georgia, and west to the Mississippi Valley. River banks and damp places. Common. Also cultivated.

Etymology. — *Clematis*, the generic name, is from the Greek word κλῆμα, a tendril, on account of the climbing habit of this genus. *Verticillaris*, Latin, is due to the mode in which the leaves are borne, in whorls. *Viorna*, Latin, *via*, the way, and *orno*, adorn — beautifier of the way. *Pitcheri*, for Dr. Pitcher, who first found it in these limits. *Cylindrica*, round, is due to the cylindrical shape of the calyx. *Virginiana*, Virginian Clematis.

Use. — Nearly all the species of this genus are ornamental, and are cultivated for their beauty.

Order II. BERBERIDACEÆ.

Shrubs or perennial herbs. Leaves alternate and exstipulate. Flowers regular, mostly 3-merous and hypogynous; sepals and petals imbricated in the bud. Stamens equal in number to the petals, and opposite to them. Anthers mostly opening by valves, hinged at the top. (In the Podophyllum, by slits.) One pistil; style short, or wanting. Fruit a berry or capsule. Seeds numerous.

No. of genera, 19. Species, 100.

BERBERIS VULGARIS (Barberry).

BERBERIS, L. (Barberry.) Leaves 1-9 foliate. Sepals roundish, 6 in number, enveloped by 2-6 bractlets. Petals 6, each with a short claw, above which, on the inside, are two glandular spots. Stamens 6, irritable; stigma circular and flattened. Fruit a sour berry, 1-several seeded. Seeds erect, with a crustaceous covering. Shrubs. Wood yellow. Flowers in nodding racemes, sometimes drooping. Fruit a sour berry.

B. vulgaris, L. (Barberry.) Leaves few on the new shoots of the season, usually merely branched spines from whose axils the leaves of the next season arise in rosettes, of obovate oblong, bristled, toothed, drooping, many-flowered racemes; petals entire; berries oblong, scarlet. Thickets and near dwellings. Eastern New England. May to June. From Europe.

Geography. — Native of Europe, naturalized in New England. *B. canadensis*, native in the Alleghenies, a curious and interesting plant.

Etymology. — Name from the Arabic, *Berbĕrys.*

Use. — A favorite ornamental shrub. The fruit is preserved, and the inner bark is held to be medicinal.

PODOPHYLLUM, L. (May Apple. Mandrake.) Early floral envelope composed of three foliaceous bracts; 6 petaloid sepals, petals 6 to 9. Stamens double the number of petals. Anthers linear; stigma large, peltate, nearly sessile. Fruit a berry, egg-shaped, 2 inches in length, fleshy, 1-celled, many-seeded. Root-stocks creeping. Roots thick and fibrous. Perennial herb.

P. peltatum, L. (May Apple. Mandrake.) Stem 1 to 2 feet high, 2-leaved, 1-flowered; flower in the crutch, or fork of the stem, nodding, large, the two parts of the stem bearing each a 1-sided leaf, palmately lobed. Flower white. May.

Geography.—The Podophyllum is found sparingly in eastern North America, from the St. Lawrence to Florida, west on the Kansas, and north near Lake Huron. There is another species found in the Himalaya Mountains.

Etymology and History.— *Podophyllum* is derived from the Greek πούς, a foot, and φύλλον, a leaf, on account of the resemblance of the leaf to the foot of a web-footed bird. *Peltatum*, the specific name, comes from the Latin *pelta*, a shield, because the foot-stalk is attached to the blade, not at the edge.

How this plant obtained the name *May apple* is not apparent, for the fruit does not ripen in May. And what gave rise to the popular name *mandrake* is equally obscure. It is possible that it arose from the similarity of the fruit of the *Mandragora officinalis* of Syria, which is no doubt the mandrake spoken of in the Old Testament.

Podophyllum was described, figured, and its emetic properties noted by Mark Catesby, in 1731. See Hist. of Carolina.

Use.—The root of the mandrake, or Podophyllum, yields to the chemist a substance known as Podophyllin. This substance is an active purgative, and is said to promote especially the secretions of the liver and kidneys, and is largely used in bilious attacks. It is also used as an emetic and a vermifuge, and as an alterative in rheumatic affections. It has been for years the principal purgative administered by the Thompsonian practitioners, and is their substitute for calomel.

Order III. **PAPAVERACEÆ.**

Annual or perennial herbs. Juice milky, sometimes colored. Leaves alternate, simple or pinnate, lobed or toothed, exstipulate. Flowers regular, terminal, and often solitary. Sepals falling soon, usually two. Petals four, sometimes six, rarely more; spreading and imbricated in the bud. Stamens numerous, and some multiple of four. Ovary solitary; style short; stigmas 2 or more, star-shaped, upon the flat top of the ovary. Fruit pod-shaped, with two divisions, or capsular with several partitions. Principal genera, about 24.

PAPAVER, Tourn. (Poppy.) Calyx composed of 2 thin sepals, falling soon after the expansion of the flower. Corolla 4-petaled, crumpled at top. Stamens numerous, attached below the ovary, 1-celled, many-ovuled. Style short, expanding into a broad, persistent stigma. Seed-vessel varying in shape from an oblate to a prolate spheroid, surmounted by the broad, persistent stigma. Seeds kidney-shaped, pitted, very oily. Leaves alternate, lobed, or cut; flowers terminal, on long peduncles; buds drooping.

There are about a dozen well-marked species, and as they are grown from seed they sport freely, producing many varieties. The P. somniferum, and one of its varieties, called the black poppy (on account of its black seeds), are the plants that produce the opium of commerce. The P. somniferum is called the white poppy, on account of its white seeds. Nos. 1, 2, and 4 are found in and near cultivated grounds in the United States.

1. **P. Rhœas**, L. (Corn Poppy of Great Britain.) Stem hairy, 2 feet high, many-flowered. Leaves glaucous, pinnatifid, incised. Flowers large, showy, scarlet, sometimes variegated and double, seed-vessel globose or oval.

2. **P. dubium**, L. (Long-leaved Poppy.) Stem slender, 2 feet high, clothed with spreading hairs. Leaves pinnate; leaflets deep-toothed and cut. Flowers light-red, sometimes scarlet, smaller than the P. somniferum. Flower-stalk slender and hairy; sepals hairy; seed-vessels club-shaped.

3. **P. somniferum**, Linn. (Opium Poppy. Garden Poppy.) Stem 1½ to 3 feet high, erect, cylindrical, glaucous-green, smooth below, with scattered soft hairs near the summit, branching. Leaves alternate, clasping, deeply toothed or cut, whole plant smooth and glaucous. Flowers large, terminal, bluish-purple or white; petals 4, edged at base with a purple border. Stamens many, attached to the receptacle; filaments bristle-like; anthers oblong, blunt, flattened, and erect. Fruit globose, 1 inch to 1½ inches in diameter, crowned with the broad, persistent, radiating stigma; the margin of stigma deflexed. Seed-vessel 1-celled, with wedge-like processes extending half-way to the center, to which the seeds are attached. Seeds kidney-shaped, whitish, numerous. It is said that Linnæus counted 32,000 in one cell, or seed-vessel. The seed-vessel, when wounded, discharges a thick, creamy juice.

PAPAVER RHŒAS (Corn Poppy).

4. **P. orientale**, L. Stem 3 feet high, rough, terminated with one rich scarlet-colored flower. Leaves pinnate, rough; leaflets serrate. Seed-vessel subglobose and smooth.

Geography. — Gardens and cultivated grounds, for ornament. (Adv. from Levant.) The geographical home of the poppy is the southern edge of the north temperate zone; but it has spread towards both the north and south. It fruits as far north as the forty-fifth parallel, but produces most largely south of the thirty-fifth degree. The great opium districts are the valley of the Ganges, Asiatic Turkey, Persia, and Egypt. The best is produced in Asia Minor. In late years the cultivation has so extended in China as to cause great alarm to the government.

Etymology. — *Papaver* is said to be derived from the Celtic word *papa*, a name applied to a soft food fed to infants, and contracted into pap in English. The seeds of the poppy were boiled in pap, to induce sleep in the infant, hence the association. *Somniferum* is from the Latin words *somnus*, sleep, and *fero*, bear; hence sleep-bearing, alluding to the sleep-producing property of opium. Poppy, the common name, is a corruption of *papa*.

History. — The medicinal properties of the poppy were known to the Romans, in the days of Hippocrates, about 400 years B. C. Vergil also speaks of the poppy. History informs us that opium was first prepared at Thebes, and called *Thebaicum*. The juice of the whole plant was used by the ancients, and called *Meconium*, which means the juice of the poppy.

Mode of Cultivation. — In Turkey the poppy is sown on well-prepared land in rows, and thinned out so as to stand from 12 to 18 inches apart, and kept free from weeds. When the capsules are nearly full-grown, incisions are made in them in the evening, with a guarded knife, which cuts through the cuticle only. The milky liquid appears on the surface the following morning, and is removed with a spoon-like instrument, and placed in an earthen vessel, from which it is poured into a shallow, open brass dish, which is tilted on the side to allow the watery part of the liquid to drain off. It is then exposed to evaporation, and daily turned until it is sufficiently hardened to be kneaded into balls, which are placed upon slats in large rooms, to dry. Here they are tended and turned by boys till they are dry enough to pack, when they are sent to Smyrna, where they are inspected and graded, packed into tin-lined cases, and sent to market.

The value of the Turkish opium depends upon the amount of morphia it contains. Opium brought to the United States is assayed before it is exposed for sale, and if found to contain less than nine per cent of morphia, is rejected. The cultivation of the poppy for opium has been attempted in the United States in Vermont, Virginia, Tennessee, and California, also in France and England, and opium far richer in morphia than the Asiatic opium has been produced, but on account of the high price of labor its cultivation proved unprofitable.

Chemistry. — Opium, the product of the poppy, is a very complex substance, containing a large number of bases in combination with sulphuric and meconic acids. Morphine, whose formula is $C_{17}H_{19}NO_3$, and Narcotine, whose formula is $C_{22}H_{23}NO_7$, are the most abundant and important.

The best known of the others are: —

Codeine, whose formula is $C_{18}H_{21}NO_3$
Thebaine, " " " $C_{19}H_{21}NO_3$
Papaverine, " " " $C_{21}H_{21}NO_4$
Narceine, " " " $C_{23}H_{29}NO_4$

So far as known, the medicinal properties of opium reside in these six substances, and principally in the first two, Morphine and Narcotine.

Use. — As a medicine opium is administered to relieve pain, to promote sleep, to allay irritation of the nervous system, and to relax the muscles in spasmodic affections. It diminishes secretions, and for this purpose is largely used. So important has it become that it has passed into a proverb that a physician without opium is like a soldier without weapons. The most important preparations of opium are laudanum, which is an alcoholic tincture, paregoric, which is a compound of opium, benzoic acid, honey, oil of anise, and dilute alcohol. Soothing syrups, given to keep children quiet, contain much opium, and sometimes cause sickness, and occasionally insanity.

The seed yields an oil only inferior to the best olive oil, and is used both as a substitute and an adulterant. Opium is used as a luxury for an intoxicant, either taken in small doses internally or smoked, when it is mixed with the hashish, or gum of the hemp, and with grateful spices. If indulged in to excess, it enfeebles the mind and enervates the body, and is said to shorten the life of the offspring of the debauchee. It enables people to bear fatigue without food, and travellers in Turkey, Syria, and India carry it with them for that purpose. Even horses are sustained, in the East, under its influence. It is eaten, not smoked, in Persia and India; smoking it is a recent Chinese invention. In Amoy, China, fifteen out of every twenty adults smoke it.

Statistics.—Great quantities of opium are carried into China and vicinity. That from India alone is over 14,000,000 pounds annually. Large quantities go overland from Persia and Turkey, but this is only about one fifth of the amount consumed there, as they produce four fifths of what they use, making an annual consumption of 71,001,840 pounds, at a cost of $280,000,000.

Order IV. **CRUCIFERÆ**.

Sepals 4. Petals 4, hypogynous, arranged opposite to each other in pairs, forming a cross. Stamens 6, 4 long and 2 shorter. Flowers perfect, usually in a terminal raceme, white or yellow. Ovary sessile, usually 2-celled. Stigmas 2. Fruit a pod, one to many seeded, seeds commonly yielding oil. Mostly herbs, sometimes woody. Juice watery, frequently acid, anti-scorbutic, and never poisonous. Stem cylindrical, or angular. Leaves simple, alternate, occasionally opposite, entire, lobed, or dissected; upper ones sometimes eared, lower ones often runcinate, for the most part without stipules. Genera, 172.

CAPSELLA, Medic. Seed-vessel, triangular-obcordate; valves, boat-shaped without wings. Seeds many, with incumbent cotyledons. Pod flattened contrary to partition. Flowers white. A common weed.

C. Bursa-pastoris. (Shepherd's Purse.) Root-leaves rosulate, cut-lobed; stem-leaves linear lanceolate, clasping, sagittate; raceme long. Radical leaves clustering, subpinnatifid. Waste ground about dwellings. Common weed. April to September.

Geography.—Naturalized from Europe, where it is a troublesome weed in gardens and near dwellings.

Etymology.—*Capsella* is the diminutive of Latin *capsa*, a box.

BRASSICA, L. (Turnip, Mustard.) Pod long, terete, somewhat 4-sided, terminating in a stout 1-seeded beak; valves 1–3-veined;

seeds in a single row, globular. Flowers yellow. Lower leaves lyrate or pinnatifid.

1. **B. oleracea,** L. (Cabbage.) Stem slender, much branched, appearing the second year, smooth, from two to three feet high. Leaf smooth, glaucous, twenty inches long and three to fifteen wide, the first year growing compactly, forming a more or less solid head, which is the edible part; the stem leaves are lyrate below, entire and lanceolate above. Flowers yellow, and in great profusion, terminal on the branches. Seed-vessels cylindrical, and curved. A biennial herb.

Varieties. — This plant sports with great freedom, yet there are a number of well-marked varieties that propagate with considerable constancy.

In the vicinity of the great Atlantic cities and in Europe there are about a dozen distinct varieties that have become favorites with market gardeners and amateurs, arranged under the following heads: —

The *Common* or *White Cabbage*, known as *Sugar Loaf*, *Flat Dutch*, *Drumhead*, *Savoy*, or *Wrinkled*, etc., etc.

The *Red*, or *Purple Cabbage*, used for pickling, etc.

Cauliflower, and several others, which have assumed new forms under cultivation. It seems almost a wonder that these varieties are so constant as they are.

CAPSELLA BURSA-PASTORIS (Shepherd's Purse).

BRASSICA OLERACEA (Drumhead Cabbage).

In Europe the number of varieties is very great. As all plants raised from the seed sport more or less freely, it is no wonder that the cabbage assumes so many forms.

There is a perennial variety grown in the Channel Islands, called the *cow cabbage*, or *tree cabbage*, or *Bore Cole*, which reaches the height of ten feet. The leaves are stripped off and fed to cattle, and the stalks are used for bean-poles, canes, etc.

Geography. — The cabbage arrives at perfection in cool, damp climates, but is successfully cultivated in the edge of the torrid zone during the wet, cooler season, and is found under cultivation in a broad zone all around the world, north of the twenty-fifth parallel, and has been carried to Australia and the islands of the Pacific.

Etymology. — *Brassica* is the Latinized Celtic name for cabbage, the signification of which is not apparent.

Oleracea, the specific name of the cabbage, comes from the Latin *olus*, a pot-herb.

History. — The home of this plant is middle and western Europe. It is not known when it was first used as food, but there is reason to believe it was so used very early in the history of European peoples; and it has become so great a favorite that its spread throughout the world is limited only by civilization. Wherever colonization has occurred, climate permitting, the cabbage has followed.

The ancients knew it. Theophrastus, who lived three hundred years before the Christian era, wrote of it, and Pliny also mentions it and speaks of its cultivation.

Use. — The most common use of this plant is in the character of a pot-herb, and it is universally esteemed. It is also prepared as a salad, under the name of Cold Chou, which has been corrupted into Cold Slaw. The Scotch call it Cauld Kail. In Germany and all northern European nations large quantities of cabbage are made into Sauer-kraut. It is chopped fine and packed tightly into casks, with alternate layers of salt, and being kept under heavy pressure it soon arrives at a state of fermentation.

BORE COLE (Tree Cabbage).

When it begins to ferment it is fit for use, and is removed to a cool place. It is eaten with oil or other dressings, and is a very important article of food in all northern Europe.

Sauer-kraut soup, with rye-bread and occasionally a little pork, is the daily food of the Russian peasant. Cabbage is also an important food for cattle, and especially for milch cows.

2. **B. alba**, Gray. (White Mustard.) Stem 2 to 5 feet high, stouter than No. 1, much branched. Leaves petioled, lyrate, or subpinnate; terminal segment large, 3-lobed. Flowers yellow, in racemes; petals larger than in No. 1, and seeds fewer. June to August. Fruits in August.

3. **B. nigra**, Koch. (Black Mustard.) Stem from 3 to 6 feet high, diffusely branched, smooth or hairy. Leaves petioled, and variously lobed and toothed; green above and lighter beneath. Flowers in slender racemes, greenish-yellow. Seeds dark-brown, sharp to the taste. Annual. June, July. Fruits in August.

4. **B. juncea**, Hooker and Thompson. A coarser species, the seeds of which are rich in oil, yielding about 20 per cent of their weight.

Geography. — The B. juncea is largely grown in India, whence the seeds are exported to England.

Etymology. — The specific name, *nigra*, is Latin for black, due to the black seed. *Alba*, Latin for white, refers to the white seed. *Mustard*, the popular name, grew out of the circumstance that mustard was prepared for the table by mixing it with new wine, called *must*.

History. — When or where mustard was first cultivated is not known. It is spoken of in the Scriptures, but it is now believed that the plant referred to was Salvadora Persica, allied to the olive, whose fruit has the taste and pungency of the mustard-seed.

The common mustard, Sinapis, is mentioned by Theophrastus, showing that it was known to the ancient Greeks and Romans, three hundred years before the beginning of the Christian era; hence it has been under cultivation and in use more than two thousand years. It has been cultivated throughout the ages of the Christian era, and was known as a medicine as well as a condiment for food. During the latter part of the Christian era, especially, it has been used more largely in western Europe and the British islands as a dressing for food than for medical purposes. A pleasant oil is obtained from the seed, used for a dressing for food and for making fine soaps.

Preparation. — When the seeds are ground and mixed with warm water, fermentation takes place, and furnishes a very pungent essential oil.

Table mustard is prepared by mixing and grinding together the seeds of B. nigra and B. alba, and is frequently ground into a paste in its own oil. The species from which the sweet oil of mustard is obtained is the B. juncea (Hook.), largely raised in India and Russia, for the oil. The seeds yield by pressure about 20 per cent of their weight of oil.

The pungency of prepared mustard is due to the presence of an essential oil which does not exist in the seed, but is generated by the powdered seed when mixed with warm water, and arises from a fermentation due to the presence of two substances, known as *myrosin* and *sinapin*. This oil is the most pungent substance known, causing strangulation when breathed. It is not present in the white seed, but a mixture of the white and black produces it in greater abundance than the black alone, and it is found that the mixture of both kinds of seeds makes the best mustard for table use.

Use. — The oil is used in dressings, for salads, etc., and for soap-making. The seeds swallowed whole act as a tonic and stimulant; in larger doses, as a laxative. The flour, mixed with warm water, acts as a quick emetic. The seed is ground into flour, in which form it is mixed with vinegar or oil, or both, into a paste for table use, as a condiment for meats. The seeds of B. alba are used whole for flavoring fancy pickles. It is also employed in an entire state, to preserve cider in a sweet condition. The flour is used for a poultice, as a counter-irritant in inflammations, and as a remedy for stomach disorders and nervous affections.

In England mustard is much sown as a crop for forage and for green manuring. When sowed at the rate of about 12 lbs. to the acre it gives an abundant crop of succulent forage, which is cut before the seeds begin to mature, and fed to cattle, sheep, and swine.

5. **B. campestris**, L. (Field Turnip.) Stem slender, appearing the second year, 18 to 30 inches high, much branched, smooth. Lower leaves lyrate, 3 to 7 inches long; lobes toothed, somewhat hairy and glaucous underneath, clasping and terminating in an abrupt acumination. Calyx closed. Corolla yellow,

half-inch in diameter, spreading. Seed-vessel 1 to 2 inches long. Root somewhat in the shape of an inverted cone, or spindle-shaped, fleshy, 3 to 10 inches in diameter, and 6 to 12 inches long, terminating in a slender, tapering radicle, besprinkled with fibrous rootlets. Biennial herb.

Under B. campestris there are several forms. The most important is the sub-species:

Napa-brassica. L. var. **rutabaga.** (Rutabaga. Swedish Turnip. Russia Turnip.) Root subglobose, flesh yellowish.

6. **B. rapa,** L. var. **depressa.** (Flat Turnip. Red Top Turnip. Strap-leaved Turnip.) Stem and leaves as above in No. 5, except that the leaves are frequently narrow, long, and linear; the root flattened at the poles, or flat above and convex beneath; radicle long and slender. A favorite variety for summer use.

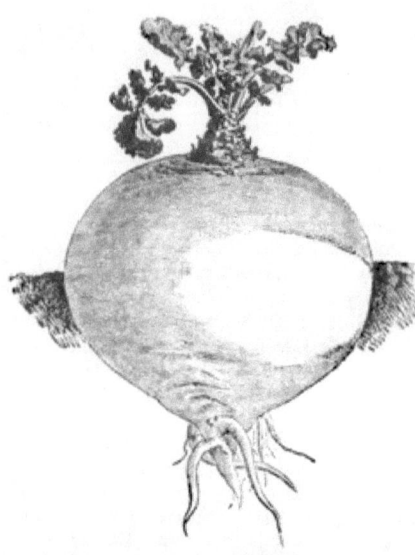

BRASSICA CAMPESTRIS (Field Turnip).

The turnip sports freely, forming many varieties. The Swedish turnip varies in color and size, but very little in quality. It is yellow or white. A recent sport is called the White Stone.

There are other varieties and forms of considerable constancy, but those already described are the most important.

Geography. — The turnip is found under cultivation in Greece, Egypt, and Syria, but especially in middle and northern Europe, and is said to be indigenous to Sweden, Russia, and Siberia. It was introduced by European colonists into North America, and is largely grown all over the middle and northern United States and southern Canada. It has also been introduced by the British into India and Australia.

Etymology. — *Campestris,* the specific name, is from the Latin *campester,* inhabiting an open place, or field. The specific name, *Rapa,* is from the Latin *rapa,* a turnip. *Turnip* or *Turnep* is from the old French

BRASSICA RAPA (Flat Turnip).

tour, turned, or round, and the Anglo-Saxon *nepe*, *white*, signifying round and white. *Swedish* and *Russia* are names due to the countries where these varieties have originated or where they are largely grown. *Flat top* is named from its form.

History. — When the turnip was first cultivated, or where it was first used for food, history does not reveal. The Romans were acquainted with it. Pliny relates that single specimens sometimes weighed forty pounds. This seems remarkable, since the turnip does not at the present day grow well in warm, dry climates, neither does it arrive at perfection in very cold regions. The British Isles and the low countries of western Europe are best suited to it. It also grows well in the middle parts of the north temperate zone in North America, to which it was taken by English colonists. Theophrastus and Dioscorides both speak of the turnip. It was taken to Britain by the Romans, most likely under Agricola.

But little attention seems to have been given to it till the early part of the 17th century, when it was an important crop in England. It requires rich, deep soil, and a damp, *cool* climate, but grows well in damp, *warm* countries. De Candolle thinks the bulk of evidence shows that it is a native of western Europe, or possibly of Siberia.

Use. — The turnip is one of the most valuable of all the root vegetables; it is prepared for the table in many ways, and is largely employed for feeding stock. Cattle, sheep, and horses are fond of it, and it is especially valuable for milch cows.

It has been sliced, dried, ground into powder, and used to adulterate flour for making bread.

Davy analyzed the turnip, and found it to contain 3 per cent of nutritive matter.

7. **B. napus.** L. (Rape.) This species differs from B. campestris in the form of its root, which is long and slender, and usually carrot-shaped; seed-vessels spreading, foliage more abundant; otherwise as B. campestris.

Geography. — The geographical range of the rape is the same as that of the turnip and cabbage. Its home is no doubt in the regions of middle and northern Europe and Siberia. It is found wild in Sweden also, but may have escaped from cultivation.

Etymology. — *Napus* is the ancient Latin name for turnip, and signifies a shape which tapers both ways from the middle.

History. — How long the rape has been in use as an economic plant is not known. Dioscorides and Pliny both speak of it under different names, and De Candolle thinks that it and the turnip have been cultivated for more than 4000 years.

Use. — In northern Europe and especially in Russia it is raised in large crops for its seed, of which an oil is made for lubricating purposes, and also for illuminating. When carefully refined it is employed for culinary and table uses. At the world's exposition in Philadelphia in 1876, a very fine rape oil was on exhibition in the Russian agricultural department. The refuse is pressed into cakes called rape cake, and fed to cattle and poultry. The rape seed is frequently mixed with turnip, cabbage, and other cruciferous seeds. To obtain the oil, the seeds are ground or beaten into paste, put into hempen bags, and placed between grooved planks standing in an upright position, so arranged as to be nearer together at the bottom than at the top. The planks are then forced together by wedges, which forces the oil out. This oil constitutes a very important article of food among the Russian peasantry.

DESCRIPTIVE BOTANY.

COCHLEARIA, Tournefort. (Horse Radish.) Calyx equal at the base. Sepals short and slightly spreading. Petals entire, larger than the sepals, with short claws. Style short, occasionally long. Pod globular or egg-shaped; valves convex, nerve dorsal. Seeds usually few, obovate, flattened, arranged in two rows in each cell. Leaves large, oblong, lower ones pinnate, those on the upper part of the stem entire. Root perennial.

1. **C. armoracia**, L. (Horse Radish.) Stem 2 to 4 feet in height, striate, much branched; branches ascending. Leaves large and thick, radical ones 12 to 18 inches long; on the lower part of the stem either pinnate, crenate, or toothed, on long stalks 6 to 12 inches in length; upper leaves on short petioles, sometimes sessile, entire, elliptical, or strap-shaped. Flowers in corymbs, white. Seed-vessel egg-shaped, seldom ripening seed. Root large, fleshy, and edible, and of very rapid growth. Flowers in June.

2. **C. officinalis**, L. (Scurvy Grass) The leaf has the shape of a spoon, hence the generic name. Sometimes cultivated in the United States. There are other species, but they are not cultivated.

Geography. — The geographical range of the Cochlearia is the middle and southern edges of the north temperate zone, extending from Great Britain to western Asia, and northeast America.

Etymology. — *Cochlearia*, the generic name, is from the Latin *cochlear*, a spoon, due to the shape of the leaf of the C. officinalis. *Armoracia*, the specific name, is derived from Armorica, a province in France where the Horse Radish was thought to be native; but it has been ascertained that Armorica is not the home of this plant, hence the name did not arise in that way, and the derivation is obscure. *Horse Radish*, the common name, signifies a *strong Radish*, due to the very pungent taste of the root. *Officinalis* means "of the shops," or "useful."

COCHLEARIA ARMORACIA
(Horse Radish).

History. — It is not certain that this plant was known to the ancients. Pliny, in the first century, applied the name to another plant. It was taken to Great Britain before Cæsar's invasion, and has become naturalized, and is growing freely and propagating itself, without cultivation, throughout temperate Europe. It was brought to northeastern America by colonists, where it is cultivated, and is also naturalized. It is an important plant to the market gardener.

Use. — The medical properties of the Horse Radish are stimulant, diaphoretic, diuretic, and anti-scorbutic, and when applied externally, rubefacient. It is administered in paralysis, rheumatism, dropsy, and in other complaints to excite the secretions.

As a food, the root is esteemed for flavoring and preparing sauces for meats and fish. It is prepared by grating the root and preserving it with vinegar. It is adulterated with grated turnip, which renders it more mild and palatable.

CRUCIFERÆ.

ISATIS, L. (Woad.) Pod or silicle oval or elliptical, flat, one-seeded; valves boat-shaped, subdehiscent; cotyledons accumbent. Biennial.

1. **I. tinctoria**, L. (Woad.) Stem 4 feet high, half an inch in diameter, much branched. Leaves thick, light-green, oval, subclasping, ears rounded, radical leaves petioled, and 10 to 12 inches long and 6 wide, abundant, giving the plant a coarse appearance. Flowers small, yellow, and in terminal panicles. Appearing in July.

Geography. — Woad is indigenous throughout the continent of Europe, and Great Britain. It is cultivated in England, Sweden, Germany, Switzerland, and in the Azores and Canary Islands.

Etymology and History. — *Isatis* is derived from the Greek word ἰσάζω, make smooth or even, because it was supposed to remove roughness of the skin. It has been called *glastum*, from the Celtic word *glas*, blue. The ancient Britons, at the time of the invasion by Julius Cæsar, adorned their bodies by painting the forms of animals and other objects on them with woad, hence the Romans gave them the name of Picts, or pictured men.

ISATIS TINCTORIA (Woad).

Woad imparts a permanent blue color, the shade depending upon the quantity, and manner of using it. It is capable of giving a very dark blue, approaching a blue-black.

Use. — Notwithstanding the inroads that indigo has made upon the use of woad, on account of the permanency of the color imparted by woad it still holds an important place among coloring substances. Dyers are in the habit of mixing it with indigo; the dyes are said to coalesce, and strengthen each other. Its use is rapidly dying out in England.

NASTURTIUM, R. Br. Sepals 4, regular, and equal at base. Petals regular, white, seed-vessel or silique tapering, cylindrical, short, and curved upwards. Seeds small, irregularly arranged in a double row. Leaves alternate. Herbs which delight in wet places, or in the edges of the waters of slowly flowing streams.

1. **N. officinale**, R. Br. (Water Cress.) Stem perennial, 6 to 18 inches long, branched, prostrate, and assurgent. Leaves pinnately divided; leaflets very inconstant, ranging in number from 3 to 5 pairs, and a terminal one, rounded, usually entire, and glabrous, occasionally sinuately toothed. Flowers in June. Fruits in July.

Geography. — The geographical distribution of this plant is very wide, as is the case with all economic plants which follow colonization. It grows all over Europe, in Palestine, Hindustan, Japan, the islands of the Atlantic and the Pacific, wherever European colonies have been established.

Etymology. — *Nasturtium* is derived from the Latin *nasus*, the nose, and *tortus*, a *twisting*, said to be due to the effect which its pungency has upon the nose when taken in large mouthfuls. *Officinalis*, Latin, signifies that the plant belongs to commerce, or the shops. *Water Cress*, the common name, comes from the Latin *cresco*, grow, and " water " alludes to its habit of growing in the water.

History.—The home of the Nasturtium is Europe and northern Asia. Where or when it was first introduced into the catalogue of table vegetables is not known. It was noticed by Dioscorides and Pliny, and therefore must have been in use in Italy before the Christian era. It has been seen near the mouth of the Columbia River, in North America. It was brought to eastern North America probably by colonists from Holland, where it is grown to a considerable extent.

NASTURTIUM OFFICINALE (Water Cress).

Use.—It is eaten as a salad, and used to garnish dishes of meats and of fish. It is served always without cooking, eaten with salt, or a vinegar dressing, and is a popular salad-plant with oysters and game.

NOTE.—The nasturtium of the garden is Tropæolum, which belongs to the geranium family, and is an immigrant from Peru. It has showy yellow flowers, and is cultivated both for ornament and its fruit, which is curiously curved and ribbed. The fruit is plucked before it is mature, and pickled. It has a pungency similar to the Nasturtium, hence its common name.

ORDER V. CAPPARIDACEÆ.

Sepals 4–8, free or sometimes cohering. Petals hypogynous or perigynous, 4–8, sometimes wanting. Stamens 6 or many, hypogynous or perigynous. Ovaries generally stipulate, 1-celled, or spuriously 2–8 celled, with parietal placentæ, ovules curved. Flowers usually perfect. Herbs, shrubs, or small trees. Leaves simple, or digitate and spiny. The Capparis is a small shrub. Number of genera, 23.

CAPPARIS, L. Calyx divided into 4 spreading, ovate sepals. Petals 4, alternate with the sepals, wedge-strap shape, longer than the sepals, spreading or recurved. Stamens numerous, inserted on a subconical receptacle, free, anthers attached lengthwise along their backs to the filaments, and the cells turned out. Pistil one, longer than the stamens. Shrub.

1. **C. spinosa,** L. (Caper.) Stem from 2 to 3 feet long, straggling or prostrate, very much branched, bark white. Leaves ovate or orbicular, thick,

glaucous, and deciduous, 2 crooked spines appearing at the base of the leaf-stalk. Flowers on long stalks, white, numerous and axillary, large and showy, but without fragrance. Petals much longer than the sepals, wedge strap-shape, spreading or recurved. Stamens sometimes 60 in number, about as long as the petals. Fruit a leathery, obovoid, succulent capsule, berry-like; seeds numerous, kidney-shaped. Shrub. Flowers from June to August.

2. **C. soldada** is found in central Africa, bearing a fruit resembling the currant, and is eaten fresh from the bush, and also dried.

3. **C. ferruginea** has a rusty, narrow leaf. It is found in the West Indies. The berries have a pungent, mustard-like taste, and on that account the plant is called the mustard shrub.

Geography. — The Capparis grows and fruits in southern Europe, and all the countries of the Mediterranean. The market is supplied from the plantations and wild copses of Sicily and Malta, and other islands of the Mediterranean, and from the south of France. It loves the rocks, and its habit is well pictured in the following quotation:

"This beautiful plant is rooted in many a crevice of the palace of the Cæsars at Rome; it spreads its green, glossy leaves and starry white flowers, with their long, purple anthers, over the ruins of that once stirring place, the Colosseum, and clothes the arches of the temple of peace with festoons which adorn without hiding their

CAPPARIS SPINOSA (Caper).

beauty; the ancient tombs of the Campagna are frequently hung with it; the rocks of Naples are favorable to it; and it has fixed itself not only on the mouldering cliffs of Malta, but in the narrow crevices of the fortification."

Etymology. — *Capparis* is Latin, through the Greek καππαρις, from the Arabic *Kabar*, the Persian name of this plant, the meaning of which is not known. *Spinosa* is Latin, meaning prickly, and is due to the thorns that appear at the base of the leaf-stalks. *Caper*, the popular name, is a corruption of *capparis*, the generic name.

History. — The caper was well known to the Greeks and Romans, and was mentioned by the early writers on natural history, especially Theophrastus, Dioscorides, and Pliny. The capparis is among the plants known to the ancient Hebrews as thorns. The flower-buds were preserved in salt or vinegar, and offered to guests just before dinner as an excitant to the appetite. It is claimed by some authors that this is the plant called in the Scriptures Hyssop, and the one used to sprinkle the doorposts and lintels with blood, in Egypt; also the plant mentioned in connection with the wisdom of Solomon, when it is stated that he knew all the plants, from the "cedar of Lebanon unto the Hyssop that springeth out of the wall:" but this has been disputed, and claims have been made for a genus of the mint family.

The flower-buds are collected by women and children (whose hands and wrists are torn by the sharp thorns by which every leaf-stalk is guarded), placed in salt or vinegar, and in Italy the fruit in an unripe state is also gathered and prepared just as the undeveloped flower-buds are. It was carried to the south of France by Greek colonists, and has been grown largely since that time near Marseilles. It grows best in rocky places or among ruins. Gerard says it refuses to be domesticated.

Use.— The medicinal qualities of the caper are stimulant, anti-scorbutic, aperient, diuretic, and stomachic.

In Italy, the unripe fruit, which is very pungent, is eaten not only as a pickle but as a salad, and is also cooked with meat. The undeveloped flower-buds are sent abroad, and are used in this country to flavor sauces and dressings for boiled meats.

The fruit of the C. soldada resembles currants, and is eaten in the same manner. The fruit of the C. ferruginea has the taste of mustard, and is largely used as a condiment for meats.

ORDER VI. **VIOLACEÆ.**

Flowers perfect, irregular, axillary, mostly solitary, 2 bracts, usually at the base or near the middle of pedicel. Sepals 5, mostly free, persistent. Petals 5, hypogynous, alternate with the sepals, unequal, lower one dissimilar and prolonged into a hollow spur, below the insertion. Stamens 5, inserted on the bottom of the calyx; filaments short, dilated, usually free. Ovary free, sessile; style simple. Fruit a capsule, many-seeded, with parietal placentæ. Seeds ovoid, or globose; testa membranous. Leaves alternate, stipulate. Herb.

No. of genera, about 21.

VIOLA, L. (Violet.) Sepals 5, unequal, and eared at base. Petals 5, unequal, the broad one spurred at base, 2 lateral ones equal. Stamens 5, approaching; filaments free; anthers connate; capsule 1-celled, 3-valved; seeds attached at the middle of the valves. Pedicels angular, solitary, 1-flowered, curved at the summit. Flowers nodding in an inverted position. Perennial herbs.

1. **V. blanda**, Willd. (Delicate Violet.) Leaves cordate or kidney-shaped, crenate, toothed or entire, early ones orbicular, flat and thin, head of sinus rounded. Flowers white, odorous, and small; sepals ovate; petals ovate, obtuse, frequently striped with purple, slightly bearded. Stigma depressed, margined. April and May.

Geography. — Found in wet places from Canada to Pennsylvania.

2 **V. Canadensis**, L. (Canada Violet.) Stem 9 to 18 inches high, smooth, slender. Leaves prominently heart-shaped, and acuminate or pointed, irregularly serrate, lower ones on long petioles; stipules large, ovate-lanceolate,

VIOLACEÆ.

entire. Flowers large, blue without, nearly white within; upper petals marked with blue lines, side ones bearded; spur short; stigma short, and without beak; sepals lanceolate. May to August.

Geography. — British America to Carolina, west to Colorado, Montana, and Wyoming. Rich woods, not rare.

3. **V. canina**, L. (**V. Muhlenbergii**, Torr.) (Dog Violet.) Var. Sylvestris, Regel. Stem leafy, 2 to 8 inches high, many from the same root, sending off creeping branches. Leaves at the root kidney-shaped or orbicular heart-shaped, upper ones acuminate; all crenate, thin, veins prominent, stipules large, lanceolate, fringe-toothed. Flowers pale-purplish, showy; petals obovate, obtuse, lateral ones bearded; spur tapering, half as long as petals; stigma rostrate. May to June.

Geography. — Eastern North America, from Labrador to Florida, and west to Colorado. In damp meadows, edges of swamps, and fringes of damp woods.

VIOLA BLANDA (Delicate Violet).

4. **V. hastata**, Mx. (Halberd-leaved Violet.) Stem slender, erect, simple, nearly smooth, leafy above, 6 to 10 inches high. Leaves on long petioles, cordate, lanceolate, or hastate, acuminate, dentate; lobes obtuse, stipules minute, ovate. Flowers yellow, peduncles shorter than the leaves; lower petal broader, 3-sub-lobed, lateral ones slightly bearded. May.

Geography. — Tenn. to Florida, mountains of Penn. and northern Ohio. In pine woods, not common.

5. **V. lanceolata**, L. (Lance-leaved Violet.) Quite smooth, lanceolate, tapering into a long petiole, obscurely toothed, or entire. Leaves generally a little longer than the scapes, 4-6 inches high. Flowers white, inodorous, striped with purple lines; spur short. Sepals lanceolate; petals beardless. April to June.

Geography. — From Canada, throughout eastern U. S., in damp places.

6. **V. odorata**, L. (Sweet Violet. English Violet.) Leaves heart-shaped, crenate, sparingly hairy, stipules lance-shaped and toothed. Flower-stalks taller than the leaves. Flowers purple and fragrant. Sepals obtuse; lateral petals with a hairy line.

There are several varieties, based upon the color and size of the flowers. —
 a. Purple Sweet Violet.
 b. White Sweet Violet.
 c. Blue Sweet Violet.

By cultivation all these frequently become double; they are great favorites with florists.

Geography. — Indigenous throughout Europe; found also in some parts of China, Japan, and India; it has escaped from cultivation, and is frequent in the fields near the great cities of the United States.

VIOLA ODORATA (Sweet Violet).

7. **V. palmata,** L. (Hand-shaped Violet.) Leaves varying from broad cordate to reniform, repand toothed, sparingly cucullate at base. Whole plant slightly pubescent. Early leaves purple underneath. Growing in dry grounds and open woods. 6 to 10 inches high.

Form No. 2. Early leaves broad cordate, or reniform, somewhat fleshy, on short petioles, under side frequently purple, serrate toothed, usually 2 or 3 in number, rarely many; later leaves usually 2–4 in number, on long petioles, 3-lobed, the middle lobe sometimes lanceolate, occasionally with parallel sides, and terminating in a blunt angle, with lateral lobes hatchet-shaped, with the margins sometimes serrate toothed, sometimes deeply cut into 2 or 3 divisions. The whole leaf is frequently divided into narrow parts, approaching V. pedata. Again, the whole margin will be made up of divisions, varying in number from 6 to 12, and from an eighth to half an inch in width, the middle ones generally the broadest, and the incisions extending half-way into the blade. The early leaves are usually smooth, the later ones covered with pubescence. Flowers apetalous and frequently subterranean.

Geography. — Dry grounds and open woods. May to August. 6 to 12 inches high.

The author watched this plant closely throughout five successive seasons in the same localities, and it seems to depart from the distinctive characters of V. cucullata as its distance from damp ground increases. He placed specimens with divided leaves in the lawn of the Freehold (N. J.) Institute, in damp, rich soil, and in the course of four years they were free from pubescence, the leaves entire, and in every way identical with V. cucullata growing within ten feet of it. He also saw specimens which had been transplanted into a dry, gravelly, rather sterile border, in Flushing, on Long Island, and they retained their pubescence and divided leaves.

8. **V. cucullata,** Ait. (Common Blue Violet.) Plant 6 to 12 inches high, flower-stalks frequently as long or longer than the petioles. Leaves glabrous, cordate, rolled in at the base, serrate-crenate, or remotely toothed, those appearing first frequently kidney-shaped, and purple underneath near the base. Flowers blue, large, late ones apetalous and subterranean. Sepals linear, lanceolate; upper one smooth, the others bearded; lateral ones obovate; spur short and rounded. April to July.

Geography. — The distribution of this species is very broad. It is found in the temperate zone quite across the continent of North America.

Var. striata, Willis. (Streaked or Spotted Violet.) Four to eight inches high. Leaves cordate, frequently reniform, early ones entire or crenate, purple underneath near the base, and glabrous, later ones becoming more and more clothed with hairs, and taking on a lobed form as the season advances, or as the plant creeps up dry hillsides from damp and lower grounds. Flowers few or many, pure white, marked with purple lines; sometimes sprinkled with purple dots, or splashed with large, irregular, or ragged purple spots, but the lines are always present. Petals very irregular as to size, even in the same plant; sometimes very small, and sometimes irregularly cut, toothed, or even fringed; later flowers cleistogamous.

Geography. — Found sparingly near Freehold, N. J., and more frequently in the vales and on the acclivities among the Gneissic hills about White Plains, N. Y.; also in the northeastern parts of New Jersey.

VIOLACEÆ.

Var. reniformis, Willis, is quite distinct and constant. Leaves very broad, cordate, or prominently reniform; frequently 4 to 5 inches wide, and an inch from base to apex; sometimes with a deep, broad sinus at the apex, seldom flowering; flowers subterranean, leaf-stalks 12 to 15 inches long, growing in rich, damp, shady places.

Var. cordata, Walt. Leaves prostrate, round, cordate, smooth or clothed with soft hairs, small. Dry hills and open woodlands. (V. villosa and V. cordata, Walt. and V. sorosis, Willd.)

VIOLA SAGITTATA (Arrow-leaved Violet).

9. **V. palustris**, L. (Meadow Violet, or Marsh Violet.) Leaves cordate, or kidney-shaped, obscurely crenate; stipules broadly ovate, and acuminate. Rhizomes creeping and scaly. Sepals ovate, obtuse. Petals small, pale blue; spur short and blunt; peduncles longer than the leaves. Plant 2 to 3 inches high. June.

Geography. — Tops of White Mountains in New Hampshire, also mountains of Colorado and Utah; identical with the European species.

10. **V. pedata**, L. (Bird-foot Violet.) Leaves pedate, smooth, 5–9-parted, lobes linear, lanceolate, obtuse or acute, 1–2-toothed or 3-lobed at the apex, tapering downwards, stipules lacerated. Flowers large, pale-blue; petals rounded at the extremities, beardless; spur short; stigma large, obliquely truncate; beak obscure. Scapes 2 to 5 inches high, several from the same root. May to June.

Var. bicolor, Pursh. Varies from the above description, in having the two upper petals deep violet, presenting a velvety appearance; the others light-blue, with yellow at their bases, resembling the V. tricolor, or pansy.

Geography. — Canada to Florida, and west to Ill. In southern exposures of sandy woodsides.

11. **V. primulæfolia**, L. (Primrose-leaved Violet.) Smooth, varying from cordate, broad ovate, to lanceolate, tapering into a winged petiole, slightly repand or crenate; when growing in dry places sparingly pubescent; sepals lanceolate; stigma beaked. Flowers white, sometimes striped with purple streaks, slightly odorous; petals slightly bearded, especially the lateral ones. April to July. Wet meadows; growing with V. lanceolata and V. blanda, and seems to be a connecting link between them. Specimens sometimes seem to possess the characteristics of both. Dr. Beck suggests that it may be identical with V. lanceolata, but the author's observations do not lead him to that conclusion. It seems more like a variety of V. blanda; its flowers are odorous like V. blanda, and its foliage more nearly approaches that of V. blanda than of V. lanceolata, 4 to 6 inches high.

Geography. — Found in the Atlantic States, and west to Tennessee; in damp grounds.

12. **V. pubescens**, Ait. (Downy Yellow Violet.) Stem 6 to 12 inches high, somewhat angular, erect, softly pubescent. Leaves broadly heart-shaped, dentate, acuminate; stipules large, ovate, sparingly toothed. Flowers middle-sized, yellow, lateral; petals slightly bearded, lower ones striped with dark purple; spur very short. Peduncles shorter than the leaves, axillary, solitary, furnished with 2 awl-shaped bracts. May.

Var. scabriuscula, Torr. and Gray. Smaller, less pubescent, brighter green, stem frequently prostrate, 3 to 9 inches high, and branching near the root.

Geography. — Canada, eastern United States to Georgia, and west to Missouri. Frequent in dry, stony, open woods throughout these limits.

13. **V. rostrata**, Pursh. (Long-spurred Violet.) Stem diffuse, erect, 4 to 8 inches high, smooth. Leaves smooth, thin, roundish, heart-shaped below, cordate-lanceolate or sub-triangular and acute above; lower ones crenate-toothed, upper ones sub-serrate. Stipules large, lanceolate, serrate ciliate. Flowers large, pale-blue; petals obovate, beardless; spur slender and very long. June.

Geography. — Eastern North America, Canada to Virginia, and south and west in the Alleghanies. Found sparingly on shaded hillsides throughout these limits.

14. **V. rotundifolia**, Mx. (Round-leaved Violet.) Early leaves orbicular or kidney-shaped, later ones longer than broad, heart-shaped, sparingly toothed, slightly crenate, 1 to 2 inches wide, 2 to 4 inches long; stalks pubescent, about as long as the blade. Flowers pale-yellow, middle-sized; side petals bearded, marked with dark lines, sometimes notched at the summit. Stalks 1 to 3 inches long, generally smooth, occasionally pubescent, sometimes bracted in the middle; spur short.

Geography. — Found sparingly in damp ground from New England to Tennessee.

15. **V. sagittata**, Ait. (Arrow-leaved Violet.) Leaf entirely smooth, when growing in damp soil; slightly pubescent when growing in dry soil. Sub-

linear, lanceolate; sometimes triangular, oblong heart-shaped, arrow-shaped, or halberd-shaped; sparingly toothed or cut toothed at the base; 4 to 10 inches high. Flower deep blue; petals obovate, bearded, and emarginate; spur short and thick; sepals lanceolate, acute. April to July.

Var. **ovata**. Leaves oblong-ovate, crenate, frequently repand toothed near the base; pubescent; stipules ciliate; flowers large and dark. Growing in dry, open woods; sandy soil. 2 to 4 inches high. April, May. (V. ovata, Nutt.)

Geography. — Found throughout eastern North America from Canada to Florida, and west to the Mississippi valley. Dry or moist open grounds.

16. **V. Selkirkii**, Pursh, Goldie. (Great-spurred Violet.) Leaves numerous, orbicular, heart-shaped, slightly hairy on the upper side, crenately toothed; sinus deep, sometimes broad, at other times nearly closed. Root-stock fibrous rooted. Flowers small, pale blue; spur very large; petals beardless, upper one marked with blue lines. Plant about 2 inches high. May.

Geography. — Found in Canada, Mass., and N. Y. Chautauqua Co. (Judge Clinton), Lake Superior (Robbins), rare.

Viola tricolor (Pansy).

17. **V. striata**, Ait. (Pale Violet. Striped Violet.) Stem assurgent, angular, or half-round, smooth, 6 to 12 inches high. Leaves alternate, heart-shaped, frequently acuminate, crenate, serrate; petioles 1 to 2 inches long; stipules large, oblong-lanceolate, strongly fringe-toothed. Flowers large, yellowish-white; side petals densely bearded; lower one striped with dark purple; spur thickish, shorter than the petals; stigma recurved. May.

Geography. — Eastern North America, from Canada, south and west. Low grounds. Found sparingly throughout these limits.

18. **V. tricolor**, L. (Pansy. Heartsease.) Stem angular, much branched, 6 to 20 inches high, leafy. Leaves oblong-ovate, lower leaves cordate, remotely toothed, or sub-crenate; stipules pinnatifid or lyrate, end lobes as long as the leaves. Flowers variable in size; two upper petals purple; side ones white; the lower one striate at base; all yellow at base. Spur short and thick. April to September.

Escaped from gardens, sparingly naturalized in fields near old dwellings.

Var. **arvensis**, DC. (**V. tenella**, Muhl.), (V. tricolor, L.). Flowers a little smaller; petals only as long as sepals. The whole plant usually smaller.

Geography. — Sparingly naturalized in New York, and south to Georgia. Around dwellings, in dry, sandy soil. Brought to North America by English colonists.

NOTE. — This violet was brought to the notice of florists in England about 90 years ago, by Mary Bennet, daughter of the Earl of Tankerville, who, aided by her father's gardener, produced several seedling varieties, the flowers of which were greatly enlarged and beautified. From this beginning the plant has been changed into the pansy of the present day.

Etymology. — Viola is from the Greek *Ιον*, from the following myth: Jupiter loved *Ιω*, the daughter of Inachus, first king of Argos, and on account of the jealousy of Juno he transformed *Ιω* into a beautiful white heifer, and immediately the earth brought forth the *violet* for her food; hence its name, the cow plant. It is also derived, by some, from the Latin *vitula*, a heifer, by eliding the *t* and changing the *u* into *o*, making viola, the cow plant, or heifer plant.

History. — It is not to be wondered at that this beautiful little flower should have attracted the attention of the ancients. The early writers on plants mention the violet on account of its beauty and delicacy, and it has been lauded by both poets and painters. In the language of flowers, the violet represents faithfulness.

It was the favorite flower of the Empress Josephine. On the day before their marriage Napoleon Bonaparte sent her a bouquet of violets, after which it became the court flower.

It was the rallying sign of the Emperor's partisans, on his return from Elba. It is related that two days before he set out on his exile journey, he, while walking alone in the garden of Fontainebleau, asked a little child to give him a bunch of violets he had gathered. These he showed to some of his officers, remarking that he considered the flower an emblem of modesty which he proposed to imitate. The next morning, a private of his old guard saw him collecting violets, and said to him, "Sire, they will be more plentiful here next year." To which Napoleon replied, "Do you think I shall be here next year?" The soldier said, "Your Majesty will permit the storm to pass." Napoleon asked, "Do your comrades think so?" "Nearly all, Sire," was the answer. Napoleon said, "Let them think, but not say so." The soldier repeated the conversation to his fellows, and it was then agreed to speak of him always as Father Violet. After this, men throughout France began to talk of the violets of the coming spring, and of a certain Corporal Violet, who would perhaps come in the spring. Ladies who longed for his coming wore violets in their bonnets. In fact, treason lurked everywhere beneath a bunch of violets, and tiny pictures of the Emperor were concealed among the leaves and flowers in every buttonhole bouquet, and bunches of the flower were painted so as to reveal his profile.

Use. — The medicinal qualities of this plant are said to be curative in lung complaints, rheumatism, and catarrh. Most of the violets contain an emetic principle, called violine, especially in their roots. The flowers are laxative and the sirup of violets is used as a laxative for infants. The sirup is also used occasionally as a test for acids and alkalies. The roots of some species produce false ipecacuanha. The flower is used by dyers to produce the color known as the Azure of Athens, and the delicate odor of the sweet violet is a most popular perfume. The florist finds it among the flowers that are largely sought after for ornamental purposes.

Order VII. BIXINEÆ.

Sepals distinct, or united at base, imbricate in the bud, 2-6 in number. Corolla polypetalous or absent; petals as many as the sepals. Stamens hypogynous. Flowers usually perfect, regular, axillary or terminal, either solitary or fascicled, sometimes racemose or panicled. Leaves alternate, simple toothed, occasionally palmately lobed. Trees and shrubs.

Genera, 29.

BIXA ORELLANA (Annatto).

BIXA, L. (Annatto.) Calyx fleshy. Sepals 5, spatulate, eared near the base. Stamens numerous, some multiple of 5. Style filiform. Fruit 1-celled, in an oblong, bristled pod, somewhat like a chestnut, but longer.

1. **B. orellana**, L. (Annatto.) Stem 8 to 12 feet high, branching. Leaves deep-green above, pale beneath, 4 inches long, broad cordate at the base, and tapering and pointed at the apex. Inflorescence a loose panicle. Flowers pink.

Geography. — The Bixa grows in tropical America, and has been introduced by Europeans into southern Europe, Burmah, the Philippine Islands, and Hindustan. It grows freely in all regions of no frost.

Etymology. — *Bixa* is the South American name for this tree, and the common name *Annotto* is quite as obscure as to its signification.

History. — It was noticed that the natives of tropical America painted or stained their skin with a bright yellow dye. On inquiry it was learned that the material was obtained from the pulp in which the seeds of the Bixa were imbedded.

The pulp is washed or soaked off the seeds, the water is then removed by evaporation, and the residuum is made into cakes, in which form it is introduced into the market wrapped in leaves. The soft wood is used by the natives of tropical America to obtain fire by friction.

Use. — The fiber of the stem furnishes excellent material for cordage.

The coloring matter is said to be of a fire color, and when mixed with cheese or butter it imparts a rich, creamy yellow; it is largely used in Holland and England as well as in America for that purpose. It is also used to color varnish, and it is mixed with chocolate to enliven the color and to improve the flavor. The roots are used in soups. The seeds are cordial, and a drink made from them is said to be remedial in allaying fevers.

Marts. — The annual import of annatto into England is about 300,000 pounds, valued at about $65,000.

Order VIII. TERNSTRŒMIACEÆ.

Sepals 5, occasionally 4–6 or 7, free or slightly connate at the base, imbricated in the bud. Petals 5, free, hypogynous, imbricated or twisted. Stamens sometimes equal in number to the petals, but usually indefinite. Flowers perfect, regular, axillary, solitary, or fascicled, sometimes in terminal racemes or panicles. Ovary 3–5-celled. Ovules pendulous or ascending. Fruit indehiscent or capsular. Leaves alternate, occasionally opposite, frequently fascicled, at summit of the branchlets. Trees or shrubs.

Number of genera, 32.

THEA, L. (Tea.) Calyx 5-parted, sepals short and scale-like. Corolla much longer than the calyx, white. Stamens many. Style 3-parted. Several flowers appear in the axils of the leaves. Capsule 3-celled. Shrubs.

1. **T. viridis**, L. Stem in a natural state grows to the height of 15 to 20 feet, but is dwarfed under cultivation, by stripping the leaves, and seldom reaches a height above 5 feet, diffusely branched. Leaves lanceolate, entire at the base, serrate, with blunt teeth towards the apex, alternate. Flowers white. When grown in the Middle States, it flowers in winter under glass.

2. **T. Bohea**, L. corresponds with T. viridis except that the flower has many petals, and is most likely a variety of it, as the only striking difference is in the numerous petals.

Varieties. — The tea plant is grown from seed, and sports freely; hence there are many varieties, differing from each other chiefly in the form of the leaf.

Geography. — The native country of the tea plant is generally supposed to be China, but it grows well in a belt included between the parallels of 25° and 35° throughout Asia, and is more prolific on hillsides than in the bottom lands. It is cultivated further north by the Chinese, but its quality and productiveness are best in the above-named belt.

Tea is cultivated in Kangra, Gurhwal, and in Assam, Cachar, Sylhet, Chittagong, Darjeeling, and Chota-Nagpore.

In Hindustan the cultivation has greatly increased, and is still increasing. The yield per acre ranges from 100 to 200 pounds. The conditions of successful tea culture are, first, a low, undulating, hilly country, where the valleys have good drainage; second, a climate warm, moist, and of uniform temperature. Assam presents a most favorable region, the temperature seldom rising above 95° in the daytime nor falling below 60° at night, while the rainfall is remarkably uniform throughout the year, being about 12 inches monthly. Wherever these conditions are approached, tea may be successfully cultivated. Japan, Australia, Jamaica, Brazil, and parts of North America all possess localities favorable to tea culture, and if labor sufficiently skilled and cheap were obtainable, these countries would be independent of tea importations. In 1836 the attention of the Indian Government was called to Assam by Dr. Royle, the botanist, as a suitable locality for the cultivation of the plant, which had been found there in a wild state.

THEA VIRIDIS (Tea).

Etymology. — The name *Thea* is derived from the Chinese word, which is pronounced *Teha*, the meaning of which is obscure.

History. — It is not known when this plant was first used to furnish a beverage, but it is well established that it has been an article of traffic for more than fifteen hundred years. It was cultivated and used in the Chinese Empire in the fourth century, and in Japan in the ninth century. Early in the seventeenth century dried green-tea leaves were presented to a Russian embassy in China, and forced on them against their protestations. When brought to Moscow the tea met with very great favor. It did not make its appearance in Europe until about the middle of the seventeenth century.

It was brought to Europe by the Dutch East India Company, and introduced into England from Holland by Lord Arlington. In 1664 the East India Company presented the Queen of England with two pounds of tea. It cost at first about $25 per pound. For a long time, because of its great price, its use was confined to the wealthy, and even in the early part of the present century it was sold in France only by druggists.

Chemistry. — The characteristic substance found in tea is *theine*, whose formula is $C_8H_{10}N_4O_2$.

Preparation. — The varieties of tea are due to different methods of preparation. The first gathering of the season is the best, and the last, which consists of large leaves, of an inferior flavor, is the worst. Black tea is exposed to the atmosphere for a considerable time; in this exposure an oxidation takes place, which produces chemical changes greatly modifying the tannin, theine, volatile oil, etc., but the green teas are not exposed to the action of the air in the same way, and the same chemical changes do not occur. They are roasted without fermenting, and are afterwards rolled and dried. Hence the different effect of the green teas upon the nervous system. Pekoe is green tea scented by flowers of the fragrant olive and other plants.

Use. — The leaves are steeped in boiling water, and the decoction is used as a beverage. This beverage has an exhilarating effect upon the system, due to a chemical substance found in it known as theine (see *Chemistry*), which is an alkaloid. It also yields a large percentage of tannic acid, with essential oil. When first introduced into Europe it was looked upon with disfavor and suspicion, and the origin of a number of diseases was traced to its use; but it has overcome all obstacles, and is now the daily beverage of more than 600,000,000 people, who consume over 2,306,500,000 pounds annually, and this quantity is constantly increasing.

Statistics. — The quantity of tea used in the world is amazingly large. In Great Britain alone (mostly in England) about 163,000,000 pounds are consumed annually. In one year 52,424,545 pounds were brought into the eastern ports of the U. S., besides what came from China and Japan to California.

The following table shows the comparative consumption among the great tea-drinking peoples: —

China consumes	2,000,000,000 pounds.
Great Britain	163,000,000 "
United States	52,000,000 "
Russia	26,000,000 "
Holland	10,000,000 "
Dominion of Canada	9,000,000 "
France	6,500,000 "
North Germany	21,000,000 "
Victoria and other British Colonies in the Pacific	11,000,000 "
From Japan more than 4,000,000 pounds are exported	4,000,000 "
and far more is consumed at home, but allowing the same for home consumption in Japan	4,000,000 "
The world's annual consumption amounts to	2,306,500,000 pounds.

Propagation. — The propagation is by seeds. The seeds must be planted as soon as they are ripe, in a moist soil, and as soon as the plants are three inches high, they must be pricked out as the gardener puts out his cabbage or lettuce; when they are six to eight inches high they may be reset in a nursery, six to twelve inches apart, and kept free from weeds. After six to ten months in the nursery they may be planted in the orchard or plantation.

The picking is done by women and children, who twitch off the young leaves and terminal buds with the thumb and finger. They are then carried to the house or shed, where they are spread out on mats, then roasted in pans, rolled in the hands, and dried over a charcoal fire, when they are ready for packing.

This is a brief description of one process. Several methods are in use to accomplish the same end.

Marts. — Canton is the great tea-exporting market for China. Most of the best teas taken into Russia are carried overland. Teas are also shipped from other Chinese ports besides Canton. Imports into Great Britain are chiefly landed at Liverpool; into the U. S., at New York, Boston, and San Francisco. The marketable character of each variety of tea depends upon its purity, time of harvesting, and the perfection of preparation or curing. When these three things are perfect, tea discharges a certain aroma and possesses a peculiar taste. The taste and aroma are so delicate that tea merchants do not trust their own judgment, but employ professional tasters, who command high salaries. These tasters suffer in health on account of breathing and absorbing a volatile oil given off by the tea while in an infused state.

Order IX. MALVACEÆ.

Herbs or shrubs. Flowers regular; sepals 5, united at the base, valvate in the bud; petals 5, hypogynous, convolute in the bud; stamens numerous, monadelphous, and hypogynous; anthers kidney-shaped, 1-celled; pistils several, distinct or united; stigmas various. Leaves alternate and stipulate. Fruit, several-celled capsules, or made up of 1-seeded carpels; embryo of the seed curved.

GOSSYPIUM, L. (Cotton Plant). Calyx cup-like, 5-toothed, encircled by a 3-leaved involucre, the cordate leaflets united at the base, incisely toothed; petals 5, large; styles united; stigmas 3–5; capsules 3–5-celled, many-seeded; seeds brown, immersed in soft, wool-like, white, fibrous hairs, which is the cotton of commerce. Herbs and shrubs.

1. **G. herbaceum, L.** (Herb Cotton.) Stem 5 feet high, clothed with stiff hairs above. Leaves large, cordate, 3–5-lobed below, 3-lobed above, somewhat in form of the grape leaf, with mucronate lobes; leaf-stalk as long as the blade. Flower-stalk longer than the petioles, flowers axillary, yellow, with a reddish center, showy, 3 inches in diameter. Herb.

2. **G. Barbadense, L.** (Sea-Island Cotton.) Leaf has 3 glands on the under side of the midrib. Seed black, cotton very white, fibers long.

3. **G. arboreum, L.** (Tree Cotton). Stem arborescent, 15 to 20 feet high, branching. Leaves 5-lobed, not so broad as those of G. herbaceum. General shape lanceolate; petioles hirsute. Flowers red and showy.

The species of Gossypium are numerous, those described above, with their varieties, are the most important that are under cultivation. As the cotton-plant is propagated from seed it is liable to sport, and a great number of forms or varieties have arisen, differing from the parent in strength, length, or color of the fiber. The cotton fiber of commerce consists of the long silky hairs with which the seeds are clothed; these hairs are tubular, unjointed, flattened, and slightly twisted. When ripe, the seed is gathered, and the hairs and the seed are separated by a machine called a gin; the cotton is then packed in bales for the market.

Geography. — The geographical distribution of the cotton-plant is mostly confined to tropical and subtropical countries, though it has some varieties that have gradually become acclimated to regions of light frost. It is cultivated in a broad belt all around the globe.

GOSSYPIUM ARBOREUM (Tree Cotton).

The cotton-plant will fruit well in the same latitude with the sugar-cane. The East Indies, China, the Asiatic islands, Greece, and the islands of the Eastern Mediterranean, the countries of the Levant, Asia Minor, Northern and Western Africa, Australia, and the isles of the Pacific, the West Indies, Southern United States, Venezuela, British Guiana, and Brazil, are the centers of cultivation.

Etymology. — *Gossypium* is from *Goz*, an Arabian word, signifying *silky*. The specific name, *herbaceum*, signifies herb-like, and *Barbadense* is for Barbadoes. *Arboreum* means tree-like. The word *cotton* is from a Syriac word meaning *fine, delicate.*

History. — It is not known when or where the cotton-plant first began to minister to man's comfort; it is a reasonable inference that it was among the first, if not the very first, of the fibrous plants to attract attention.

MALVACEÆ.

The flowers and the bursting pods are showy and very beautiful, and must always have been objects of admiration. Cotton was unknown to the ancient Egyptians, as no cotton clothes or wrappings have been found in the mummy pits. The seeds of tree-cotton were found by Rosellini in an Egyptian tomb. Herodotus speaks of a plant in India which produced a finer and better quality of wool than that of sheep, of which the natives made their clothing. Five centuries after Herodotus, Pliny describes the cotton-plant, and states that it was under cultivation in Egypt, and that the fiber was used to make the fabrics worn by the priests. Arabian travellers who visited China during the ninth century state that the Chinese did not at that time use cotton fabrics, such as were used in Southern Europe, Northern Africa, and the countries of the Levant, but instead used silk. It is believed that Alexander the Great, about 325 B.C., carried the cotton-seed to the Levant from India, where he found it growing in the country between the forks of the Indus. Another account gives the Arabs credit for its introduction into Egypt, whence it spread into Asia Minor and the islands of the Eastern Mediterranean.

The raw material did not become an article of commerce until many years after the occupation of India by the British; but manufactured cotton goods were imported into Great Britain as early as 1666 from Bombay and other ports of Hindustan. The Dutch, English, and Portuguese merchants all dealt largely in cotton fabrics made in Southern Asia. The first importation of raw cotton into England from the East Indies occurred in 1798. The profit on cotton when it first entered into commerce was five hundred per cent. Napoleon I. during his reign cut off all trade with neighboring nations, and one of the results was an attempt to bring cotton under cultivation in Italy, Southern France, and the island of Corsica.

It has been claimed that a cotton-plant has been found in Mexico entirely different from the Asiatic varieties, growing without cultivation, and that the Mexicans and the Peruvians wore cotton clothing when conquered by the Spaniards, soon after the discovery of the new world. The plant grown now in America was introduced in early colonial times, but did not reach any commercial importance till the beginning of the present century, when about two thousand pounds were shipped to England; from that time the quantity rapidly increased until the outbreak of the Civil War.

In the year 1860, 2,160,000,000 pounds were exported. The quality of the American production is so far superior to all others that it brings in the open market a much higher price than the cotton of India. Next to the United States, India takes the greatest quantity to Great Britain.

Before the power loom and spinners were brought into use, China and India made the cotton fabrics and prints of the world; but now England exchanges the woven fabrics with those countries for the raw material.

Use.—The wool or fiber of the cotton-plant is now wrought into every sort of fabric that enters into the clothing of civilized peoples of tropical and subtropical countries, and that constitutes the *under* garments of people of higher latitudes, and it forms a large part of the attire of females throughout the civilized world. Gun-cotton, a highly explosive substance, is produced by soaking cotton in nitric and sulphuric acids. Gun-cotton treated with sulphuric **ether** gives collodion.

Of the seeds an *oil* is made which rivals the best olive oil for culinary purposes. The seeds, ground and pressed into masses, are sold under the name of oil-cake, and used to feed poultry and cattle, for which purpose they are highly **valued.**

Cotton, treated with antiseptics, forms a very important article of modern surgery. From the roots a fluid extract is made, much used in the Southern United States by irregular practitioners.

Statistics.—The amount of cotton consumed in the world is not easily ascertained. There are about 3,000,000,000 pounds exported annually from Southern Asia, and 2,000,000,000 pounds from the United States, and about the same amount is consumed in those countries; so that an estimate of 12,000,000,000 pounds for the whole world would not be too high. This quantity, at ten cents per pound, amounts to the enormous sum of $1,200,000,000 for the value of the raw material.

Order X. STERCULIACEÆ.

Trees or shrubs, agreeing with Malvaceæ, except that the anthers are extrorse and 2-celled; petals sometimes wanting; capsules united into a 2-5-celled ovary.

THEOBROMA, L. (Cocoa.) Calyx spreading, sepals 5. Petals 5. cordate at the base, extending into a strap. Stamens united at the base, extending upwards into 10 divisions, each alternate one terminated by 2 anthers each, the other divisions sterile. Pistil filiform, divided into a 5-parted stigma. Fruit a 5-angled, elongated, warty capsule, in the form of a cucumber, 5-7 inches long, and 3 inches in diameter, containing from 25-40 seeds. Seeds about three eighths of an inch long and two eighths wide, imbedded in pulp.

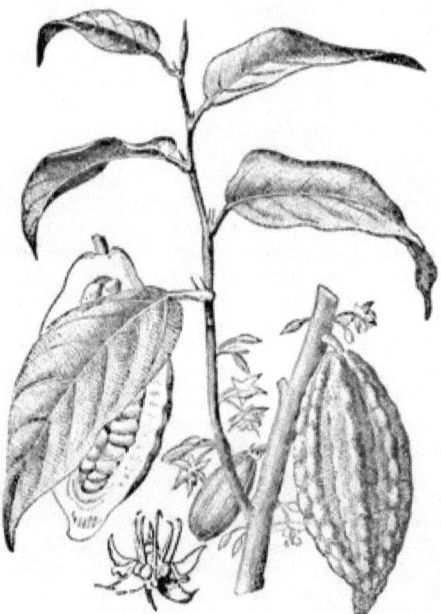

THEOBROMA COCOA (Cocoa).

1. T. **cocoa**, Wallich. Stem upright, much-branched; branches commence about 4 or 5 feet from the ground, and form a symmetrical head. Bark light-brown, smooth; the whole from 10 to 15 feet in height, presenting a beautiful appearance. Leaves oblong, lanceolate, 4 inches in length and 2 in width, entire, dark-green, and evergreen. Flowers, leaves, and fruit after the third year of the plant's life are always present. Fruit harvested twice a year, June and December.

Besides the above-described species, there are some 10 others, all of which are the same description, with the exception of some slight variation in the size and shape of the fruit or leaf, the most important of which are:—

2. **T. angustifolia**, DC. Narrow-leaved.

3. **T. ovatifolia**, DC. A Mexican species called Soconosco, with ovate leaves.

4. **T. bicolor**, Humboldt. A species found in Colombia, S. A., with variegated leaves.

5. **T. Guianensis**, Aublet. A species indigenous to Guiana, S. A.

6. **T. microcarpa**, Mart. A West India species, with small seeds.

7. **T. sylvestris**, Martius. A species found in the selvas of Brazil. As the plant is produced from seed, varieties are constantly occurring.

Geography. — The Theobroma is a native of America, and is indigenous to tropical and subtropical regions, or at least to regions of no frost. It is found in Brazil and all parts of South America north of Brazil, the West India islands, and Mexico. It has been introduced into the Philippine islands and other parts of the Indian Archipelago, but the great supply is produced in the new world.

The Spaniards introduced its culture into southern Europe about the year 1560. Some years ago the British introduced the cultivation of the cocoa into India with success.

Etymology. — *Theobroma* is derived from the Greek Θεός, God, and βρῶμα, food, hence food of the gods. *Cocoa* is supposed to be a contraction of the Portuguese word *macoco*, monkey, applied to the tree on account of the resemblance of the end of the fruit to the face of a monkey.

History. — Cocoa was first brought to the notice of Europeans at the time of the invasion and conquest of Mexico by the Spaniards in 1520. The Mexicans prepare it with spices, as chocolate is now prepared. That prepared for the emperor was flavored with vanilla. When Cortes, the conqueror of Mexico, sent the valuable products of the New World to Charles V., cocoa was sent as one of the choicest. The Spaniards took it to Europe and introduced it into commerce. The medical men found that it possessed curative properties, and a monograph was written by Hoffman in which he entitled it Potus Chocolait. He states that Cardinal Richelieu was cured of a general wasting of the body by its use.

Though used in Spain and Portugal early in the sixteenth century, it was not known in England until more than a hundred years afterwards. The earliest mention of its use was in a periodical known as Needham's Mercurius Politicus, 1659. The mode of preparing it was introduced into England by Dr. Sloane.

A. De Candolle states that, when the Spaniards explored and conquered Mexico, the fruit of the Theobroma was so highly prized that the seeds were used for money. In 1674 the Spaniards carried the plant to the Philippine Islands, where its cultivation became wonderfully successful.

Chemistry. — Various results have been obtained by chemists in the analysis of the bean of the Theobroma, arising no doubt from different conditions or varieties of the bean. In one thousand parts it yields —

Fat (Cocoa-Butter)	510
Albuminoids	210
Starch or Sugar	220
Mineral substances	40
Theobromine	20
	1000

FORMULA OF THEOBROMINE. — C_7, H_8, N_4, O_2. This is the substance that affects the nervous system somewhat as theine and caffeine do.

Comparative analysis with milk (approximate): —

	Milk.	Cocoa.
Fat	35	510
Albuminoids and Caseine	50	210
Starch or Sugar	40	220
Theobromine	0	20
Other substances	15	40
Water	860	0
	1000	1000

This shows that the cocoa bean contains materials to sustain life. Among the mineral substances of Theobroma is Phosphate of Lime.

Use. — The dried and split cotyledons of Theobroma seeds are cocoa nibs, and when ground and made into a paste, they form chocolate. The seeds, when roasted and ground, are cocoa; this when mixed with starch and finely ground, is soluble cocoa.

Like tea and coffee, cocoa is one of the most important and useful articles of domestic economy, and in the formation of warm table beverages stands first among the peoples of Southern Europe, South America, and Southern North America, and forms an article of consideration in commerce in the marts of all the civilized world. The natives of the countries where it grows break the capsule or seed-vessel and suck out the pulp, which has a glutinous, sweetish taste. It is also preserved, and vinegar, spirits, liquors, and jellies are made of it; but the most important part is the seeds, which are roasted, ground, and made into a beverage, as coffee is prepared, and are used in the same way.

Statistics. — The production of cocoa in the New World amounts to about 60,000,000 pounds, worth $7,000,000. In the absence of statistics, the quantities produced in Europe and Asia cannot be arrived at.

Marts. — The markets of South America are La Guayra, Puerto Cabello, Maracaibo, Campano; of North America, Vera Cruz; of the West Indies, Martinique and Guadaloupe.

ORDER XI TILIACEÆ.

Calyx valvate, and falling early; sepals 5; petals 4–5, or as many as sepals, rarely gamopetalous, hypogynous, twisted, imbricated, or valvate in the bud. Stamens some multiple of the number of the petals, free or in bundles: anthers 2-celled. Flowers perfect, with rare exceptions, regular, axillary, or terminal, solitary, or in few-flowered cymes or panicles. Leaves alternate, occasionally nearly opposite, simple, entire, or palmately lobed; sometimes crenulate or dentate, frequently coriaceous; veins prominently reticulate beneath:

TILIACEÆ.

stipules 2. Fruit 2-10-celled, varying from 1 cell by suppression to 10 by false septa.

Trees and shrubs, seldom herbs. No. of genera, 40.

CORCHORUS, L. (Yellow Jute.) Calyx 5-parted, sepals valvate, petals alternating with the sepals; stamens two or three times as many as the petals, nearly all fertile, or a few of the interior ones without anthers, filaments free, anthers 2-valved, opening by longitudinal slits. Style simple; apex broadened, hollow, truncate, and toothed, or crenulate. Ovary 2-5-valved, cells frequently incomplete, the placenta falling short of the center, many-ovuled. Capsule short, subglobular, or elongated, varying to ovate-oblong or subcylindrical, velvety, or clothed with soft bristle-like processes or hairs. Seeds numerous. Leaves simple or compound, alternate, serrate, stipulate. Flowers solitary, or in small terminal and bracteate cymes.

CORCHORUS OLITORIUS (Jute).

1. **C. capsularis**, W. (Jute.) Stem 5 to 10 feet in height, branched. Leaves alternate, acuminate, serrate, palmately compound; leaflets 6, lanceolate, tapering to the base. Flowers terminal and at the ends of the branches, in loose compound racemes, small, yellowish. Capsule globose, wrinkled, 5-celled seeds, few in a cell.

2. **C. olitorius**, W. (Jute.) Stem 5 to 6 feet in height, erect. Leaves alternate, ovate, acuminate, serrate, the lower teeth terminating in thread-like processes. Flowers small, yellow, on a peduncle; sepals 5; petals 5. Capsules subcylindrical, resembling the seed-vessel of a cruciferous plant, 10-ribbed, 5-celled, 5-valved. Seeds numerous.

3. **C. siliquosa**, L., whose fruit resembles a silique, is found in Florida; but the first two species described above furnish the fiber known as jute.

Geography. — Southern belt of the north temperate zone and the tropics.

There are about forty species, all of which are natives of tropical and subtropical countries.

Etymology. — *Corchorus* is said to be derived from the Greek κόρη, the pupil of the eye, and κορέω, cleanse, alluding to the purgative properties of some of the species. (How it applies to the eye is not apparent.) *Capsularis* alludes to the peculiar form of the seed-vessel of this species, and *olitorius* means a garden-plant. Jute is from Sanskrit *djot*, the Indian name for the plant, the meaning of which is obscure.

History. — The home of the C. capsularis is Java and Ceylon, and undoubtedly other Asiatic islands. C. olitorius is a native of western India, and has spread by cultivation to Asia Minor, Africa, etc. It is certain that the Greeks used this plant as a pot-herb, and by many other nations around the shores of the Mediterranean this use of it was common.

DESCRIPTIVE BOTANY.

Cultivation. — A hot, moist climate, an abundant rainfall, and a rich, alluvial soil appear to be the conditions most favorable for the successful cultivation of the jute plants. The land must be well tilled and abundantly manured. The time for sowing the seed in India extends from about the middle of March to the end of May; the seed is sown broadcast in the prepared ground, the young plants are thinned out to 6 inches apart, and the ground is carefully weeded. The stalks are ready for cutting down between the middle of August and the middle of October.

Use. — The fiber of the jute is long, hard, coarse, and glossy, but much inferior to hemp and flax in point of strength. It is cultivated in Southern and Western Asia and in the Grecian Archipelago, and in Central and Northern Africa. In the Levant the C. olitorius is grown for a pot-herb, and eaten for a vegetable with meats. The fiber is obtained by steeping the plant and rotting it, in the manner described for preparing hemp; the fiber is hard and woody. When used to manufacture wearing-apparel, it is worked through a machine and treated with fish oil, which in a measure remedies the evil.

It is used in the manufacture of coarse cloth, gunny bags, sails, and cordage for vessels. It is employed in the adulteration of hemp, and of late years has been applied to the adulteration of silk.

The pulverized bark is an active purgative, and is employed by the Hindus for that purpose. The Hindus also manufacture paper of jute. Theatrical wigs, switches, water-falls, fronts, backs, and bangs are also made of the fiber of this plant.

Statistics. — Jute ranks in commercial and economic importance, as a textile, next to flax, cotton, and hemp. Great Britain imports large quantities annually, the greatest supply, over 500,000,000 pounds, coming from India. Dundee, in Scotland, is the center of jute manufacture.

ORDER XII. LINACEÆ.

Calyx of 5, rarely 4, parts, imbricate in the bud. Petals as numerous as parts of calyx, hypogynous, twisted. Stamens equal in number to petals, and alternate with them. Ovary 5–4 celled; cells 2-ovuled; or ovary spuriously 10–8-celled. Styles 3–5, free. Flowers perfect, regular, in terminal racemes or corymbs. Fruit a globular capsule. Seed compressed. Stem herbaceous, sub-woody.

Herbs. Number of genera, about 14.

LINUM, L. (Flax.) Calyx 5-parted, persistent. Petals, stamens, and styles in 5's, alternating with each other. Seed-vessel 5-celled, each cell partly or entirely separated into 2 cells by a false partition, thus forming 10 imperfect cells; true cells mostly 2-seeded; seeds flattened. Leaves sessile, entire, simple, alternate, occasionally opposite, exstipulate. Herbs. Annual.

1. **L. usitatissimum**, L. (Common flax.) Stem about 3 feet high, slender, tapering, smooth. Leaves alternate, linear-lanceolate, acute, an inch to an inch and a half long. Flowers in a corymbose panicle; sepals ovate, acute; margins membranaceous. Petals subcrenate, large, blue, with a purple tinge, caducous. Seeds compressed, ovate, smooth, and shining, mucilaginous. Annual.

LINACEÆ.

2. L. angustifolium, L.   (Narrow-leaved Flax.) Herb, bears the above description, with narrower leaves. Perennial.

These two species and their varieties furnish the flax of commerce.

3. L. Virginianum, L. (Wild Flax), is a beautiful plant, indigenous to eastern North America; bears the above description except that the sepals are mucronate and the seed-vessel depressed. Hills and edges of woods. Common.

There are nearly a hundred species of the Linum, the inner barks of all of which in a greater or less degree possess the strong, fibrous character of the L. usitatissimum and the L. angustifolium.

Geography. — No plant known to domestic economy has a wider geographical range than the flax. It is found growing luxuriantly in the subtropical bottom lands of Hindustan, in southern Egypt, and throughout Europe to the shores of the Baltic, in southern and middle Russia, and northeastern America.

Etymology. — *Linum* is from the Greek word λίνον, thread. The specific name *usitatissimum* signifies most used, or useful, and *angustifolium* has reference to the narrow leaves of this species. *Virginianum* signifies Virginian flax, or flax found in Virginia, a name once applied to a much larger extent of territory than that occupied by the present State. *Flax* is the Anglo-Saxon name for the plant, and signifies to plait, or weave, referring to the use made of the fibrous bark of the plant.

History. — The home of the varieties of flax under cultivation is believed to be the countries of the Levant and the Mediterranean. Flax was used by the people of western Asia at a period prior to the founding of Babylon. It was well known to the ancient Egyptians. The mummies were wrapped in linen cloth, usually of very fine texture. In the Swiss lake dwellings evidence is found that the L. angustifolium had been cultivated. The stems of the plant used were cut, indicating a perennial species. Similar discoveries have been made in the peat-bogs of Lombardy. The species grown in ancient Egypt was L. usitatissimum. In the Scriptures fine linen is frequently mentioned. In the temple of Minerva at Lindus there was kept a linen corslet of fine workmanship, which had been worn by Amasis, an Egyptian king who reigned 600 years before Christ, each thread of which was composed of 360 filaments. The garment was decorated with cotton and gold. At the beginning of the Christian era linen fabrics were in use not only in Palestine and Egypt, but also in Europe.

LINUM USITATISSIMUM
(Common Flax).

Pliny, in speaking of the flax of Spain, says that it was superior to that produced in other countries, showing thereby that it was under cultivation throughout the south of Europe at that time. It was introduced into England soon after the Norman rule began, but never became a profitable crop there.

Cultivation and Preparation. — Flax requires a damp, moderately fertile soil. It is sown broadcast, about three bushels to the acre, and rolled or harrowed; and just before the seed is ripe, it is pulled, made up into bundles, and sunk under water, to dissolve the gummy substance between the bark and the stalk. After it is removed from the water it is spread out to dry; it is then passed between grooved rollers to break the woody parts of the stem, and beaten with a large, broadsword-like wooden knife to separate the fiber from the broken woody parts. The whole of this preparation at the present day is better performed by machinery.

Use. — The use of flax is well known. It furnishes material for the coarse sails of our shipping, for the cordage with which they are rigged, and for the celebrated Irish linens, and the delicate laces which adorn the ladies' toilet. The ancient Romans did not, as the Italians do now, make great use of linen as wearing-apparel, but they used it for cordage and for the sails of their vessels.

In early colonial times it was raised in the British colonies of North America, and every family prepared, spun, and wove the linen necessary for the beds, table, and underwear for the family. A coarse fabric was made called tow-cloth, which was used for pantaloons and farmers' and teamsters' frocks.

Of the seed is made linseed oil, used in mixing paints. It rapidly oxidizes, and causes a hard, glossy surface. It is an ingredient in printers' ink and in the manufacture of oil cloths. The seeds when boiled are used for drinks for throat and bronchial troubles. A liniment made of lime-water and linseed oil is applied to burns, with great success.

Marts. — Flax for linen fabrics is produced more largely in Belgium than in any other European country. The finest flax in the world is raised in Flanders, where the material for the celebrated Brussels lace is produced. The flax used for Brussels lace is sold for $500-$900 per ton, yielding a greater return per acre than the price of the land upon which it grows would amount to. Flax is also the chief staple of the north of Ireland, Belfast being the metropolis of the linen trade. The great markets for linen fabrics are the large towns of these countries.

Seed for linseed oil is furnished to the world by Russia, Holland, America, and the East Indies.

ERYTHROXYLON. L. Sepals 5. Petals 5, imbricate, each furnished with a double or plaited scale on the inner side of the base. Stamens 10, rarely 12. Leaves alternate. Flowers axillary. Fruit drupaceous. Shrub.

E. coca, Lam. (Coca.) Stem 6 to 8 feet high, ramified into a symmetrical head. Leaf dark-green above, paler beneath; thin, entire, ovate; tapering at each end; strongly veined, two lateral lines extending from the base to the apex parallel with the midrib. Flowers small, white, solitary, on short pedicels. Stamens united at the base. Ovary 3-celled, 2 cells empty, fertile cell 1-seeded.

Geography. — Tropical and subtropical. Found in the northwestern parts of South America, in Bolivia, Peru, Ecuador, and Colombia; also in northern parts of Brazil.

Etymology. — *Erythroxylon*, the generic name, is derived from the Greek words ἐρυθρός, red, and ξύλον, wood, redwood, due to the color of the wood. *Coca* is the Indian name. In the northwestern parts of Brazil it is called *Spadic.* The signification of neither of these names is known.

History. — The practice of chewing the leaves of the coca existed in Peru at the time of the Spanish conquest. How long previously it had been used we have no means of knowing.

ZYGOPHYLLACEÆ. 77

Use. — The dried leaves are placed in the side of the mouth with a little ashes or slaked lime, and chewed until exhausted, the operation being repeated several times during the day. It acts as an excitant to the salivary glands. The saliva, having extracted the properties of the leaves, is swallowed, and produces a pleasurable stimulating effect upon the system, neutralizing the pangs of hunger and thirst, and wonderfully increasing the power of endurance of physical exertion, enabling travellers and burden-bearers to endure wonderful fatigue.

It is believed by some that the chewers of coca are speedily injured in both mind and body by effects similar to those produced by alcohol and opium habits.

Dr. H. H. Rusby, who has recently returned from a visit to Bolivia and other parts of tropical South America, where he has spent some time in the examination of the botanical characters and medicinal properties of the trees and plants of those regions, is eminently qualified to speak of their wonderful qualities. From his article on the coca we gather that the most important medical properties residing in the plant are found in the leaf, and are stimulant, carminative, anæsthetic, and supporting.

Dr. Rusby has had unusual opportunities for watching and studying the effects of the use of the coca, and his observations do not lead him to the conclusion that the constant use of the leaf by the natives either impairs their health or enfeebles their intellect.

ORDER XIII. ZYGOPHYLLACEÆ.

Calyx of 4-5 parts, imbricate in the bud. Petals hypogynous, imbricate or contorted. Stamens double the number of petals, hypogynous; filaments with a scale inside. Ovary several-celled. Flowers perfect, regular or irregular, axillary, 1-2 from an axil. Leaves opposite, pinnate; petioles stipulate, sometimes with spines. Branches frequently divaricate and jointed. Fruit a loculicidal capsule. Herbs, shrubs, and trees.

Number of genera, 17.

GUAIACUM OFFICINALE
Lignum-vitæ).

GUAIACUM, L. (Lignum-vitæ). Calyx 5-parted. Petals 5. Stamens 5 or 10. Fruit a pentagonal capsule, with 5 cells: a single seed in each cell. Tropical and sub-tropical trees.

G. officinale, L. (Lignum-vitæ. Guaiacum Wood.) Stem crooked, 40 feet high and 18 inches in diameter, branching irregularly. Bark gray, with green spots. Root very large, descending very deep into the earth. Leaves compound, with 3 pairs of ovate, blunt leaflets. Flowers axillary clusters, blue, handsome, like hepatica. Sap wood yellow; heart wood greenish-brown; close grain, sinks in water; a gum resin exudes from wounds in the bark, which is the Gum Guaiac of the Materia Medica. The resin yields to the chemist HO, $C_{12}H_7O_5$, Guaiacic Acid.

Geography.—This tree is exclusively American; it has a narrow geographical distribution, indigenous in the West India Islands, but thus far has not been found upon the mainland. Another species, G. sanctum, has been detected in Florida.

Etymology.—The name *Guaiacum* is from the native name, *Guaiac*, which undoubtedly refers to some property of the plant. *Officinale* signifies, of the shops, referring to its sale and use. *Lignum-vitæ*, the wood of life, is named from its medicinal properties, as it is said to preserve life.

History.—The Guaiac is a resinous substance which flows from wounds in the bark, and hardens on exposure to the air and sun heat. It is friable, and of a greenish-red color. The bark and wood also are charged with this resin, and are therefore medicinal. The leaves, fruit, and flowers are all purgative. As it is an American tree its use is recent, but soon after the discovery of America, in 1508, it was obtained from the natives of Haiti by Gonsalvo Ferrand and taken to Europe.

Use.—The wood is very hard, takes a fine polish, and is so heavy and close-grained as to sink in water. It is a favorite wood in the hands of the turner; it is manufactured into rulers, pulleys, ships' blocks, bearings in steam machinery, mortars, pestles, bowls, and vases.

The resin is stimulant and very diffusive in the system, and affects the skin, kidneys, and the intestinal canal very actively. In large doses it is cathartic. It is administered in the form of pills and tinctures.

Order XIV. RUTACEÆ.

Flowers inferior or perigynous. Sepals and petals imbricate, 4–5. Stamens double the number of petals, or equal (in the citrus, numerous and indefinite), inserted on the receptacle or a surrounding disk. Leaves alternate, sprinkled with pellucid dots, containing a bitter, aromatic oil. Fruit in the orange family a berry. Shrubs and small trees.

Genera, 83. Species, 450.

CITRUS, L. Calyx 5-sepaled. Petals 5–10, white, fleshy, fragrant. Stamens numerous and some multiple of 5. Filaments flat at the base and united in sets. Anthers versatile. Style 1. Ovary many-celled, ripening into a pulpy berry. Shrubs and small trees.

1. **C. aurantium**, L. (Sweet Orange.) Stem about 6 to 8 feet to the point where the head begins to form, much branched; branchlets armed with spines, forming a symmetrical tree from 15 to 25 feet in height. Leaves ovate, tapering to a point; petiole winged; blade leathery, dark-green above, lighter beneath, evergreen, articulated to the petiole. Flower white, with a very delicate, fragrant odor. Fruit a berry, globular, or flattened at the poles, bright yellow, two to four inches in diameter, composed of a juicy, edible pulp, divided into 10–13 cells, each cell with 1–3 seeds, the whole inclosed in a bright golden, tough rind. In a state of cultivation the spines on the branchlets are usually wanting.

Var. **sanguinea**. (Blood Orange.) This form has become constant, and differs from the typical plant in the color of the pulp in the fruit, which varies from a light blood-red to a deep dark-red, the branches being usually, but not always, without spines. The author has consulted intelligent fruit-growers and dealers, who all say that the characteristics are perpetuated by budding, and that as the trees grow older the fruit shows a tendency to return to the normal type.

Botanists and naturalists have thus far treated it as an accident, unexplainable, or a *lusus naturæ*, thus leaving the nurseryman and fruit-grower to solve the mystery by experiment and observation.

This form is more frequently taken on when the pomegranate stock is used, but occurs occasionally with regular stocks, and does not always appear when the pomegranate stock is used.

Varieties. — As the orange is propagated from seed, it departs in form from the parent, and forms varieties. In a very complete natural history of the orange family, published by Risso, an eminent scholar of Nice, one hundred and sixty-nine sorts are described, with characters sufficiently distinct to make varieties; these are grouped under eight species. Under the first, C. aurantium, the author arranges forty-three varieties, differing as to qualities, form, or size of the fruit.

An orange-tree in full bearing presents an object of surpassing beauty to the landscape. Conceive a tree with a well-formed, symmetrical head, the branches clothed with a dark-green foliage, besprinkled with delicate white flowers, and dotted all over with bright golden-colored fruit, and you have an object whose beauty is simply enchanting.

CITRUS AURANTIUM (Sweet Orange).

Geography. — The geographical distribution of the orange is very wide. It is to be found in all the regions of no frost, where agriculture and horticulture are practiced. The Arabs carried it from India into western Asia, northern Africa, and southern Europe. The China orange was carried through Persia, Syria, and along the northern coasts of the Mediterranean to southern Europe. The bitter, or Seville orange, went by way of Arabia, along the southern shores of the Mediterranean, to northern Africa and over into Spain. Though the orange-tree has a very broad geographical range, it is a tropical and subtropical plant. It does not ripen its fruit well except where the temperature has a mean above 60° Farenheit. China, southern Japan, India, western Asia, southern Spain, Sicily, northern Africa, Australia, Brazil, West India Islands, Florida and southern California, and the Azores Islands are the chief growing regions.

Etymology. — The generic name, *Citrus*, is derived by some from κίτριον, supposed to be a corruption of κέδρος, a cedar-tree, because the orange, like the cedar, is evergreen. This is a very improbable etymology.

It is held by others that the name is due to the city of Citron in Judea. The specific name, *aurantium*, arises naturally from its golden color, the *Golden Citrus*. The common name, orange, is a corruption of the Latin word *aureum*, golden. The ancient Romans were not acquainted with the orange.

History. — The date of its introduction into Europe is not known with certainty, but it is believed that it came to Spain with the Moors. It was taken into Portugal early in the sixteenth century. Sir Walter Raleigh, who lived in the latter part of the same century, has the credit of introducing the orange into England, in the southern part of which, with careful protection, it flowers, but does not mature fruit. One account of the introduction of the orange into Europe makes it due to the Crusades, which occurred during the twelfth and thirteenth centuries.

The home of the orange has been a subject of considerable research. One account makes it a native of India, and states that it was carried from southern India to Syria and the countries of the Levant by Alexander the Great on his return from the invasion of that country, and that it was thence taken into Europe during the period of the Crusades.

It requires no great stretch of the imagination to picture the foot-sore, famished pilgrim reposing in the shade of the orange-tree, while he cools his parched tongue with its golden fruit, whose enchanting beauty is as startling to his astonished vision as the fragrant fluids of its juicy pulp are delightful to his palate.

Gallesio maintains that the orange was not among the fruits mentioned by Nearchus as seen by Alexander in his invasion of India, and hence infers that it could not at that time have been known in the countries through which he passed. It has been stated as probable that the Arabs carried it from India (where they found it east of the Ganges) to southwestern Asia and northeastern Africa, whence it has emigrated with civilized man into all countries where the climate favors its growth.

A. De Candolle believes its origin to have been in China, and thinks the bitter orange of central Asia may be the ancestral stock, and the sweet orange the offspring, having been obtained from seed in China and Cochin-China.

The orange-tree attains a great age. It has been known to reach the age of 600 to 700 years. In Cordova, the ancient Moorish capital, the broad avenues are skirted by old orange trees 30 feet high, whose heads are frequently 30 to 40 feet in diameter. The larger trees bear from 12,000 to 16,000 oranges in a single crop, and the fruit of the current year is frequently mingled with the flowers for the following crop.

Use and Products. — The orange is a favorite dessert, but is not preserved to any great extent. It is recommended by medical men as a stomachic, taken before breakfast. The bitter orange is very largely used for marmalade (from its rind), to flavor sauces for puddings, etc.

The wood of the orange-tree is close-grained and takes a fine polish, and is used by turners and wood engravers. The stems of young trees when about an inch in diameter are highly prized for walking sticks.

The peel of the sweet orange yields by pressure an *essential oil* of great value for perfumers' use and for flavoring confectionery. It is known in Europe as Essence de Portugal, and in the United States as Oil of Sweet Orange. All that reaches this country is the product of the island of Sicily. It is shipped from Messina in copper canisters of twenty and forty pounds each.

Essential oils, etc., from *C. vulgaris*, Riss. (Bitter Orange). The fruit of this tree is the bitter, or Seville orange. From the peel is extracted the Essence of Bigaradia, called, in America, Oil of Bitter Orange. It is used for flavoring liqueurs and bitters. From the flowers, the Oil of Neroli Bigaradia is obtained. The best is made at Grasse, in the southeast corner of France. It is the chief ingredient of Eau de Cologne. This distilled Oil of Orange-flowers was known to Porta in the sixteenth century. It obtained its present name from the Princess of Neroli, who used it for perfuming gloves towards the end of the seventeenth century.

Orange-flower Water is consumed very largely in Europe in cookery, and sugar saturated with it is eaten by the French and Italians. It is obtained by distilling the flowers with water.

Oil of Petit-Grain, as its name signifies, was originally made by distilling the immature oranges when about the size of large peas, but the oil now known by that name is obtained from the leaves by distillation. It is used as a cheap substitute for Neroli.

The sweet orange yields from the flowers and leaves Oils of Neroli and Petit-Grain, but they are scant in quantity and poor in quality. They are known commercially as Neroli Portugal and Petit-Grain Portugal.

The fruit of a variety of the C. aurantium, known as Curaçoa orange, yields an essential oil of a peculiar flavor; the fruit of C. myrtifolius, the mandarin orange, another.

Statistics. — North America is largely supplied with oranges from Jamaica, the Bahamas, and from the Mediterranean; but the extensive and rapidly increasing cultivation of the tree in Florida and southern California is beginning to supersede the foreign importation.

The climate of Florida is remarkably adapted for orange culture, and orangeries are becoming yearly more numerous and more extensive. In the other Gulf States this industry is pursued to some extent, and in California the orange groves are very productive.

Oil of orange-peel and oil of orange-flower are imported into the United States in great quantities.

Large numbers of oranges are exported to Great Britain from the Azores, from Portugal, and from Spain, Sicily, and other Mediterranean countries.

France consumes great quantities of oranges, a large percentage of them being exported from Algeria.

The loss by decay on European oranges is 37 in each 100; on those from American ports 33 in each 100 (the voyage being shorter).

2. **C. decumana**. (Shaddock.) Tree from 15 to 20 feet in height, forming spreading head, branches armed with prickles. Leaves downy underneath, ovate, somewhat acute, occasionally blunt; wings of the petioles as broad as the leaves and heart-shaped at the base. Flower white; stamens 10-30; petals 5; sepals 5. Fruit a berry of a dull greenish-yellow, form an oblate spheroid, from 4 to 8 inches in diameter. It is said that in Japan it grows to the size of a child's head, weighing 14 pounds. The pulp of those brought to the American market is bitter, though in the varieties grown in

China and Japan it is sweet. As it is propagated from seed the varieties are numerous.

Geography. — The zone of the Shaddock is the same as that of the orange and lemon. It is cultivated in China, Japan, and the islands of the Pacific, and has been carried by Europeans to the West Indies and southern Europe.

Etymology. — The application of the name *decumana*, which signifies "by tens," is not apparent. The common name, *Shaddock*, was given because it was taken from China to the West Indies by Captain Shaddock.

History. — According to the best authorities the home of the Shaddock is the islands south of Asia. It is cultivated in China, where it is called sweet ball, on account of the sweet taste of the pulp. The varieties brought to our market are bitter. The trees are raised from seed, and the Chinese engraft or propagate upon them buds and cuttings of the sweet varieties.

The tree is a beautiful evergreen, with dark, shining leaves always adorned with fruit, and part of the time with both flowers and fruit, — the fruit in various stages of advancement making the most charming object imaginable for the lawn, in all regions of no frost.

The fruit forms an important article of commerce.

3. **C. Limonum**, Risso. (Lemon.) Stem 12 to 15 feet high, branched into a symmetrical head, branches armed. Leaf ovoid or elliptical, dark green on both sides, leathery, entire, petioles winged. Flower white and fragrant. Fruit a golden yellow, from 2 to 4 inches in diameter, and lengthened at the poles, making a prolate spheroidal berry, ending in a short, teat-like process; pulp is divided into cells or compartments from 10 to 20 in number, each containing one or more seeds. In the cultivated varieties the seeds are frequently few in number, many cells being vacant.

Varieties. — Like the orange it sports freely, and there are 38 distinct varieties, all of which possess some prominent quality which commends them to the grower, — the differences being in size, shape, thickness of skin, or intensity of acidity.

Geography. — The zone of the lemon is the same as that of the orange. Most of the lemons that enter into commerce are grown in the south of Europe. The island of Sicily produces excellent lemons, and sends nearly all it produces to the U. S. Large quantities are now grown in Florida.

Etymology. — The specific name of the lemon is from *Lymoun*, the Arabic name for lemon. The common name, lemon, is a corruption of the specific name, Limonum.

History. — The home of the lemon is India and western Asia. It is said to have been known to the ancient Jews. Theophrastus and Dioscorides both speak of it, but it was not known to any extent in Europe before the Crusades, at which time it is said to have been carried into western Europe from Syria, where it had been brought by the Arabs from beyond the Ganges. Writers of the twelfth century inform us that it was common in Italy and Egypt. It has also been stated that it was under cultivation in Italy in the third century.

4. **C. Limetta**, Risso. (Lime.) Small tree, 8 to 10 feet in height, trunk crooked or zigzag, forming a dense, spreading head 6 to 8 feet in diameter, branches armed with sharp, strong spines, or prickles; some of the petioles winged. Leaves ovate-orbicular, serrate-toothed or entire, dark green above, lighter underneath. Flowers white; petals 5. Stamens sometimes 30 in number. Fruit a globular berry, protruding at top, an inch and a half in

diameter, dull yellow; skin thin; pulp sharply acid, divided like other species of the genus into a number of cells, containing 1 or more seeds each.

Varieties. — Less attention has been given to the cultivation of the lime than to that of the orange, hence the varieties are less numerous than of any other species of this genus.

Geography. — The geographical zone of the lime is subtropical and tropical. It arrives at the greatest height of fruit-bearing in a temperature not below 70° Fahrenheit. It grows well in southern India, and in the West India Islands, in southern Europe, and in northern Africa.

Etymology. — The specific name of the lime, *Limetta*, is said to be derived from the city of *Lima*, near which it was largely cultivated, and the common, or English name, *lime*, is an abbreviation of the botanic name. This history of the name is not to be depended upon.

History. — The home of the lime is the same as that of the other species of this genus, India east of the Indus, whence it was carried into western Asia, southern Europe, and northern Africa by the Arabs; but as it is not edible, its culture for fruit was not prosecuted. Much attention has been paid to its cultivation of late in the British West Indies. It grows and fruits well in moderately fertile soil, needs but little care, and lives to a great age.

Use. — The lime is used for its juice, which is the material for the manufacture of citric acid. It is also useful for setting or fixing dyes. The juice is more acid than the juice of the lemon, though by some it is considered more palatable. It is an important article of commerce between Portugal and England, countries of central Europe, and the United States. In the West Indies it is cultivated for hedging as well as for its fruit.

Statistics. — Lime juice is manufactured in large quantities in the British West Indies. Nearly 13,000 gallons are annually exported from the island of Domingo alone. The ports of southern Spain export large quantities of the fruit as well as of the juice. It is consumed in France, Germany, England, and the United States.

5. **C. Medica,** Risso. (Citron.) Fruit oblong, 6 inches long; rind thick.

Order XV. **MELIACEÆ.**

The characters of this order are like those of Rutaceæ. Leaves alternate and pinnate. Stamens united, forming a tube, the introrse anthers sessile on its top. Leaves rarely dotted. Mostly trees.

Number of genera, 37.

SWIETENIA, L. Calyx 5-cleft. Petals 7. Stamens 10, united into a tube with 10 teeth, inclosing the anthers. Style short. Stigma 5-rayed. Leaves alternate, even-pinnate. Leaflets opposite, entire, ovate-lanceolate, unequal at the base. Flowers greenish-yellow, in axillary panicles. Fruit woody, pear-shaped, 2 to 4 inches in diameter, 5-celled, 5-valved; seeds numerous and winged, imbricated in two rows. Large tree.

S. mahogani, L. (Mahogany Tree.) Stem 80 to 100 feet high, and 5 to 8 feet in diameter. Irregularly branched. Leaves compound, with four pairs of leaflets, green, shining, and about two and a half inches long. Flowers small, in a thyrse, yellowish-white.

This is the only species of the genus.

Geography. — The mahogany is tropical or subtropical. It is indigenous to the West Indies, the Bahamas, Central America, and to southern Florida. It has been planted and is successfully growing in southern British India.

SWIETENIA MAHOGANI (Mahogany).

Etymology. — *Swietenia* was the name given by Jaquin, in honor of Gerard L. B. Van Swieten. *Mahogani* is the name by which the tree is known to the aborigines of South and Central America, but the signification of this is unknown.

History. — This beautiful wood was introduced to notice in the latter part of the sixteenth century. A few planks were sent to Dr. Gibbon of London

MAHOGANY.

by his brother, a sea-captain, sailing to the West Indies. At the time he received them he was erecting a dwelling, and gave the planks to his joiner to finish some part of the house; but the workmen refused to use them on account of the hardness of the wood. The cabinet-maker was then ordered to construct a candle-box of a part of one of the planks, which, when finished, so far exceeded in fineness all Gibbon's other furniture that it became an object of notice and wonder, and was placed upon exhibition. Soon after this it became a favorite material for the construction of furniture, and very speedily found its way into the dwellings of the wealthy classes.

Mahogany is to this day the favorite wood for cabinet ware. The tree was first noticed as a native of the territory of the United States by Dr. Muhlenberg. It is found on the Florida Keys, reaching the height of 50 to 90 feet. That which furnishes the fine curls and apparent interlacing of the fibers, used for veneers, grows on rocky hillsides, and is from that part of the tree where the branches join the trunk. It is said to attain a great age. Sir William Hooker counted 200 rings in a block, but the rings may not each have denoted a year.

Use.—Mahogany is one of the best woods known to the cabinet-maker. The finest is sawn into thin slices and used for veneers. It is worked into chairs, tables, cabinets, desks, bureaus, bedsteads, and other furniture, and is much more common in Europe than with us. Its value depends upon its hardness, and its ability to take a high polish. It is not liable to shrink and warp. It was formerly used in naval architecture, and ranked by the English as second to the celebrated English oak for that purpose. On the Bay shore in Central America the wood of the mahogany, which is coarser grained and softer, is used for inferior purposes, such as common tables, wainscoting, flooring, the making of cigar-boxes, etc., and is called Bay Wood.

There is a bitter principle residing in the bark which is efficacious in remittent fevers, and the bark is used by the natives in the same way and for the same purposes as the cinchona bark.

Statistics.—Great quantities are imported annually by England and by the United States. A single log has been known to sell for $5,000; this when sliced up into material for veneers was made to cover an immense surface.

Order XVI. ILICINEÆ.

Sepals 4–6, imbricated in the bud, small. Corolla 4–6 cleft, hypogynous, imbricated in the bud. Stamens inserted in the throat of the corolla, alternate with the segments; anthers adnate. Flowers perfect, small, solitary, or grouped in the axils of the leaves. Fruit a drupe. Leaves alternate or opposite, simple, leathery, glabrous, shining, and without stipules. Trees and shrubs.

Number of genera, 3.

ILEX, L. (Holly.) Calyx with 4 or 5 teeth. Corolla 4–5-cleft, wheel-shaped. Stamens 4–5, alternating with the segments of the corolla. Ovary sessile, with 4 stigmas. Berry with 4 or 5 one-seeded nuts. Small tree.

I. Paraguayensis, Lamb. (Paraguay Tea, or Yerba Maté.) Stem from 5 to 15 feet in height and 3 to 6 inches in diameter. Much branched. Leaves alternate, oval, crenate, glossy, leathery, 4 inches long, evergreen, dark green above, paler underneath. Fruit a berry, bright red, small and smooth.

There are many species to this genus, and as it is propagated from the seed it has many varieties. The I. Paraguayensis is very constant, and is the only species whose leaves furnish tea.

Geography. — This plant is indigenous to Paraguay, and forms entire forests, extending over large tracts throughout the central, eastern, and northern parts of the republic. It is also found under cultivation, but no observations have been made to justify a statement as to the comparative value of the cultivated plants. The Jesuits, previous to their expulsion from Paraguay in 1867, gave great attention to its cultivation, and instructed the natives in its preparation. It grows also in Paranagua, but that found in Paraguay yields four times the strength. The Government monopolizes the sale, buying cheap and selling at a high price, thus securing a large revenue.

ILEX PARAGUAYENSIS (Paraguay Tea).

Statistics. — Europeans do not usually relish its peculiarly bitter taste. The inhabitants of South America, on the other hand, prize it highly, and have come to consider it, not as an article of luxury, but as one of necessity. In harvesting, the work is so carelessly and slovenly done that the plant in a wild state is rapidly undergoing destruction. Its increasing demand will bring about a more careful method of harvesting. In Parana about 10,000,000 pounds are annually produced. About 6,000 persons are employed in Brazil in the preparation of Maté; and about 3,000,000 pounds are shipped annually at Itaguy, a town on the Uruguay.

Etymology. — *Ilex* received its generic name from the resemblance of its leaf to that of the Quercus Ilex. The specific name is from the country where it is found indigenous (Paraguay). *Holly*, the common name of the genus, is a corruption of the word *holy*. Holly bush, or Holly tree, — that is, *holy-tree*, because the evergreen leaves of the Ilex opaca are used for decoration at the holy time of Christmas.

Chemistry. — The leaf of the I. Paraguayensis yields to chemical analysis about the same percentage of Theine as the leaves of the Thea viridis.

Preparation and Use. — The small branches are cut or broken from the tree; they are thrown upon a wire gauze over a fire and thus dried; the leaves and twigs are then bruised or ground in a rude mill, which reduces the whole to powder. It is then packed in green bullock-hides, making packages of about 200 pounds each, and thus sent to market.

Marts. — It is carried to Itaguy, on the Uruguay, and to the ports of Porto Alegre, Uruguay, and Paraguay. From the last three ports about 13,000,000 pounds are annually exported.

Order XVII. **RHAMNACEÆ**.

Flowers perfect, or defective, regular, perigynous. Sepals, petals, and stamens 4–5, frequently apetalous; stamens alternate with the valvate sepals, and opposite to the petals. Fruit a 1-seeded capsule. Trees and shrubs.

Number of genera, 37.

RHAMNUS, L. (Buckthorn.) Calyx pitcher-shaped, from 4–5-cleft. Petals notched, sometimes wanting. No. of styles, 2–4, partially united. Drupe with 2–4 cartilaginous nuts. Leaves opposite or alternate. Flowers in clusters, axillary. Shrubs and small trees.

R. infectorius, L. (Yellow Berries.) Stem made up of ramifications commencing to branch at the root, subprocumbent; the assurgent, spiny branches rising to the height of 2 feet. Leaves ovate-lanceolate, serrulate, smoothish, and deciduous. Flowers diœcious; petals in both sexes greenish-yellow. Fruit 3-seeded, black when ripe, but harvested when green, to be used for a dye by calico printers.

RHAMNUS INFECTORIUS (Yellow Berries).

Geography. — This plant grows as far north as England, though it properly belongs to a more southern climate. Its native haunts are southern Persia and southern countries of the Levant. The fruit is known in commerce also as *Persian berries*.

R. chlorophorus and R. utilis are natives of China. Their fruit yields a fine green dye for silk, which is imported into Europe under the name of Chinese Green Indigo.

Etymology. — *Rhamnus* is derived from the Celtic word *ram*, a bunch or tuft of branches. The specific name comes from the Latin word *inficio*, color, tinge, or paint. *Buckthorn*, the common name of the genus, is not easily explained. Indeed it is difficult to comprehend why the name is applied to the plant. Some suppose it is due to its crooked stems, which resemble the horns of the buck. It is also known under the name of Yellow Berries, on account of the color produced by the fruit.

History. — The home of the Rhamnus infectorius, the *Staining buckthorn*, is Persia, Syria, and southern Europe. It is cultivated in Turkey and the

Levant, and in the south of France. The largest and best berries come from Persia.

Use. — The berries furnish a yellow dye, which is employed in the dyeing of morocco in Turkey, and in calico printing. It also produces, by the aid of chemicals, the Sap green and the Dutch pink.

Marts. — The Persian berries come from Aleppo and Smyrna. Some are also shipped from France and Turkey. These ports ship to England annually 1,200,000 pounds, but more are used in the countries where they are produced.

CEANOTHUS, L. (New Jersey Tea. Red Root.) Calyx a semi-globular tube, with 5 segments. Petals clawed, rolled in at the edges, bending down. Stamens with ovate, 2-celled anthers. Styles 3, diverging. Stigmas papillate. Fruit tricoccous. Small shrub, 1 to 3 feet high.

CEANOTHUS AMERICANUS
(New Jersey Tea).

C. Americanus, L. (New Jersey Tea.) Stem 1 to 3 feet high, 1 inch in diameter. Leaf ovate-acuminate, serrate, pubescent beneath, alternate, stipulate. Flowers in a thyrse, axillary.

Geography. — Geographical limits, eastern North America.

Etymology. — Ceanothus, is derived from κεάνωθος, a name given by Theophrastus to indicate a plant with spines, from κέω, prick, split, or cleave. It does not apply well to the American plant, which is without thorns. The specific name explains itself. The common name, *Red Root*, is given on account of the color of the roots, and *New Jersey tea* is a name given because it is said that the leaves were used in New Jersey, during the time of the Revolution, for *tea.*

Use. — The young leaves are collected and dried in the shade, when they are said to furnish a beverage superior in flavor, and resembling the China teas. It has become an article of local commerce in some parts of Pennsylvania.*

ORDER XVIII. **AMPELIDEÆ.**

Woody, climbing by tendrils. Calyx small; petals 5; stamens 5; stigmas sessile and capitate; filaments distinct, or slightly cohering at the base. Ovarium 2-celled; fruit a pulpy berry; seeds 4-5, bony; leaves simple, alternate, stipuled, palmately veined; tendrils opposite the leaves. Trees or shrubs, usually climbing by tendrils. Small number of genera. (The term *vine* is used widely in America for every climbing or trailing plant.)

VITIS, L. (Grape Vine.) Petals deciduous, spreading, or attached at the top. Calyx very short, entire, or obscurely 5-toothed.

* The young leaves of the Rubus strigosus (common red raspberry) when dried are used for tea, and are said to be far superior to the common grades of the imported article.

AMPELIDEÆ.

Although the genus Vitis comprises more than 200 species, mostly natives of tropical and subtropical regions, less than half a dozen species have any economic value.

1. V. labrusca, L. (Fox Grape.) Stem trailing or climbing into trees, reaching sometimes the enormous height of 100 feet, branching freely. When in the borders of woods, it sends out lateral branches, covering the shrubbery and small trees for long distances. The main vine sometimes attains a diameter of 6 inches or more; bark rough, sloughing off in strips. Leaves very large, 3-lobed; when young, white and downy beneath. Flowers in compound panicles, green, and opposite the leaves. Fruit globular, three-quarters of an inch in diameter.

Species. — There are 6 other species growing in North America, but most of the varieties now under cultivation in the United States, amounting to several hundred, are seedlings of the V. labrusca or its offspring.

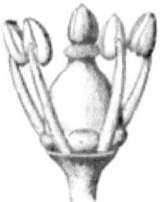

VITIS LABRUSCA (Fox Grape).

2. V. vinifera, L. (European Wine Grape.) Habit of stem as V. labrusca. Leaves cordate, 5-lobed, deeply sinuate, serrate, downy underneath. Flowers greenish-yellow, odorous. Fruit green, red, or black. This is the wine grape of Europe, from which hundreds of forms have been obtained. All the varieties of the grapes raised under glass in America are seedlings of this vine or of its progeny. There are also hardy varieties from it grown in the United States. Zante Currants, the name of which is supposed to be a corruption of Corinth, from the city of that name whence they were formerly shipped, are a small variety of the V. vinifera. They form a considerable article of commerce.

Geography. — The zone of the grape extends from about 21 degrees of north latitude to 48. In the old world this belt trends east from the British Isles and Portugal to Persia, but the fruit ripens to greatest perfection along the middle and southern parts of the belt. In the new world the zone does not extend above the 43d degree north latitude, reaching south to 33 degrees, trending east and west from the shores of the Atlantic to the Pacific coast. The centers of growth are the middle Atlantic states and California.

VITIS VINIFERA (European Wine Grape).

In late years the grape has been largely cultivated in Australia.

Etymology. — The generic name, *Vitis*, is derived from the Celtic word *Gwid*, signifying "the best of trees." This origin is disputed, and it is said to be derived from an Aryan root, WI, signifying to bind, or twine. The specific name, *labrusca*, is a Latin word, signifying "wild," and the name *vinifera* is from two Latin words, *vinum*, wine, and *fero*, bear, whence *wine-bearing* plant. The common name, *grape*, is from the old French *rappe*, a bunch or cluster.

History. — The grape was known to man at a very early period of his existence. Mention of this fruit comes down to us among the first records of history. Pictures left by the ancient Egyptians, and remains by the lake-dwellers of Italy and Switzerland, from the bronze age, reveal to us that the grape was known to those peoples. It was cultivated in Egypt as early as 4000 B. C.; not in China until 122 B. C. The Egyptians and the Greeks both have a mythological history of the introduction of the grape.

The ancient Scriptures frequently speak of the grape. The weight of history points to western Asia as the home of the species known as the Vitis vinifera, but there seems little doubt that it is also native to Europe. It is supposed to have been introduced into England by the Romans, yet it was not known there during the time of Agricola, who was governor in the reign of Claudius Cæsar, about 78 years after the beginning of the Christian era. But among the earliest conveyances of estates under the Saxon rule vineyards are named. And in the reign of William the Norman vineyards are among the valuables enumerated in the Dooms-Day Book. It is also recorded that in the twelfth century there was a vineyard attached to every monastery in the south of England.

The grape is indigenous to America, and much attention has of late years been given to its cultivation. In the United States it has become a very large and important industry, and considerable wine is made in Ohio, Indiana, New York, and in California.

The best wines and brandies are made in southern Europe. Southern Germany, Spain, Portugal, Italy, and Madeira are wine-making countries; but France is the great wine-producing nation. More than 500,000,000 gallons are produced annually in France alone. In Spain and Portugal much attention is paid to the production of raisins, which are the sweet grapes of these countries dried. Along the southern shores of Spain and Greece the grape forms sugar in abundance, but does not easily enter into vinous fermentation, hence the fruit is dried; and the raisin crop is one of great economic and commercial importance.

Cultivation. — The vine requires a high summer temperature and a prolonged period in which to ripen its fruit. When these conditions exist it can be profitably cultivated, even though the winter temperature be very low. In the Alps it is profitably cultivated to an altitude of 1870 feet, and in the north of Piedmont as high as 3180 feet.

Use. — The grape is among the most delicious and highly prized fruits of the temperate zone. It is a great favorite as a dessert, and for a table ornament has no equal; it is also preserved, and is a favorite jelly material. The raisin, or grape in a dried state, is also a favorite dessert. In central and southern France and Italy the chief use made of the grape is for the manufacture of wine. The common wines are cheap, but the finer sorts cost several dollars a bottle.

Argols is the commercial name for the crude deposit from wine, which forms on the bottom and sides of wine casks, from which Cream of Tartar, or Bitartrate of Potassa, and Tartaric Acid are obtained.

Diseases of the Vine. — The organic diseases which affect the vine may be divided into two categories, — those caused by parasitic fungi, and those caused by insects. In 1849 the vine in France, southern Germany, Italy, and Spain was attacked by a mildew fungus, *Oidum Tuckeri*, which in many localities destroyed the crop. After some years this blight disappeared, and a new pest made its appearance, in the character of a tiny insect, the *Phylloxera vastatrix*, which lives and propagates upon the roots of the vine, and in a very short time destroys the plant. There are certain symptoms of the disease caused by this insect by means of which an infected spot may be readily recognized. The vines become stunted and bear few leaves, and those are small. When the disease reaches an advanced stage, the leaves are discolored yellow or reddish, with the edges turned back or withered. The grapes are arrested in growth, and their skin is wrinkled.

Plants grown in California possess the power to resist this enemy better than those grown in Europe, and are frequently used for stocks upon which to graft the European plants.

Marts. — The great markets for export are Marseilles, Bordeaux, and Havre, in France; Genoa and Naples in Italy, — but mostly Malaga and Lisbon. The import markets are Liverpool and London, in England; the north German ports, in northern Europe; New York and Boston, in the United States; and Quebec, in Canada.

Order XIX. SAPINDACEÆ.

Flowers unsymmetrical. Stamens sometimes twice as many as calyx-lobes or sepals, usually fewer, or equal, alternating with the petals in the Maple family (sub-order *Acerineæ*). Flowers usually polygamous or diœcious, sometimes without petals. Ovary 2-lobed and 2-celled, 2 ovules in a cell, maturing one seed in each cell. Fruit winged, cotyledons crumpled in the embryo. Leaves opposite, exstipulate, lobed. Trees and shrubs.

Genera, 73.

ACER, L. (Maple.) Calyx 5-parted. Petals 5, sometimes 4-12. Stamens 4-12. Anthers 2-lobed, seeds 2, sometimes 3, in a 2-winged vessel or samara united at the base, wings diverging. Leaves simple, opposite, lobed. Trees. Flowers axillary, in corymbs.

A. saccharinum, Marsh. (Sugar Maple. Rock Maple. Bird's-eye Maple.) Stem 50 to 80 feet high, 1 to 2 feet in diameter; branches erect; head symmetrical. Leaves cordate, smooth, glaucous beneath, green above, 5-lobed; lobes acuminate, coarsely toothed or sublobed. Flowers small, yellowish, on long, slender peduncles. Samaras brown when ripe. Flowers in May. Fruit ripe in September.

Var. **nigrum** (Mx. f.). Britt. (Black Maple.) A. nigrum, Mx. f. (Sugar Tree.)

This species or variety is like the A. saccharinum, with darker leaves, sometimes nearly peltate; bark rough. The sap is as rich in sugar as the sap of A. saccharinum.

The leaves of both these species are about 5 inches wide, and from 5 to 7 inches long when the tree is young, shorter as the tree grows older, palmately or unevenly divided into 5 lobes; edges coarsely toothed.

Geography. — The geographical range of the sugar tree is not great. It does not flourish south of 38° N. latitude except in high mountains. It abounds in the northern parts of the United States and in southern British America.

DESCRIPTIVE BOTANY.

Etymology. — *Acer* is a Latin word, signifying sharp, and is supposed to have been applied to the maple tree because it was, on account of its hardness, used for spears. *Saccharinum*, the specific name, is from the Latin word *saccharum*, sugar, due to the sugar-bearing sap. *Nigrum*, name of the variety, is a Latin word signifying black, due to the dark foliage.

History. — When or where the sap of the maple was first used for the manufacture of sugar is not known; but we have no record that sugar was made from this tree till after the colonization of northeastern America. It is therefore probable that its manufacture was begun by the early settlers of the French and British colonies of this continent. At the present time about 10,000,000 pounds are exported from Canada; allowing 5,000,000 pounds for home consumption would make the amount produced about 15,000,000 pounds. In the United States the production is about 30,000,000 pounds, which makes an aggregate production of 45,000,000 pounds. The sap flows from the tree through wounds made in the trunk near the ground, into which are inserted

ACER SACCHARINUM (Sugar Maple).

tubes; it is caught in pails or tubs and placed in large pans, in which it is evaporated by heat to a syrup. A tree will yield from 2 to 4 pounds yearly, and will continue to do so for 40 years without suffering injury. The trees are tapped early in spring, when the sap is ascending. The boiled sap is used as molasses under the name of maple molasses or syrup. By further evaporation, straining, and refining by boiling with it lime, milk, and eggs, a white sugar is produced of a very delicate flavor.

Use. — Maple Sugar is used for the same purposes as the cane sugar, and when purified by the ordinary modes of refining, it has much the same character; but when used without refining, it has a smoky taste, which is grateful to most palates.

The sap of the Sugar Maple has been for more than a century used for the manufacture of sugar.

The wood of the Sugar or Rock Maple is also of very great value. Wheelwrights use it for axles of carriages. It constitutes a large part of the material used for school furniture, bedsteads, table-legs, and chairs. It also furnishes to the cabinet-maker the beautiful curled, or bird's-eye maple, and is excellent for fuel.

Order XX. ANACARDIACEÆ.

Flowers perfect, diœcious, or monœcious, regular; small, spiked, or panicled. Sepals 3-5, united at the base. Petals 3-5, sometimes absent, imbricated. Stamens alternate with the petals, and same number, perigynous. Ovary 1-celled, free. Stigmas 3. Fruit a berry or drupe, 1-seeded. Trees and shrubs. Number of genera, about 16.

RHUS, L. (Sumach.) Sepals connected below, small and persistent, 5 in number; petals free, ovate, spreading from the margin of a rounded disk; stamens 5-10 or wanting, inserted on the disk; styles 3, sometimes united; stigmas subcapitate. Fruit, a small dry nut, hard and globose, 1-celled. Shrubs with alternate, compound, or simple leaves.

1. **R. glabra**, L. (Common Sumach, Smooth Sumach.) Stem 3 to 12 feet high, much-branched, forming a flattish top; branch and leaf-stalks smooth. Leaves of 8-15 pairs of sessile leaflets, and a terminal one which is petioled; upper side dark green, under side lighter. Flowers yellowish-green, frequently abortive, in densely crowded panicles. Fruit a little drupe, covered with a crimson down which is charged with malic acid, sour but agreeable to the taste. Flowers appear in June and July, fruit in autumn. The color of the leaves in autumn is a rich crimson.

2. **R. typhina**, L. (Stag-horn Sumach.) Stem reaches the height of 20 feet; leaflets serrate; otherwise as in R. glabra.

3. **R. copallina**, L. (Mountain Sumach.) Stem from 5 to 12 feet in height, much-branched; leaflets 4-10 pairs, with a terminal one unequal at the base; the common petiole margined with a wing between each pair of leaflets; otherwise like R. glabra.

4. **R. venenata**, DC. (Poison Sumach.) Trunk 10 to 15 feet high, tree-like, 3 to 5 inches in diameter, branching so as to make a spreading top. Leaves of 3 to 6 pairs of leaflets, with a terminal one, deep green, shining above. Flowers in panicles, small and green. Fruit a drupe, the size of a pea; juice poisonous, producing an eruption of the skin, accompanied by swelling.

5. **R. Toxicodendron**, L. (Poison Ivy. Poison Oak.) Stem trailing or climbing, vine-like, 10 to 50 feet long, fastening itself to the trunk of trees by rootlets. Leaves green, shining, in threes, terminal leaflet pointed. Flowers racemed in axillary panicles, greenish. Fruit a dull-white berry; juice poisonous, and forms an indelible ink. When growing without support, it assumes the form of a little tree.

6. **R. Cotinus**, L. (Venetian Sumach. Smoke Tree.) Stem 6 to 8 feet high, irregularly and stragglingly branched. Leaves alternate, simple, obovate, entire, conspicuously veined, veins nearly at right angles to the midrib, stiff and translucent, on long petioles. Flowers very small, purplish, in loose panicles, pedicles of abortive flowers lengthen and become hairy after blooming; groups of these feathery pedicles give the plant at a distance the appearance of a fleecy cloud. Fruit white.

7. **R. coriaria**, W. (Hide-tanning Rhus.) Stem 6 to 10 feet in height, dividing near the root into an irregular ramification; bark hairy, and of a brown color. Leaves compound alternate, in 7 or 8 pairs of leaflets, and a terminal one yellowish-green and hairy on the under side. Flowers in terminal, loose panicles, greenish-white.

Geography. — All the above-described species are indigenous in North America, except R. coriaria. The American species are common throughout the northeastern parts of North America, from Canada to the Gulf States. The Cotinus is found in Arkansas, and is identical with the European species, which is indigenous from the Levant to Western Europe. The Coriaria is indigenous in Syria, but has been introduced into Sicily, Italy, and Turkey, also into Spain and Portugal, where it is carefully and extensively cultivated.

Rhus Cotinus (Venetian Sumach).

The bark of all the American species is highly charged with tannin. The Coriaria is especially rich in this material, and is so highly prized as to have found a market in America.

Etymology. — *Rhus* is from the Greek word ῥοῦς, an old name, the signification of which is not known. It is also supposed to have been derived from the Celtic word *rhudd*, signifying "red," due to the color of the fruit. *Sumach* is supposed to come from the Arabic *summage*, a shrub. *Glabra* signifies "smooth," due to the smoothness of the leaves. *Typhina*, giant, on account of the size. *Copallina*, connected, from the winged petiole by which the leaflets are united. *Venenata*, from *venenum*, poison. *Toxicodendron*, from the Greek τοξικόν, poison, and δένδρον, a tree, hence poison-tree. *Cotinus*, ancient name, signification obscure. *Coriaria*, from *corium*, a hide, referring to the use of the bark and leaves of this species in tanning hides.

Use. — The bark and leaves of most of the species of Rhus are charged with tannin of a superior quality. The R. glabra in America and R. coriaria in Europe are especially rich in this material, which is used in making the fine moroccos.

The fruit of the R. glabra is used by the Thompsonian practitioners as a remedy for canker, sore mouth and throat. The wood and fruit of the other species are used for dyes and inks. The juice of R. Toxicodendron produces an indelible ink. R. venenata is very poisonous, causing an inflamed eruption of the skin. R. Toxicodendron produces similar effects of a milder character.

Marts. — About ten million pounds are carried from the continent to Great Britain annually. It sells for four dollars a hundred.

ANACARDIUM, W. (Cashew Nut.) Calyx 5-toothed; corolla 5-parted; stamens 5, styles 3. Fruit a kidney-shaped or heart-shaped nut, on the end of a pear-shaped, fleshy peduncle, which is edible.

A. occidentale, W. Trunk branching a few feet from the ground, ramifying into a beautiful second-class tree. Leaves elliptical, green, leathery, alternate, obcordate, or deeply emarginate. Flowers in a loose corymbose panicle, red and fragrant. Fruit of the size and somewhat of the shape of a rabbit's kidney.

Geography. — The Anacardium is a native of the tropical regions of both Asia and America; flourishes in Jamaica, and is cultivated for its fruit, and also used in planted grounds for ornament in that island.

Etymology. — *Anacardium*, the generic name, is

ANACARDIUM OCCIDENTALE (Cashew Nut).

from the Greek ἀνά, like, and καρδία, heart, heart-shaped, due to the form of the fruit. *Occidentale* is the Latin word for west, or belonging to the western continent.

Use. — The fleshy stem or the apple is eaten as it is plucked from the tree; it has a slight acid taste, and an agreeable flavor. The juice produces a delicate wine; the wine distilled produces a liquor far superior to rum, used for disease of the kidneys, and for a beverage, in mixing punches, etc. The dried and broken kernels are used for flavoring Madeira wine.

The cotyledons are inclosed by a double covering, or by two separate shells; between these shells a thick oily substance forms, which is inflammable. It is also very caustic and blisters the skin. For this reason it has been applied by practitioners for eating away corns, ulcers, ringworms, and even cancers.

The kernels, when fresh, are eaten raw; they are also used for making puddings, and they form an ingredient in custards, etc. When older, the nut is roasted and eaten as chestnuts are; it is also roasted and ground with cocoa in the manufacture of chocolate. By tapping, a milky juice is also obtained, which makes an indelible black ink. A gum, which possesses the character of gum arabic, is also obtained by wounds made in the bark.

DESCRIPTIVE BOTANY.

Order XXI. **LEGUMINOSÆ.**

Flowers irregular or regular, perfect. Sepals 5, more or less united, unequal, the odd one anterior. Petals 5, odd one posterior. Stamens distinct, or 9 united, and one (the posterior one) free. Ovary superior, single and simple. Style and stigma simple. Fruit a legume. Leaves stipulate, usually compound. Herbs, shrubs, or trees. Number of genera, about 400.

INDIGOFERA, L. (Indigo.) Calyx with 5 acute segments; vexillum roundish and emarginate; keel with a spur on each side, at length reflexed; legume 2-valved, and 1- to many-seeded. Shrubs.

I. tinctoria, L. (Indigo Plant.) Stem 2 to 3 feet high, and from a quarter to half an inch in diameter, subligneous, branching. Leaves compound, consisting of about 6 pairs of longish bluish-green leaflets, and a terminal soft one, darker above; flowers in axillary racemes, papilionaceous and pale-red. July and August.

Species. — There are about 150 species of this genus, most of which yield indigo, but the species that are found to be most productive are —
1. **I. tinctoria**, L. (Coloring Indigofera.)
2. **I. argentea**, N. (Silver-leaved Indigofera.)
3. **I. Caroliniana**, L. (Carolina Indigofera.)

Besides the 150 species of A. De Candolle, there are as many or more *varieties*, most of which yield this dye.

The following plants also contain indigo:
Nerium tinctorium, Rottb. (Apocynaceæ.)
Tephrosia tinctoria, L. (Leguminosæ.)
Tephrosia apollinea, L.
Tephrosia toxicaria.
Polygala tinctoria, Persoon. (Polygalaceæ.)
Polygonum Chinense, L. (Polygonaceæ.) China.
Polygonum tinctorium, L. China and Japan.
Polygonum barbatum, L.
Polygonum perfoliatum, L.
Polygonum aviculare, L.
Wrightia tinctoria, R. Brown. (Apocynaceæ.) India.
Amorpha fruticosa, L. (Leguminosæ.)
Baptisia tinctoria, R. Br. (Leguminosæ.)
Marsdenia tinctoria. (Asclepiadaceæ.) India.
Randia aculeata. (Cinchonaceæ.) West Indies.

INDIGOFERA TINCTORIA (Indigo).

Geography. — The indigo-bearing plants flourish in a hot climate. The Indigofera tinctoria is a native of India, and is cultivated between 20° and 30° of north latitude. It is grown in Java, the East Indies, Northern Africa, the West Indies, and Central America. That from India and Central America is the most valuable.

Etymology. — *Indigofera* is derived from the Indian word *indigo*, and the Latin word *fero*, bear or carry, signifying indigo-bearing. The specific name comes from the Latin word *tinctorius*, coloring.

History. — We do not know when indigo was first used as a dye. For some time after it was introduced into Europe it was believed to be a mineral substance. Marco Polo, the earliest traveller into India and China, gave an account of the plant that produces the indigo, and the methods of preparing it. After the discovery of America the plant was found in the warm parts of the new world, and it was also learned that the ancient Mexicans were acquainted with it as a dye. In 1747 the Indigofera Caroliniana was discovered in Carolina, and large quantities of indigo were manufactured there and sent to England, but the cultivation in the United States is not now prosecuted to any great extent. Soon after the discovery of the indigo-plant in America, the French began to produce it at Goree, an island on the west coast of Africa.

Cultivation. — The seed is sown in drills eighteen inches apart. The ground should be damp. The seed soon germinates, and in two months begins to flower, at which time it is fit to harvest. Great care is exercised to cut it at the exact time, to prevent damage from the rains. As soon as harvested it is carefully placed in a vat; the vat is then filled with water, and a heavy frame is laid upon the plants to keep them under water. After fermentation, the liquor is drawn off into another tank. It is then violently agitated by dropping heavy blocks into it, or heavy buckets whose bottoms and sides are perforated with many holes; this separates the fecula, or grain, as it is called, from the liquid. It is then drawn into a third vat, where by evaporation it is freed from the liquid, and the indigo is left; and before it is quite dry it is cut into small cakes, in which form it is sent to market.

The supply of indigo is subject to many contingencies, which is the cause of great fluctuation in the price; hence it is frequently the basis of commercial speculation.

Use. — The food value of the plants of this order is very great, due to the large amount of nitrogen stored in the seeds. Peas yield 23 per cent. Many species furnish important dyeing substances. Indigo is a most important substance in the hands of the dyer. It has a strong affinity for fibrous texture, whether animal or vegetable, and imparts, without a mordant, a permanent and beautiful blue. It yields to the chemist a substance known as Indigotin ($C_{16}H_5NO_2$), which is the coloring matter.

Marts. — Indigo is shipped from most of the ports of British India and the Eastern Archipelago. In America, the ports are Vera Cruz in Mexico, Belize in Yucatan, Truxillo in Honduras, and San Juan in Costa Rica, Kingston in Jamaica, and the ports of New Granada. The United States is supplied by Mexico, Central America, and South America direct.

ASTRAGALUS, Tourn. (Milk Vetch.) Calyx tubular, inflated, 5-toothed; teeth short, nearly equal. Petals long-clawed; standard ovate, or fiddle-shaped; wings unequally oblong; limb sometimes eared above the base; keel a little shorter than the wings. Stamens 10, 9 connate into a sheath, cleft above, 1 free. Ovary sessile; ovules numerous, in two series. Style slender, straight or curved. Stigma small and terminal. Legume sessile or stipitate, with its sutures turned in so that it is sometimes 2-celled. Small shrubs or herbs, variable in form.

A. gummifer, Labill. (Gum Tragacanth.) Shrub, 2 feet high. Stems short, naked; branches numerous and straggling; bark reddish-gray, slightly rough, marked with scars of fallen leaves; young twigs woolly. Leaves numerous, spreading in all directions, two and a half inches long, pinnate; rachis hard, stiff, smooth, yellow, acutely pointed, furnished at the base with broad, membranous, acute, glabrous, rusty, clasping stipules, cut at the edges; leaflets opposite or alternate, nearly sessile, very small, obovate, acute, entire, glabrous, both sides grayish-green, veined, articulated with the rachis, soon falling off. Flowers small, sessile, solitary or two to three together in the axils of the lower leaves, each with a membranous, acute bract as long as the calyx. Calyx cut to the base into 5 equal, very narrow, acute segments, clothed with silky, white hairs, persistent. Petals papilionaceous, a little longer than the calyx, pale yellow, and persistent; wings a little shorter than the standard, with a long linear claw; keel-petals nearly as long as the wings. Stamens 10, upper one free, 9 united into a sheath, which is attached to the petals at the base. Ovary villous; style long and filiform; stigma minute. Pod small and kidney-shaped, smooth, and pale brown.

This is a very large genus. Most of the woody and spiny species produce the tragacanth gums, but this species is prominent among those that produce it, and the first that was accurately described. The species A. tragacantha, from which the gum takes its name, does not yield the drug.

Geography. — The Astragalus gummifer and other gum-bearing species are subtropical plants, and do not produce the gum unless they grow in a warm climate. The gum which supplies the market is produced in Persia and the region south of the Black Sea, Greece and the islands of the eastern Mediterranean, also in Syria.

ASTRAGALUS GUMMIFER
(Gum Tragacanth).

Etymology. — *Astragalus*, the generic name of this plant, is from the Greek ἀστράγαλος, vertebra, an allusion to the crowded and apparently jointed appearance of the beans or seeds in the pods of some of the species of this large genus. *Gummifer* is from the Latin *gummis*, gum, and *fero*, bear, hence gum-bearing. *Tragacanth*, the name of the gum, is from the Greek τράγος, a goat, ἄκανθα, beard, hence a goat's thorn, this name being an allusion to the slender spines with which the branchlets of the A. tragacantha are armed, and which bear a slight resemblance to a goat's beard, which is somewhat like a thorn in shape.

History. — When or by whom this drug was first used is not known. The ancients were acquainted with it. Theophrastus, who wrote more than three hundred years before the commencement of the Christian era, mentions it.

Preparation. — The mode of collecting the gum is to remove the earth from the crown of the root, and then make wounds in the bark, from which exudes a whitish gummy sap that hardens in flakes, when it is removed. This

LEGUMINOSÆ. 99

is the fine flake gum of commerce. There are also small lumps constantly appearing on the stem and branches, which are picked off. These kinds are mixed, but are afterwards separated into several varieties, according to quality, for the markets of Europe and America. The gum consists of two substances, *Arabin*, which resembles gum Arabic, which is readily soluble in water, and *Tragacanthin*, which water causes to swell but will not dissolve. Gum tragacanth forms a mucilaginous jelly, with fifty times its weight of water.

Use.— In medicine, gum tragacanth is used as a demulcent, and a medium to aid in suspending liquid medicines. It is also used for a paste or cement; for suspending inks and dyes; for preparing fabrics for dyeing, and for stiffening crapes. Shoemakers use it for a paste to fasten the linings in shoes. It is used by confectioners and pharmacists to furnish adhesiveness to materials of which lozenges are made.

ARACHIS, L. (Peanut.) Calyx of the staminate flower with a slender tube; limb 2-lipped, upper lip 4-toothed. Corolla resupinate. Stamens 9, united in a tube. Pistillate flowers without calyx, corolla, or stamens. Ovary on a slender peduncle, which lengthens downwards, and forces the fertilized pistil into the ground, where the legume matures. Legume oblong, obtuse at each end, somewhat cylindrical, 1, 2, or 3-seeded; seeds ovoid. Flowers axillary, lower ones fertile, upper ones sterile.

A. hypogæa, Willd. (Peanut. Ground Nut. Ground Pea. Monkey Nut.) Stem 9 to 18 inches long, prostrate, branching, and hairy.

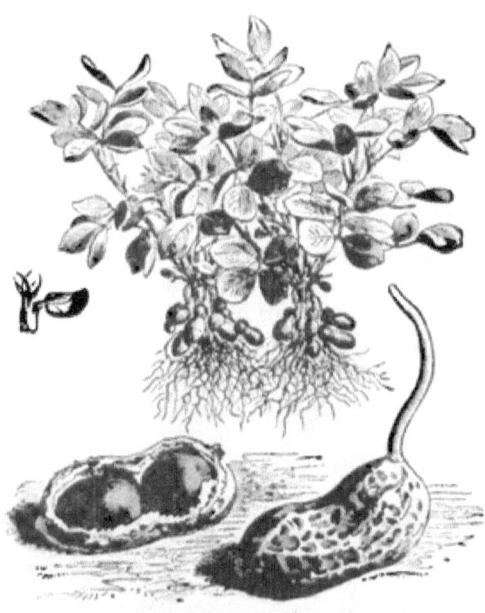

ARACHIS HYPOGÆA (Peanut).

Leaves composed of 2 pairs of leaflets, which are an inch to an inch and a half long, obovate, entire, mucronate at the apex, bordered by a hairy nerve, sub-sessile; common petiole 1 to 2 inches long, channeled above, and hairy. Flowers axillary, orange-yellow, appearing in July. Fruit ripens in latter part of September. While the process of ripening is going on the pod sinks under ground.

Of this plant there are seven species, only one of which seems to be under cultivation. Of this one there are several varieties, differing from each other in size and delicacy of flavor, one very conspicuous for its large fruit, which is about two inches long and nearly an inch in diameter. The ordinary nut of commerce is about an inch to an inch and a quarter in length and half an inch or less in diameter. A third variety is barely more than half an inch in length, and correspondingly small in diameter, but of very delicate flavor.

Geography. — The peanut is a subtropical plant. It grows and fruits in the southern United States, but will not fruit in regions of severe frost. It is an important crop in southern and central Virginia, and the Carolinas and Tennessee.

Etymology. — *Arachis*, the generic name, is derived by some from the Greek words α, without, and ράχις, the backbone, which signifies, in this application, "without a stem," which is incorrect; hence this derivation is obscure. Others derive it from ἄραχος, a name applied to a kind of vetch by Theophrastus. *Hypogaea*, the specific name, is from the Greek words ὑπό, under, and γῆ, the earth, *i. e.*, underground, due to the mode of ripening the fruit. *Peanut* is named from the fact that the plant appears like the pea while growing. *Ground Nut* and *Ground Pea* are thus named because they ripen under ground. *Monkey Nut* receives its name from the fact that monkeys are fond of it.

History. — De Candolle believes the peanut to be an American plant. It is found in the tombs of the ancient Peruvians. It was introduced into Africa from Brazil by the Spaniards, to feed the slaves on the passage over the ocean, and it spread by commerce into northern Africa, southern Europe, India, China, Japan, and North America.

Use. — The peanut is a very popular nut among children and young people in general. It is used sometimes as a dessert. It yields an excellent sweet oil, which is substituted for olive oil in the arts and for culinary purposes. In China the oil is used for illuminating, and also for lubricating. Its roots are sometimes used as a substitute for liquorice.

In Virginia it is ground into flour and used for making pastry and biscuits, and is said to be superior to wheat, on account of its rich, delicate flavor.

Statistics and Commercial Importance. — The annual yield of the peanut in Virginia is about 2,100,000 bushels; in Tennessee, 250,000 bushels; in North Carolina, 135,000 bushels. Besides these, large quantities are imported from South America and Africa.

LENS, Moench. (Lentil.) Calyx 5-parted, parts narrow, lanceolate. Corolla small, nearly concealed by the long segments of the calyx, varying in color from white to lilac, or pale blue. Style erect. Stigma smooth. Pods short and blunt, thin, smooth, and 2-seeded; seeds in form of a circular double-convex lens. Annual.

1. **L. esculenta**, Mœnch. Stem weak, about 10 to 18 inches high, branching. Leaves pinnate, composed of 6 to 8 pairs of elliptical leaflets, the main leafstock terminating in a branched tendril; lower leaves without tendrils. Fruit a short pod, with 1 to 2 seeds, about two eighths of an inch in diameter, yellowish-brown. Flowers June to July. Fruits August.

There are several varieties of the lentil, three of which are usually under cultivation. The Lens esculenta is the most common, and the most esteemed.

Var. **lutea**, the yellow lentil, is said to be a favorite with the French.

Var. **Provence** is a large, coarse sort, grown for stock.

Geography. — The zone of the Lens is the middle and southern parts of the north temperate zone, Syria, Egypt, southern and central Europe, and

LEGUMINOSÆ. 101

Hindustan. It escapes from cultivation and becomes a troublesome weed, and it has been suggested that it was the plant mentioned in the parable of the *Tares*.

Etymology. — *Lens*, the generic name, is the old Latin name of the plant, the signification of which is very obscure. *Esculenta* is Latin for "eatable," hence eatable lens or lentil. *Lentil*, the common name, is from the Latin *lenticula*, a little lens.

History. — When or where this food plant was first cultivated it is very difficult to determine. It is believed by the best authorities to have been grown in western Asia, and along the shores of the Mediterranean, as far west as Italy, in prehistoric times, and thence introduced into Egypt, after which it spread over Europe, and crept eastward into India. It is spoken of by the ancient writers on Botany, and was no doubt the material employed by Jacob to prepare his pottage with which he purchased his brother's birthright. It is at the present day an important food in Palestine. It is occasionally cultivated in the eastern United States, but is not very profitable. It is to be found on sale in the Italian and German groceries of our large cities.

Use. — It is prepared as beans are, boiled with or without meat, it is also baked with pork or mutton, and is made into soups, and used to thicken gravies. It is largely used by the Arabs in a parched state, while on their marches. It was in early times the only food of large armies while on the march, but is greatly inferior in quality and delicacy to either the pea or the bean. Its meal is sold as a food for invalids under the name "Revalenta."

PISUM SATIVUM (Garden Pea).

PISUM, L. (Pea.) Calyx with leafy segments, 5 in number, 2 upper ones shorter. Petals 5, upper one broad and turned back. Stamens 9 and 1. Style flattened and ridged, velvety on the upper edge. Pod oblong; seeds globular, from 5 to 10 in a pod. Annual herb.

1. **P. sativum.** (Garden Pea.) Stem 1 to 4 feet high, terete, smooth, and weak, climbing by tendrils. Leaves composed of 2 to 3 pairs of elliptical, obtuse, entire, mucronate leaflets, an inch to two inches long; the common leaf stalk strong, terete, terminating in a long branched tendril. Stipules large, ovate, somewhat sagittate, crenate, dentate at the base. Flower-stalks

axillary, 1 to 6 inches long, terminated by 1 to 2 flowers each. Style turned back; flowers white. Pod an inch and a half to three inches long, somewhat cylindrical or flattened. Flowers in June. Fruits July to August.

As the pea is grown from the seed, it sports. Gardeners, making use of this tendency, have produced a great number of varieties, differing in length of stem, size, shape, and especially in delicacy of flavor in the fruit.

The varieties are usually arranged under the heads *Short-stalked* and *Long-stalked*, and named from some real or imaginary quality, or after the propagator, or discoverer.

2. **P. arvense**, L. (Field Pea.) Differs from the P. sativum in being less delicate to the taste. Has only one flower on a flower-stalk. Flowers red. Seeds crowded in the pod, presenting the form of short, quadrangular prisms. Whole plant coarser, and more hardy, enduring heavy frosts; one variety planted in France endures the hardest frosts of winter, and fruits the following summer. Raised largely in Europe for feed for cattle and horses.

Geography. — The habitat of the pea is from the middle of the temperate zones to the edges of the tropics. It fruits well throughout central and southern Europe, Egypt, Syria, Japan, India, China, and Cochin China; but nowhere is it more productive and more largely cultivated than in southern Japan, where it constitutes a very important article of food and of internal commerce.

Etymology. — *Pisum*, the generic name, is derived from the Latin word *piso*, beat, pound, or bruise, due either to the means employed to separate the seeds from the pods, or to grind them into flour. *Arvense*, the specific name, is from the Latin, signifying "field," and *sativum* from the Latin, sow, or plant. *Pea*, the common name, is a corruption of the word *piso*, grind, or bruise.

History. — The home of the Pisum arvense is not positively known. It is found without cultivation in Italy. The P. sativum is not known at present to be wild anywhere, hence the place where it originated is difficult to discover. It has been claimed to be a variety of P. arvense, but its botanical characteristics are so distinct and constant as to throw that hypothesis into great doubt.

There are reasons to believe that it was carried into Europe by the Aryans, at a remote period of history; and it is supposed to have been indigenous in western Asia, along the foothills of the Caucasus, towards Syria, and southeastward to Persia. It was brought to North America by European colonists, and is grown in the kitchen and market garden, and in many regions in the field.

Cultivation. — The pea flourishes in a light, rich soil, and yields an abundant harvest to generous cultivation. The garden mode of cultivation is usually in drills six to eight inches apart (called double rows), with a space of three feet, and another double row. In the field it is either sown in drills and worked with a horse-hoe or a plough, or sown broadcast like the oat. Among the peasantry of Scotland, England, and the Isle of Man the P. arvense is an important field crop, furnishing food for themselves and their domestic animals.

Use. — The pea is plucked before the seed is ripe, when it is in the milky state; it is then shelled, and boiled in a small quantity of water, to preserve the delicate flavor, and served plain or with gravies; it is also cooked with a small quantity of salt meat. When ripe, it is used for soup, or ground into

LEGUMINOSÆ. 103

flour to thicken gravies and soups; it is also ground and fed to cattle and horses. It constitutes a very important article of food, and is found in every kitchen and market garden.

PHASEOLUS, L. (Bean.) Calyx with 2 bracts at the base, bell-shaped, somewhat 2-lipped, upper lip 2-parted, lower one 3-parted. Keel of the corolla beaked, and, together with the stamens and style, spirally twisted. Legume, or pod, linear curved, flattened, or somewhat cylindrical; many-seeded; seeds kidney-shaped. Annual herb.

1. **P. vulgaris**, Savi. (Pole Bean, Kidney Bean, Haricot, String Bean.) Stem 4 to 10 feet long, slender, voluble, and twining always against the sun, or towards the west or southwest. Leaves trifoliate, or a pair of leaflets and a terminal one, common petiole 1 to 6 inches long. Flowers white, in racemes, on stout peduncles, 1 to 4 inches long. Pod 3 to 6 inches long; seeds more or less kidney-shaped, whitish, dull-yellow, or mottled.

Var. **nanus**, L. (Bush Bean) is a dwarf, with a short, erect, branching stem 8 to 15 inches high. Leaflets sharp-pointed, and bracts larger than in the above; otherwise like P. vulgaris.

2. **P. lunatus**, L. (Lima Bean.) Stem as in P. vulgaris, except longer and disposed to branch more; leaflets larger, oblique or triangular, 2 to 4 inches long, common petiole 2 to 6 inches long, racemes loose, pedicels short. Flowers greenish-white, smaller bracts shorter than the calyx. Pods 2 to 3 inches long, an inch wide, curved or moon-shaped. Seeds 1-3, large, flat, greenish, or white. Flowers in July; fruits in August to September, remaining till frost appears.

PHASEOLUS VULGARIS (String Bean).

Species. — The above are the species which have given rise to all the forms now under cultivation. In the south of North America, Mexico, and the West India islands, a small black bean, called turtle soup bean, is largely cultivated, especially by the Spanish Americans.

104　DESCRIPTIVE BOTANY.

(The beans of a different species, *Vicia faba, L.*, otherwise called *Faba sativa*, are not flattened, and are smaller than the above. They are often given to horses.)

Geography. — The zone of the bean is very wide, including the tropics and the temperate zones to the 45th parallels, and even beyond in some localities.

PHASEOLUS NANUS (Bush Bean).

Etymology. — *Phaseolus*, the generic name is from the Latin *phaselus*, a little ship, due to the shape of the flower. *Vulgaris*, from the Latin, means common or usual. *Nanus*, from the Latin, signifies small or dwarf. *Lunatus*, Latin, indicates that the pod is in the shape of a new moon. *Bean*, the common name, comes to us from the old middle English, and is supposed to signify good, i. e., good food, or good for food. *Bush Bean* is named from the fact that the plant appears in the shape of a little bush. *Pole Bean* is named from the circumstance of planting poles in the hills with the seed for the stems to climb upon. *Kidney Bean* takes its name from the kidney-shaped seeds. *Lima* is supposed to be named from the fact that this species was found near the city of that name.

History. — The bean was supposed to be a native of India and Western Asia, and to have thence worked its way into southern Europe by commerce and travel; but recent inquiries have disturbed this belief. De Candolle, who is the best authority on the subject, thinks that the Phaseolus vulgaris, and P. lunatus are indigenous to South and Central America, and that the bean was carried thence to southern Europe by the Spaniards, and to western Africa by Spanish slave-traders. To reconcile these statements with the accounts of ancient writers, we must conclude that the beans of the ancients were varieties of other species, or different genera, of the Pulse family.

PHASEOLUS LUNATUS (Lima Bean).

LEGUMINOSÆ. 105

Cultivation. — The bean grows so easily in almost every variety of soil that nearly every locality produces sufficient for home consumption. Hence, though it has become one of the most important table vegetables, it has little more than a local sale.

It is planted in hills two and a half to three feet apart, or in drills three feet apart.

Use. — The bean, in its numerous forms, constitutes a very important article of food. The ripened seeds are boiled, and served plain or made into soups. A flour is also produced from the ground seeds, and employed to thicken gravies and soups. In New England, *baked beans* form a favorite dish. The green pods of the P. vulgaris, when the seeds are about half-ripened, are cut into half-inch pieces, and boiled, either with or without salt meat, and served as a vegetable. Beans are largely used at sea and in the army, for rations for the sailors and soldiers, and they are believed to afford more nutritive material than any other substance of the same bulk.

GLYCYRRHIZA, L. (Liquorice.) The description of Astragalus applies to Glycyrrhiza, except that in the latter the anther-cells are united, and the legume is continuous internally.

G. glabra. 3 to 4 feet high. Leaves pinnate, 4 to 5 pairs, and a terminal one; leaflets ovate. Flowers axillary, in racemes, whitish-violet.

Geography. — The liquorice is native to Italy and southern Europe; it also grows in the south of England, and is cultivated in Spain and Portugal.

Etymology and History. — *Glycyrrhiza*, the generic name, is from the Greek γλυκύς, sweet, ῥίζα, root, *sweetroot*, due to the well-known sweet taste of the liquorice. It was grown in England in the time of Elizabeth.

Use. — Its medicinal qualities are demulcent and emollient. It is administered in catarrh and other irritations of the mucous membrane, and for sore throat. It is largely used to sweeten tobacco. Brewers also use it to give body and to impart a sweet taste to porter and to Scotch ale. It is also mixed with purgatives, under the name of liquorice-powders, to disguise the taste of other drugs, as senna, etc.

HÆMATOXYLON, L. (Logwood.) Calyx cup-shaped, hemispherical. Sepals 5, nearly equal, imbricated; corolla papilionaceous. Stamens 10, 5 shorter. Ovary inserted in the hollow receptacle, free, short, stipitate, usually 2-ovuled. Pods furnished with lance-shaped, flattened, leaf-like beans or seeds, 1–2 in a pod. Small tree.

H. Campechianum, L. (Logwood.) Stem 20 to 30 feet in height, and 12 to 18 inches in diameter; crooked, much-branched, branchlets armed with sharp spines; sometimes appears as a shrub forming dense thorny thickets. Bark dark and rough. Leaves pinnate, with 4 or 5 pairs of irregular obcordate leaflets. Flowers yellow, in terminal spikes. Pods long, double-valved; seeds oblong, kidney-shaped, flattened.

The only species of the genus.

Geography. — The logwood-tree is native in all parts of the damp forests of Central America, being most abundant on the peninsula of Yucatan, and along the low-wooded shores of Guatemala and Honduras; but it grows well along the low banks of streams and damp grounds of the Isthmus of

Panama and the West Indies. It has been planted by the British in Burmah, where it grows to perfection.

Etymology. — *Hæmatoxylon* is from the Greek words **αἷμα**, blood, and **ξύλον**, wood, signifying blood-wood, on account of its red color. *Campechianum* is the name of the region where it was first obtained. *Logwood*, the common name, is due to the form in which it is brought to market, i. e., in short logs, four feet long and six inches in diameter. For the same reason it is called blockwood. On the continent of Europe it is called *Campeachy wood*.

History. — It grows and thrives best in damp ground. Though a quick grower, the wood is of a fine, hard texture. It was known as a dye-wood as early as the middle of the sixteenth century.

The dyers of that day prepared beautiful colors from this wood, but its chemistry not being understood, they were unable to fix them; hence its use was forbidden by a law, which was rigorously enforced. After about a hundred years the act was repealed or made void by the passage of another, in 1661, which read as follows: "The ingenious industry of the times hath taught the dyers of England the art of fixing colors made of logwood, so that by experience they are found as lasting and serviceable as the color made by any other sort of dye-wood." And logwood from that time became a popular dye. In 1675 the demand for the wood developed a great industry in cutting, preparing, shipping, and freighting the wood. The Spaniards interfered with the English, who had established a colony of choppers on the shores of the Bay of Campeachy. The English thereupon made plantations in Jamaica, but the wood produced did not yield the dye of the wood grown in its native swamps.

HÆMATOXYLON CAMPECHIANUM (Logwood).

When the tree is 10 years old it is about 20 feet high and 10 inches in diameter; it is then felled, the sap-wood chipped off, cut into pieces 3 to 4 feet in length, and shipped to Great Britain or the United States. The best wood is from Honduras, the next best from St. Domingo, and the third class from Jamaica.

Chemistry. — A blood-red crystalline substance is dispersed through the wood, and this when extracted gives the violet dye. It yields to the chemist a substance indicated by the following formula: $C_{32}H_{14}O_{12} + 2 HO$. This when isolated appears in yellow crystals, and has the taste of liquorice, and was named Hæmatoxylon by Chevreuil, a professor of chemistry in Paris, who obtained it. It is not itself a dye; but when united with certain alkaline bases, and exposed to the action of the air, it produces beautiful *red*, *purple*, and *blue* colors.

Use. — It was first used in 1646. Its medicinal properties seem to be a mild astringent and tonic, and it is administered in the form of extract or decoction for infantile cholera, chronic diarrhœa, and chronic dysentery.

Its most important use is as an ingredient in the production of dyes.

As a hedging-plant it is highly esteemed; on account of its rapid growth, its crooked habit of growth, and its strong spines, it is one of the best-known plants for fencing purposes.

It makes excellent fuel, and is very hard and heavy.

Statistics.—About 64,000 tons are annually taken into Great Britain, and nearly as much into the United States.

Marts.—The principal ports to which logwood is taken are: London, in Great Britain; Versailles, in France; and New York, in the United States.

CASSIA, L. (Senna.) Sepals 5, barely united at base. Petals 5, unequal, spreading. Stamens 5 to 10, unequal apart, frequently imperfect. Anthers opening by two chinks at the apex. Pod many-seeded, often with cross partitions. Leaves pinnate. Flowers yellow.

1. **C. acutifolia**, Delile. Stem woody, 3 feet high. Leaves alternate, pinnate, stipulate; leaflets in from 4 to 6 pairs, sessile, oval-lanceolate, acute, oblique at base, nerved, three fourths of an inch long. Flowers yellow, in axillary spikes. Fruit a pod or legume, an inch long, half an inch broad, flat, elliptical, obtuse, membranous, and smooth, divided into 6 or 7 cells, each containing 1 seed.

2. **C. obovata**, DC. Like the above, except that it is 18 inches in height. Leaf with 5 to 7 pairs of leaflets, obovate and mucronate. Legume flat, kidney-shaped, and clothed with a short inconspicuous down.

3. **C. angustifolia**, Wahl. Stem erect, smooth. Leaflets in 4 to 8 pairs, sessile, lanceolate, obscurely mucronate, smooth, downy beneath, with a wavy line along the under side of margin, one to one and a quarter inches long. Legume oblong, abrupt at base, round at apex, an inch and a half long and half an inch broad. Annual.

CASSIA MARILANDICA (Senna).

4. **C. Marilandica**, L. Stem 3 to 5 feet high. Leaflets in 8 to 10 pairs, an inch and a half long and half an inch wide, oblong, blunt, and mucronate; the main petiole has a club-shaped gland at its base. Flowers in short axillary racemes on the upper part of stem, yellow, fading to white. Anthers black, 10 in number, and unequal. Pods hanging, 3 inches long, flat, linear, hairy at first, stipules falling off.

5. **C. fistula**, L. Large tree, branching regularly, and forming a symmetrical head; wood hard and heavy. Leaves of 5 to 6 pairs of opposite leaflets; leaflets 3 to 5 inches long, ovate, pointed, undulate, smooth, on short petioles. Pod a foot long, an inch thick, cylindrical, woody, dark-brown, hanging. When the plant is disturbed by the wind the pods strike together and produce a sound which may be heard at a considerable distance. Native in Upper Egypt.

Species.—There are many species of the Cassia, but those already described are the ones from which the commercial products are obtained that enter into the American trade.

Geography. — The geographical range of most of the species of Cassia which enter into the materia medica is tropical and subtropical, extending quite around the globe.

Etymology. — The common name *Senna* is derived from the Arabic name *Sana.* The generic name is said to have been traced back to the Hebrew word *Ketzioth,* signifying "to cut;" but the application of this signification is not apparent. The specific names are Latin, with one exception, *Marilandica,* which signifies "Maryland," this being the State from which specimens of the plant were first sent to Europe. *Acutifolia,* Latin, acute- or sharp-leaved *Obovata,* obovate-leaved. *Angustifolia,* narrow-leaved. *Fistula,* a tube-shaped fruit.

History. — Senna was introduced into medical practice in the tenth century by the Arabic physician, Serapion.

The most valuable is the Alexandrian senna. It is a mixture of the leaves and pods of C. acutifolia and C. obovata and the leaves of Cynanchum oleæfolium; the mixture is prepared at Boulac, in Egypt, where it is put up in bales and sent to Alexandria.

The East India senna is obtained from C. angustifolia, in southeastern Arabia, where it grows without cultivation. The leaves and pods are gathered and sent to Bombay, whence it reaches Europe.

Tinnevelly senna is obtained from C. angustifolia, which was introduced into India from Arabia, and is now largely cultivated in the vicinity of Tinnevelly. The imported article consists of unbroken leaflets, of a fine darkgreen color.

The American senna is from the C. Marilandica, and is found throughout the Middle and Southern United States. It is collected in a wild state, dried in the shade, and the leaflets and pods are sent to market.

Use. — The active medicinal principle of senna has thus far defied the skill of the chemist. By experience the drug is found to be a safe and efficient, but not a violent purgative. It is usually administered in combination with salts of magnesia. It is used for alterative purposes in the form of confections; the pulverized leaves, the pulp of the fruit of C. fistula, with fruits and spices, are made into a mass, and are prepared in the form of lozenges.

CERATONIA, L. (Carob Tree.)

C. Siliqua Found in the countries of the Levant, bearing large pods, which are fed to cattle, and believed to be the plant referred to in the New Testament in the parable of the Prodigal Son. Pods have been found in Egyptian monuments, with a piece of wood, which microscopic examination proved to be from this tree.

DALBERGIA, L. (Blackwood.) Calyx unequally 5-toothed; vexillum obovate or orbiculate; wings oblong; keel blunt. Stamens 10 or 9, bifid-didymous. Ovary stipitate; ovules 2 or more. Pod oblong linear, compressed, thin, wingless. Leaves unequally pinnate or 1-leaved. Trees and vines.

1. **D. nigra,** Allemo. (Fine Rosewood.) Trunk 50 to 70 feet high, branched into a symmetrical head. Leaves unequally pinnate, or solitary. Flowers papilionaceous, white. Pods flat, 1-2-seeded. Wood brown, and beautifully mottled with yellow spots and veins.

LEGUMINOSÆ.

2. **D. latifolia**. Rox. (Indian Rosewood.) Trunk 80 feet in height, beautifully branched. Leaves pinnate, with few leaflets. Pods flat, few-seeded. Wood hard, heavy, close-grained, takes a fine polish, color brown to black, streaked with rich yellowish veins.

NOTE. — The name Rosewood is applied to several different trees. There is some confusion with regard to the genera that produce all the woods known by that name. The Dalbergia nigra, and D. latifolia are known to be trees from which rosewood is produced, and D. nigra is the tree whose wood discharges an odor of roses.

Geography. — Dalbergia nigra is found native in Brazil, both tropical and subtropical. The D. latifolia is indigenous to southern Asia. Though tropical, it extends north to the edge of the north temperate zone, and is extra tropical in regions of no frost.

Etymology. — *Dalbergia* was the name given to this genus to compliment Nicholas Dalberg, a Swedish botanist of distinction. *Nigra*, Latin, signifying black, due to the color of the wood. *Latifolia*, Latin, broad-leaved, from *latus*, broad, and *folium*, a leaf. *Rosewood*, the popular name is due to the odor given off by the wood when under the saw and plane of the workman. The names, *Kingwood*, *Blackwood*, are also applied to all the woods known as Rosewoods. The *Dalbergia* is known among the inhabitants of Brazil under the name *Jacaranda*, the signification of which is unknown.

DALBERGIA NIGRA (Rosewood).

Use. — It is brought to market in logs or planks. The logs are usually split in half, in order to make sure that they are sound.

The wood is highly prized for musical wind-instruments, polishing-sticks for shoemakers, piano-cases, chairs, sofas, bedsteads, bureaus, and for veneering all sorts of cabinet work.

CAESALPINIA. (Brazil Wood.)

C. crista, L., and **C. Braziliensis** furnish fine dyes and red ink. The wood takes a fine polish, and is used for violin-bows, etc.

The name *Braziliensis* does not seem to come from Brazil, for it was known before the discovery of America. It has been suggested that the discovery of the tree in Brazil may have given its name to the country.

TAMARINDUS. L. (Tamarind.) Calyx funnel-shaped, narrow, divided into 4 ovate, lanceolate, acute segments, imbricated in the bud. Petals 3, 1 posterior and the other 2 lateral, oblong, white or yellowish-white, with red veins. Stamens 3, filaments long and free. Anthers opening lengthwise. Ovary stalked, 1-celled; ovules numerous. Style long and hooked. Fruit pendulous, pod-shaped, compressed, 3 to 6 inches long, 1 inch wide, curved, nearly smooth, chocolate-brown; seeds imbedded in a firm pulp; 3 strong woody cords extending along the edge of the pulp from base to apex. Seeds 2 to 8 in a pod.

T. Indica, L. Trunk from 60 to 80 feet in height, and 2 to 4 feet in diameter; bark rough, twigs smooth or pubescent. Branches long, horizontal, forming a very large head. Leaves alternate; stipules falling; leaflets in

8 to 16 pairs, opposite, one half to three quarters of an inch long, sessile and overlapping, oblong, blunt, unequal at the base, thick-veined underneath. This is the only species of the genus; but as it is propagated from the seed, it sports and produces varieties differing in the size and quality of the fruit. The pods of the Indian and African varieties have more beans in them than those of America.

Hymenaea verrucosa, of Madagascar, and *H. combaril* of the West Indies, are allied to the Tamarindus. H. combaril is the locust-tree of the West Indies.

Trachylobium Hornemannianum, of eastern Africa, is also an ally of Tamarindus.

Geography. — The geographical zone of the tamarind is tropical and sub-tropical. It is indigenous to Africa, but it has spread by cultivation to Arabia, to southern India, Ceylon, Java, the Philippines, northern Australia, the tropical isles of the Pacific, the West India islands, and to tropical South America.

Etymology. — *Tamarindus* is from the Arabic *Tamar*, a date, and *Indus*, India, hence *Indian date*. *Indica* denotes the country in which it grows, yet it is not known to be a native of India. *Tamarind*, the common name, is the generic name Anglicized.

History. — There is reason to believe that the ancient Greeks and Romans were not acquainted with the tamarind, but it seems to have been known to the ancient Egyptians. It is mentioned in the Koran, and was well known to Arabic physicians in the middle ages.

TAMARINDUS (Tamarind).

Preparation. — In the West Indies the fruit is picked when ripe, packed in small kegs, and hot syrup is poured over it; then the vessels are closed, and it is fit for the market.

In Asia the fruit is packed in salt, and a syrup made from the fruit is poured over it.

In Africa the pulp is separated from the pod and seed and pressed into cakes, then dried in the sun.

Use. — The medicinal properties of the tamarind are laxative, cooling, and anti-febrile. It contains about one tenth of its weight of citric acid, also small quantities of acetic, malic, and tartaric acids, hence its value in producing acid drinks, for which it is largely used in the countries where it grows.

It is an important article of food among the natives of the hot countries of Asia and Africa. The seeds are roasted, and reduced to flour, of which cakes are made; they are also boiled. The Hindus make use of the leaves, flowers, bark, and seeds in the preparation of healing remedies.

In the Atlantic cities of the United States it is used as a preserve. In the famine of 1878-9 the leaves were used as food in the Deccan.

The timber produced by these trees is hard and durable, and they all produce a resinous gum known as copal, which when heated with linseed oil or spirits of turpentine, dissolves and forms the best varnish. The gum flows from wounds made in the trunk and branches. It also flows from the roots, and is found by digging about the foot of the tree. Much of this resin that comes to market is fossil, as it is frequently found where there are no trees.

Marts. — The ports of export are Aden, in southern Arabia, Alexandria, and ports of the East and West Indies.

ACACIA, Necker. Flowers regular, perfect or polygamous. Calyx 4–5-toothed. Petals united below. Stamens free or united below, numerous, longer than the corolla. Anthers small. Style thread-like. Pod sometimes two-valved, and at other times not opening when ripe; flattened or cylindrical, containing many flattened seeds. Leaves bi-pinnate; leaflets small; stipules frequently spinous. Flowers small, in globular heads or cylindrical spikes, axillary, and yellow. Shrubs or small trees, usually armed with prickles or thorns.

A. Senegal, Willd. (Gum Arabic.) Stem 20 feet high, erect. Branches irregular, crooked, and twisted, the young branchlets thickened at the nodes, which are armed with three hooked thorns. Bark smooth, grayish, or white. Leaves alternate, or appearing in bunches, or fascicles, bi-pinnate; rachis slender, tomentose, terminated with a gland, having one also at the base; pinnæ opposite, 3 to 5 pairs; leaflets opposite, 10 to 20 pairs, sessile and linear-oblong, rigid, grayish-green, one sixth of an inch long. Flowers axillary, sessile, small, in slender, cylindrical, erect spikes. Calyx bell-shaped, downy, cut into 5 acute segments, reaching half-way down. Corolla campanulate, twice as long as the calyx, divisions extending half-way down. Stamens numerous; filaments slender, erect, 3 times the length of corolla, yellowish, united at the base into a short tube, which is inserted on the base of corolla. Anthers small and roundish. Ovary on a short stalk, small, oblong. Style filiform, shorter than stamens; stigma terminal. Pod short-stalked, 3 to 4 inches long, and three fourths of an inch wide, constricted between the seeds, smooth, pale, membranous, with a strong marginal rib. No. of seeds 2–6; funiculus long; beans roundish, much flattened, brown.

ACACIA SENEGAL. (Gum Arabic).

There are over 400 species of this genus, but the pure Gum Arabic is from the *A. Senegal*, found in Kordofan.

Geography. — The geographical distribution of the acacia is very extensive. It occupies a broad belt both sides of the equator, all around the globe. Though it is for the most part a tropical and subtropical plant, it reaches far into the temperate zones. Many species grow in Australia, and some in America. Most of the species, however, are found in tropical Africa and Asia, and in the tropical Pacific Islands.

Etymology and History. — *Acacia* is from the Greek ἀκακία, a name given to a thorny plant by Dioscorides, derived from ἀκή, a sharp point. *Senegal*, the specific name, is from the district in Africa where the tree abounds. *Gum Arabic*, the popular name, is due to the circumstance that formerly the gum was carried from Aden in Egyptian ships through the Red Sea to Egypt, and thence reshipped to Europe.

At present none of the gums of commerce known as Gum Arabic are obtained from Arabia.

The pure gum used in medicine is from Kordofan, is carried down the Nile to Egypt, and is a white opaque substance which when pulverized resembles wheat flour in color. The several gums sold for gum arabic are from other species, and are usually brought to market mixed.

The gum exudes from wounds or incisions in the bark, and appears in tears from the size of a pea to that of a small hen's-egg. The different sorts are known in commerce by names which indicate the countries whence they are brought, as Mogador gum, North Africa gum, Jedda gum from Jedda in Arabia, Cape gum from the Cape of Good Hope. East India gum is carried from the east coast of Africa to Bombay, from which point it is shipped to Europe.

There is also a gum sold for gum arabic which is an Australian product, and is obtained from the Acacia pycnantha, Benth. The beautiful *A. dealbata* of Australia, frequently seen in our green-houses, yields a good gum.

The *Acacia Seyal* is the Shittim wood of Scripture, and the *Acacia Suma* is one of the sacred trees used by the Brahmins to obtain fire by friction, for their altars.

The fine white gums of commerce are known as Turkey gums. The darker, translucent, reddish gums are known commercially as Senegal gum.

Use. — The gum begins to flow in the flowering season early in December, and the harvest extends to the last of January, during which time the harvesters subsist almost entirely upon the gum.

A. Catechu of India yields by decoction a valuable tonic (Catechu), and in the hands of the dyer it forms the colors *black*, *brown*, *green*, *drab*, and *fawn*. The decoction is highly charged with tannin.

As a *medicine*, gum arabic is used largely as an emollient and demulcent; it is prescribed in stomach difficulties, dysentery, and other bowel disorders; and is used in throat troubles, and for cough mixtures.

In confectionery, it is mixed with sugar and formed into lozenges and gum-drops.

It is largely used for a cement, or sticking substance. The Egyptians employed it to suspend their water-colors in painting.

The commoner qualities are used for giving luster to crape, silk, etc., to stiffen the fibers in cloth-finishing, and in calico-printing. For labels, etc., it is usual to mix sugar or glycerine with it to prevent it from cracking.

The tree has great beauty, and is highly prized in planted grounds where it is able to endure the temperature. The wood is hard, and takes a fine polish. The bark of many of the species is highly charged with tannin, and though used in the manufacture of leather, is not a favorite for that purpose, because it imparts a stiff, brittle character to hides during the process of tanning. These barks, however, are largely imported into England. They are known in commerce as Wattle Barks.

A species of an allied genus Prosopis (P. juliflora, DC.), a native of Texas, yields an inferior gum locally substituted for gum arabic.

Order XXII. ROSACEÆ.

Flowers perfect, regular, terminal, usually in a corymb, cyme, or umbel. Sepals 5, occasionally fewer, united at the base. Petals 5, occasionally wanting. Stamens numerous, in several series, distinct or cohering together, inserted with the petals on the disk which lines the calyx-tube. Leaves alternate, stipulate. Fruit a pome, drupe, or akene. Seeds one or few in each carpel. Herbs, shrubs, or trees.

Number of genera, 71; of species, 1000.

RUBUS, L. Calyx spreading, 5-parted; petals 5 in number, falling. Stamens many, on the border of the disk; ovaries numerous, with 2 ovules, 1 abortive. Akenes pulpy, drupe-like, aggregated upon a succulent receptacle. Shrub.

1. **R. strigosus**, Mx. (Wild Red Raspberry.) Stem 3 feet high, half an inch in diameter, sparingly or diffusely branched, armed with weak prickles. Leaves pinnately 3-5-leaved; leaflets oblong-ovate, obtuse at the base, pointed at the apex, serrate, gashed, teeth unequal, sessile and hoary beneath, wrinkled. Flowers white; corolla cup-shaped, and smaller than the calyx. Fruit hemispherical, when removed from the receptacle; it is hollow, and forms a little cup. The aroma and taste are very grateful. Common in Northern United States. Flowers in June. Fruits in July and August. Root perennial. Stem biennial.

Rubus (Raspberry).

2. **R. occidentalis**, L. (Black Raspberry. Thimble Berry. Black Cap.) Stem 3 to 5 feet high, glaucous, recurved, bending to the ground, armed with strong recurved prickles. Leaf 3-foliate; leaflets acuminate, subsessile, doubly serrate, tomentose, or white downy beneath. Flowers axillary and terminal; corolla smaller than the calyx. Fruit like the last, except that it is black. Flowers in May and June; fruit ripens in July. Common where the last is found. Dr. Gray says it flourishes best in ground that has been burned over.

3. **R. Idæus**, L. (Garden Raspberry.) Stem 5 to 8 feet high, armed with strong bristles or recurved prickles. Leaf pinnate, with 3 to 5 leaflets; the leaflets broad-ovate, acuminate, unequally cut and toothed, hoary underneath; lateral ones sessile, terminal one petioled. Flowers in corymbs or panicles; petals shorter than the divisions of the calyx, white, terminal. Fruit red, like No. 1. Dr. Gray thinks it identical with the American species R. strigosus. Wood says Dr. Robbins found it in a wild state in Vermont, also in Connecticut.

R. Idæus is the plant from which all the varieties of the red raspberry have sprung, either by hybridizing or from seedlings.

The black cap varieties have arisen by similar means from the R. occidentalis.

There are about 150 varieties under cultivation in North America.

Geography. — The geographical distribution of the Rubus is very broad. It grows well in the temperate zone, between 30° and 50° latitude in North America, and the belt extends from the Atlantic to the Pacific oceans. In Europe it is found as far north as the 60th parallel, and extends to the northern parts of Africa, and from Asia Minor west to the British islands, and eastward into India. It is said to be found in Japan, but it is supposed to have been carried there by Europeans.

Etymology. — *Rubus*, the generic name of the raspberry, is derived from the Celtic word *rub*, signifying red, Latin *ruber*. The specific name *strigosus* is a Latin word, which means scraggy, or meager, relating to the small size of the plant. The specific name *occidentalis* means western. The name *Idæus*, is from Mount Ida, where it is believed this species had its origin. *Raspberry* comes from the Italian word *raspo*, rough, on account of the roughness of the stem and leaves; it is also called *raspis* in Scotland for the same reason.

The *Idæus* is the cultivated plant in Europe, and was brought to North America by European colonists.

History. — There is no record to show when the raspberry was first brought under cultivation or when it was carried into Europe, but its value as a food-plant must have drawn attention to it at a very early period of man's civilization. The seed of the raspberry is said to have been found in the hands of mummies, which points to great antiquity in its use.

Use. — The raspberry is a favorite dessert fruit. It has a delightful perfume, and a subacid taste agreeable to most palates.

It ripens just at the end of the strawberry period, and thus prolongs the early fruit season. It is used for jams, raspberry vinegar or wine, for syrup to flavor soda water and other drinks. It is largely canned and dried. A wine made from it is distilled into Raspberry Brandy.

RUBUS STRIGOSUS (Wild Red Raspberry).

Marts. — The ease with which it is cultivated enables gardeners in the vicinity of our large cities to supply the market. It is so perishable that it cannot be shipped to long distances, hence the markets must be local.

4. **R. villosus**, Ait. (High Blackberry.) Stem from 3 to 8 feet high, curved, from half an inch to an inch in diameter; young branches, and villous peduncles, grooved, and armed with strong curved prickles. Root creeping. Leaves 3-foliate, or pedately 5-foliate; stipules subulate; leaflets ovate or oblong-lanceolate, unequally serrate, villous beneath, petioles and midrib aculeate. Flowers in a raceme, abundant, white; sepals linear at their extremities;

petals longer than the sepals, obovate, spreading. Fruit ovoid-oblong or cylindrical, from half an inch to an inch in length, and half an inch in diameter, changing from green to red, and black when ripe. Flowers in May; fruits in July.

Var. **frondosus**, Gray. Leaflets incisely toothed, smooth. Flowers more corymbosed, with leafy bracts, and roundish petals.

Var. **humifusus**, Gray. Stem trailing, and smaller peduncles; few-flowered.

5. **R. fruticosus.** (High Blackberry.) A plant common in the British Isles. But little attention is paid to it, and it has not been brought under cultivation.

Varieties. — There are some 20 varieties now under cultivation, differing from each other as to the quality of the fruit and the hardiness of the plant.

The varieties of the blackberry grown in our gardens, and from which our markets are supplied, are seedlings from the R. villosus, or high blackberry, found in our fields and fence-rows all over the Northern and Middle States. The Kittatinny was found in the Kittatinny Mountains in Warren County, New Jersey, growing without cultivation. The New Rochelle blackberry was found by Mr. Lewis Secor by the roadside in the town of New Rochelle, Westchester County, New York, and was called Secor's Mammoth. Mr. Lawton, of New Rochelle, took great interest in it, and propagated it in his nursery at New Rochelle. The names Lawton and New Rochelle blackberry are both due to this circumstance.

RUBUS VILLOSUS (High Blackberry).

Geography. — The Rubus villosus or high blackberry is an American plant, and grows freely all over the Northern and Middle States, in fence-rows, pastures, and edges of woods and old fields.

Etymology. — The specific name *villosus* comes from the Latin word *villus*, wool, and signifies woolly, a name applied to this plant because it is clothed with weak, long hairs. The common name *blackberry* arises from the color of the fruit when ripe.

History. — There is but little to be said of the history of the blackberry. The villosus, the parent of all the cultivated varieties, is an American plant, and has been used as a food-plant since the settlement of the country, but has only recently become an article of commerce. Of late years it has engaged the attention of fruit growers to a great extent, and many fine varieties have been produced by hybridizing and from seedlings.

Use. — It is a favorite dessert fruit, eaten with sugar or milk without cooking; it is preserved in sugar and brandy, and is canned; it is also prepared as a jam. A syrup made from it is used as a remedy in chronic stomach and bowel difficulties, because of the astringent properties it contains.

Marts. — On account of the perishable character of this fruit the markets must be local.

FRAGARIA, (Strawberry.) Tourn. Calyx concave, deeply cleft; sepals or divisions 5 in number, with 5 alternate bractlets; petals obcordate, white and large; stamens numerous; styles numerous; akenes naked, on the surface of a subglobular, heart-shaped, or irregular pulpy eatable receptacle. Perennial stemless herb.

1. **F. Virginiana,** Duchesne. (American Strawberry.) Without stem. Leaves and flower-stalks pubescent; leaves on long radical petioles, composed of 3 dentate leaflets, lateral ones oblique, nearly sessile; flower-stalks less hairy than the petioles. Flowers in a cyme; calyx erect. Flowers in April; fruits in May, June, and July.

2. **F. vesca,** L. (English Strawberry.) Calyx spreading or reflexed. Akenes superficial, not imbedded in pits in the receptacle. Otherwise as in F. Virginiana.

By propagating from seeds and by hybridizing, many varieties have been produced. American nurserymen catalogue about 400.

Geography. — The geographical range of the strawberry is very wide; in fact, it extends around the globe. Captain Cook speaks of the fine strawberries he found in great profusion in Kamchatka and Alaska, where they are still found to grow in abundance.

FRAGARIA VESCA (English Strawberry).

Etymology. — *Fragaria*, the generic name was given to this plant by Tournefort, on account of its fragrance; it is derived from the Latin *fragrans*, a pleasant odor. The specific name *Virginiana* is from the place where it was found native; and *vesca*, small, on account of the size of the fruit. The name *strawberry* is said to have arisen from the circumstance that in England straw was spread around the plants upon the ground for the fruit to rest upon to keep it from the sand and mud.

History. — There is very little history to this fruit. No mention is made of it until the days of Henry VI. of England, the last of the reigning sovereigns of the house of Lancaster, 1453, when a poem appeared which shows that strawberries were known in London at that time.

It is also related that when Gloster was planning the murder of Hastings, he requested the Bishop of Ely to send him strawberries, and Shakespeare makes him say : —

" My lord of Ely, when I was last in Holborn,
I saw good strawberries in your garden there."

ROSACEÆ.

The strawberry requires a generous soil of light loam to bring it to perfection; and during the ripening season it needs dry, sunny days and warm nights to perfect its aroma and taste. When practicable, the fruit should be taken from the garden immediately to the table.

Use. — The strawberry is a delicious fruit for dessert or for preserving. The mode of serving is well expressed in the line: —

"A dish of ripe strawberries smothered in cream."

It is not only noted for its delicate fragrance and delightful flavor, but has a high reputation for its healthfulness. It is related that the father of botany, Linnæus, was cured of a fit of gout by eating strawberries, which, if true, would establish its sanitary or curative properties.

Marts. — Markets for the strawberry must be local, on account of the perishable character of the fruit.

PRUNUS, L. (Plum, etc.) Ovary superior. Carpel 1. Style terminal. Ovules 2, pendulous. Drupe 1-seeded. Trees or shrubs.

1. P. domestica, L. (Damson Plum.) Stem from 4 to 6 feet to the point where the head begins to form, and from 4 to 6 inches in diameter; much-branched, forming an open head about 15 feet in diameter; whole tree 10 to 20 feet in height. Leaves ovate-lanceolate, acute or obtuse, varying very much in shape, 1 to 3 inches long, and three fourths of an inch wide; petioles about 1 inch long. Flowers white, usually solitary, appearing with the leaves. Fruit very dark, varying to nearly white, clothed with a glaucous bloom. Stone smooth, more or less flattened. Flowers in April and May. Fruit ripens in August.

The number of varieties is very great; about 300 are catalogued by the nurserymen and fruit-growers, differing in shape, size, color, or taste of the fruit.

Geography. — The plum is widely distributed; it is found in all parts of the temperate zone south of 60°, throughout Europe and western Asia. It flourishes best in the northern and throughout the middle regions, and is so well spread throughout western Asia as to make it difficult to fix upon its native home. It was brought to northeastern America by European colonists.

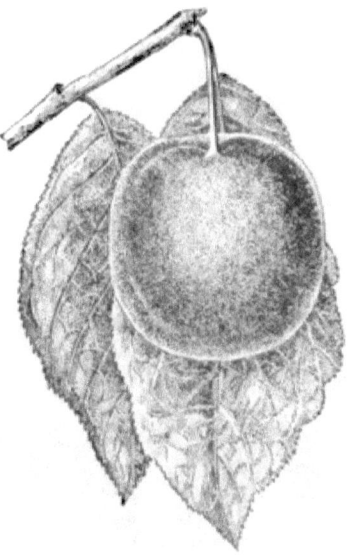

PRUNUS DOMESTICA (Damson Plum).

Etymology. — The generic name *prunus* is from the Latin, *prunus*, a plum. The specific name, *domestica*, given by Linnæus, explains itself, — the house-plum, or cultivated plum. The word *plum* is of obscure signification; no meaning is known for it.

History. — The plum is a native of northern Persia. It has been thought by some botanists to be either indigenous in Europe or well naturalized. This, however, is disputed by De Candolle, and he gives good reasons for his opinion; yet it has been found in the ruins of the Swiss lake-dwellings.

Preparation. — The prunes, so well known in our markets, are dried plums. They are sold under the name of Turkish, French, and German prunes. Those prepared in Turkey are mostly dried in the sun; but the German and French prunes are kiln-dried, and the German fruit, especially, frequently has a smoky taste.

The French prunes are prepared by first exposing them to the sun in thin layers on frames made of wicker-work. They are then placed in slightly heated ovens, removed, turned, and put back. A slight heat is kept up, and after twenty-four hours they are again withdrawn and turned; the oven is then heated to 120° Fahrenheit, the fruit is again put in and left a day, after which it is packed in boxes holding about fifty pounds each. The finest fruit is gathered, dried as described above, and carefully packed, each one put in singly in small boxes weighing from five to ten pounds each, and sent to market for dessert purposes. Some sorts are used as remedies to regulate the bowels. Large quantities of prunes of an excellent quality are now grown and prepared for market in southern California. In Bosnia, Servia, Spain, Portugal, and southern France, the industry of preparing prunes is also largely carried on.

Use. — The plum, though not so delicate as the peach and apricot, is nevertheless a delicious and favorite dessert fruit, and highly esteemed for culinary purposes. For pies, tarts, preserves, and canning it ranks high, and there is no fruit dried that enters so largely into commerce. The French and Turkish prunes are well known to every housekeeper in our cities and towns.

Because of its hardihood, the plum is one of the most valuable fruit-trees for the farmer. It is not particular as to soil, and the crop is not likely to be destroyed by spring frosts.

2. **P. avium**, W. (Ox Heart. English Cherry.) Trunk 6 to 8 feet to the point where the head begins to form, and from 10 to 18 inches in diameter; bark smooth or cracked. Branches erect, forming a compact head; entire tree from 20 to 40 feet in height. Leaves oblong-ovate, acuminate, hairy beneath, and double-toothed, about 3 inches long. Flowers in umbels, appearing with the foliage. Fruit globular, ovoid, or heart-shaped. Flowers appear in May; fruit ripens in June and July. Drupe smooth, no bloom. Stone smooth, globular.

3. **P. cerasus**, L. (Morello, or common Red or Sour Cherry.) Trunk 6 to 12 inches in diameter; head low and globular. Leaves serrate, acute. Fruit globular, red, acid, esteemed for preserving.

The cherry sports freely, and we have many varieties; American nurserymen catalogue about 500.

The French divide their varieties into three sections: Griottes, tender-fleshed; bigarreaux, hard-fleshed; guignes, small-fruited cherry.

The Romans had eight varieties during the first century.

Geography. — The cherry grows well throughout the temperate zone wherever the apple flourishes, and even further north than the apple. It has spread over northern Africa, and the Dutch and Portuguese have taken it to southeastern Africa. It was brought to America by European colonists, where great attention has been given to its cultivation. The climate of England suits the cherry, and Belgium and the British Isles produce the best cherries in the world. Cherry-trees in blossom are greatly prized by the Japanese as ornaments to their gardens.

Etymology. — The specific name, *avium*, is derived from the Latin word *avis*, a bird, and arose from the circumstance that birds are fond of this fruit. *Cherry*, the common name, is a corruption of the old Greek name *cerasus*, a name applied to this fruit because it was found growing at Cerasus, a town in Pontus.

History. — This delicious fruit is said to have been brought from Armenia to Italy by Lucullus, a victorious general, about seventy years before the commencement of the Christian era, whence it spread westward, and was no doubt carried to England in the days of Agricola. Its popularity may be inferred from the circumstance that it spread over southern and middle Europe in a very short time; for about the beginning of the second century of the Christian era it was to be found in the grounds and gardens of the wealthy throughout northern Italy, Spain, France, southern Germany, and England, — this too at a period in the world's history before agriculture or fruit-growing had attained any scientific importance.

In the southern parts of Europe and in northern Africa the cherry was called the "berry of the king."

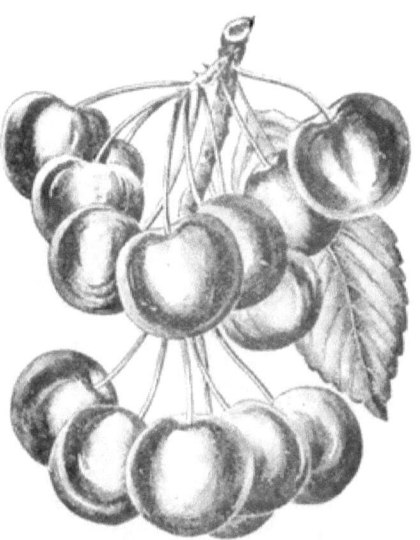

PRUNUS CERASUS (Common Sour Cherry).

For many years this fruit has been a favorite in Germany. Loudon, in his account of trees, says that in Germany and Switzerland the local governments encourage the inhabitants to plant trees, and in some neighborhoods the highway passes through avenues of cherry-trees, to the fruit of which the traveller is at liberty to help himself, provided that he takes no more than he can eat on the spot. In Moravia the highway from Brunn to Olmutz, a distance of sixty miles, passes through an avenue of cherry-trees.

The ancient home of the cherry is believed by De Candolle to be the country south and east of the Black Sea, extending from the Caspian to the Propontis. He believes also that the cherry was known to the Romans before the days of Lucullus, and that he only introduced an improved variety, which gave rise to the supposition that he had brought to Rome a new fruit.

Use. — The cherry varies in form from globular to heart-shaped, and in size from half an inch in diameter to three quarters, and from three quarters of an inch to an inch and a quarter in length. It is the most delicious fruit of its season. It is highly prized as a dessert. It excels as a fruit for pies, puddings, and tarts, is a favorite preserve in sugar or brandy, and is largely canned and dried.

4. **P. Armeniaca**, Willd. (Apricot.) Calyx 5-cleft, regular, falling; petals spreading; stamens 15 to 30. Fruit a drupe, fleshy, usually glaucous, covered with bloom; the pit or nucleus ovate, compressed, smooth; kernel shape of the shell. Small tree. Stem 4 to 5 feet to the point of ramification, and 3 to 5 inches in diameter. Branches numerous, spreading, forming an open head; whole tree 10 to 20 feet high, and about 15 in diameter through the head. Leaves ovate, acuminate, subcordate at the base, denticulate conduplicate in the bud. Flowers sessile and solitary, white, appearing before the foliage; petioles with one or more glands each. Fruit yellow. It flowers in April and fruits in July and August.

There are many varieties of this tree. Nurserymen in the United States catalogue 120, based upon some form or quality of the fruit.

Geography. — The geographical zone of the apricot is tropical and subtropical, reaching as far north in Europe as the 40th parallel, and in America to the 38th.

PRUNUS ARMENIACA (Breda Apricot).

It has become very widely spread in Asia, Armenia, Persia, China, and Japan. Its great abundance and high quality in Armenia led the early botanists to suppose it indigenous there, but it is quite as common both in China and Japan, and in the latter country the tree grows to a larger size than elsewhere. It grows well in the Southern United States, and especially in California, where it is canned in large quantities for the eastern market.

Etymology. — The specific name indicates that the Greeks and Romans received it from *Armenia*, and therefore called it *Armeniaca*. *Apricot* seems to be derived from the Latin words *præcox*, early, and *arbor*, tree, early tree, on account of the early ripening of the fruit.

History. — The apricot was carried into Europe about the beginning of the first century of the Christian era. It is supposed to be a native of China as well as of western Asia. It attains to great perfection in Persia; and on account of its delicate flavor the Persians call those grown in the vicinity of Ivan, "the seed of the sun." The French naturalist, M. Regnier, thinks the apricot is a native of Africa, and has spread by cultivation into Asia and Europe. De Candolle, in his "Origin of Cultivated Plants," gives substantial reasons for believing that the apricot is a native of China, introduced thence into western Asia, and finally into Europe, and by European colonists brought to northeast America, where considerable attention has been given to its cultivation.

Use. — The apricot is a beautiful smooth fruit, smaller than the peach, but closely resembling it in form, and only a little less delicious in flavor. It is used as a dessert, and is highly prized for preserving, either in sugar or brandy. For puddings, pies, and dumplings, it is next to the peach in favor.

Marts. — San Francisco is the great shipping mart in America for the canned fruit.

5. P. Persica, L. (Peach.) Calyx tubular, bell-shaped, 5 parted. Petals 5. Fruit either smooth or tomentose. The stone or pit rugose or wrinkled, ovate, flat, or compressed, acute, separating lengthwise when struck with a hammer, inclosing a kernel much the shape of the horny shell, charged with Prussic acid. Small tree. Stem about four feet to the point of ramification, and 4 to 8 inches in diameter, forming a symmetrical head, from 12 to 20 feet high and 10 to 15 feet in diameter. Leaves lanceolate, serrate, shining above. Flowers solitary, rose-colored, appearing in May. Fruit ripens August, September, and October.

The peach, as it is propagated from the seed, sports, and many varieties occur. Charles Downing catalogues over 400. About 80 of these are sufficiently constant to be relied upon. Of this number a few are clings, so called because the pulpy envelope adheres or clings to the stone; when the flesh comes off freely, or separates from the stone, the peach is called freestone.

Var. **lævis** (Nectarine) is an important variety, having a number of sub-varieties. It is a tree like the peach in form and size, flowers and leaves. The fruit is glabrous or downy.

Darwin considered the peach a variety of almond, and the nectarine a form of peach, and held that the same tree has borne in successive years peaches and almonds, and that *peach* trees have produced *nectarines;* also that the seeds of the *peach* have produced *nectarine trees.*

On the other hand Decaisne and A. P. De Candolle consider the peach and the almond distinct species, each possessing peculiar botanical characteristics.

PRUNUS PERSICA (Peach).

Again, the home of the almond is western Asia; that of the peach China, where it was grown centuries before the Chinese became acquainted with the almond.

The peach is not at the present day found in a wild state, though it escapes from cultivation, and propagates itself freely, especially in the southern part of the north temperate zone.

Geography.—The geographical zone of the peach is the southern half of the north temperate zone in Asia, Europe, and America.

In America, south of 40° north latitude and north of 32°, the peach is an important crop. In the United States, southern New Jersey, Pennsylvania, and the states of Delaware and Maryland are largely engaged in its cultivation.

Etymology.—*Persica*, the specific name of the *peach tree*, is due to the circumstance that the Greeks and Romans received it from Persia. The common name, *peach*, is supposed to be a corruption of the low Latin name *pesku*.

History.—It was thought for some time that Persia was the home of the peach, but De Candolle, in his "Geographical Distribution of Plants," and again in his recent "Origin of Cultivated Plants," shows conclusively that

China is the place of the peach's nativity, whence it has travelled westward to Persia, and finally to Europe.

It was introduced into Italy from Persia by the Romans, in the reign of Claudius Cæsar. It was introduced into Great Britain during the sixteenth century, and thence brought in 1680 by the settlers of Virginia to America, where it grows to great perfection. It does not ripen well in England without the protection of glass, or at least of walls. It is a considerable crop in France, especially in the south. In China and the United States it reaches the greatest perfection.

The peaches of Pekin are said to be the finest in the world. But Delaware and Maryland raise peaches whose lusciousness it is difficult to equal and impossible to excel.

It is related that in Persia the delicious flavor of the peach was supposed to be accompanied by poisonous qualities. Pliny, in his work on plants, says it was supposed that the king of Persia sent the peach into Egypt to poison the people with whom he was then at war.

It is also noteworthy that the peach-tree holds the same place in the ancient writings of the Chinese that the tree of knowledge does in the sacred Scriptures, and that the golden Hesperides apples of the heathens hold among the western nations. There are traditions of a peach-tree whose fruit possessed the power of producing immortality upon those who partook of it, but which bore fruit only once in a thousand years. There is also preserved in the early books of the Chinese, an account of a peach-tree which existed in the infancy of the nation, growing on a mountain whose approaches were guarded by a hundred demons. The fruit of this guarded tree, when taken, produced instant death.

Use.—The peach ranks with the pear as a dessert fruit, and if it were not for its perishable character it would grade far above the pear.

The season of the peach is very short, lasting from August through September and October, to November. For pies, puddings, dumplings, and preserves it has no equal.

The fermented juice of the peach is distilled, and produces a highly prized brandy.

The commercial value of the peach crop, including the large quantities that are dried and canned, is of great importance.

Marts.—New York is the point to which most of the fresh crop is brought. Baltimore in Maryland, and Dover in Delaware, are the centers of canned and dried peaches. Canning for winter use and export has become a very important industry.

6. **P. communis**, L. (Amygdalus communis.) (Almond.) Stem 10 to 12 feet high, branching into a symmetrical head, entire tree reaching the height of 25 feet. Leaves oblong-linear or lanceolate, tapering towards the base, serrate and glabrous. Flowers developing before the leaves, white or pinkish, appearing in March and April. Drupe tomentose or stone furrowed, compressed.

Var. **amara**, De Candolle (Bitter Almond), is the variety producing the bitter almonds of the market. Flowers larger, pink, tinged with rose; nut hard. A sub-variety has brittle shells.

7. **P. nana**, L. (Dwarf Almond.) Differs from the P. communis in being a shrub 2 to 3 feet in height. Flowers solitary with a colored calyx.

Varieties.—As the almond is produced from seed it sports freely, and varieties are numerous. There are about a dozen varieties under cultivation, differing in the size and quality of the fruit and the fruit envelopes.

ROSACEÆ. 123

Geography. — The geographical limit of this fruit is between 30° and 42° north latitude. The tree grows well in the latter parallel, but does not fruit freely north of 45°. It fruits well in Virginia, and as far north as central New Jersey.

Etymology. — The old name, *Amygdalus*, is from the Greek ἀμύσσω, lacerate, due to the gashes and fissures in the shell or husk. The specific name, *communis*, Latin, signifies "common," and *nana* means "dwarf." The variety name *amara* is also from the Latin, and signifies "bitter."

History. — The home of the almond is Persia and western Asia. It is also indigenous throughout the countries of the Levant, and was no doubt carried thence to northern Africa, southern Europe, and eastern Asia, and by European colonists was brought to the United States.

Use. — The sweet almond is a favorite nut, and is much esteemed as a dessert. It is largely used in confectionery and sauces. The bitter almond is used in cookery, for flavoring, and in perfumery. The nut constitutes an important article of commerce. Turbid water from the river Nile is cleared by rubbing bitter almonds on the inside of the vessels which hold it.

Oils. — Fixed or sweet oil of almonds of commerce is obtained by pressure from both sweet and bitter almonds. When bitter almonds are used, the residuum or cake is subjected to fermentation, and the volatile oil or essential oil of bitter almonds is obtained therefrom by distillation. This oil contains Prussic or hydrocyanic acid, in its concentrated form a virulent poison. Properly diluted, essential oil of almonds is a pleasant and wholesome substance for flavoring custards, puddings, etc.

PYRUS, L. (Pear, Apple.) Calyx superior to ovary, pitcher-shaped, 5-cleft. Petals 5, roundish. Stamens many. Styles 5, frequently united at the base. Fruit a fleshy pome, with 5-2 carpels, consolidated with the fleshy calyx-tube. Trees, with mostly simple leaves and free stipules.

1. **P. malus,** L. (Apple.) Stem 5 to 10 feet to the point where ramification begins to form a diffusely branched head from 10 to 20 feet in diameter, and 20 to 30 feet high. Branches slender. Leaves ovate, serrate, acute, crenate, woolly on the under surface, glabrous and shining above. Flowers in sessile corymbs, roseate, appearing with the leaves. Fruit spherical. Carpels 2-seeded. May. Fruiting from July to October.

The number of varieties of the apple is very great, and as it is propagated from the seed, it sports freely, and new varieties are constantly arising. Of late years the number has been greatly increased. In 1870 the number grown in the United States was over 1500, every one of which claims some desirable quality, and amateurs are yearly adding to the vast catalogue; yet so frequently do some of these varieties deteriorate or die out that only about 89 are regarded as constant.

Geography. — The apple-tree flourishes in the parts of the north temperate zone between the parallels of 35° and 50°. It does not fruit well south of that limit unless in elevated localities. England, France, Germany, the Netherlands, Prussia, ancient Poland, the United States, and southern Australia are the most important apple-growing countries. The varieties which have originated in America are numerous, and some of them for size and delicacy of flavor excel any yet produced in Europe.

DESCRIPTIVE BOTANY.

Etymology. — The word *pyrus* comes from the Celtic word *peren*, signifying pear. *Malus*, the specific name, is the old Latin name for apple.

The common name, *apple*, is said to come from two Sanscrit words meaning water-fruit. Others derive it from *abala*, or *ab*, a ball, and *ala*, little, a little ball-shaped fruit. The Latin word *pomum* favors the signification of a watery fruit, inasmuch as it comes from *po*, drink.

History. — This fruit was known and extensively used by the Swiss lake-dwellers. They preserved it by cutting it lengthwise and drying it in the sun.

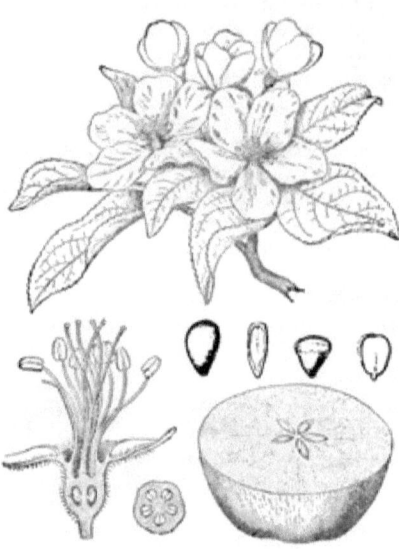

PYRUS MALUS (Apple).

The home of the apple is supposed to be Persia and the northern regions of Asia Minor. It is found without cultivation in northern India and middle China, and throughout middle Europe.

It is held by some that all our varieties are from the crab apple. While apples are spoken of in the Old Testament, it is not certain that the fruit referred to is our apple. It was taken to Rome 450 years before the Christian era, but was confined to the gardens and planted grounds of the wealthy for many years. Pliny, in his book on plants, written in the first century, speaks of the apple as a profitable crop on account of its scarcity in the vicinity of Rome.

It seems that engrafting was practiced at that time; for the same writer speaks of apples that will honor the fruit-grafter forever.

When this method of improving fruit was first used is not known. The Romans had 22 varieties, but no doubt some of these were natural.

At the time of Vergil, apples must have been common, and were no doubt an important article of food for the peasants; for the poet in the First Eclogue makes Tityrus say to Meliboeus : —

"Stay with me to-night, for I have *ripe apples*, soft chestnuts, and plenty of cheese."

The apple was brought to northeast America by European colonists, and was the most important fruit grown in British North America. The attention paid to the apple in the early days of the colonies is due, no doubt, to the superior intelligence of many of the first settlers. Its distribution in western Pennsylvania and Ohio was greatly aided by the efforts of an eccentric man who appeared in the settlements of western Pennsylvania early in the present century and busied himself about cider-mills in collecting apple-seeds from the pomace. Having placed the seeds in sacks, he loaded them upon a horse and proceeded to the unoccupied lands in Pennsylvania and Ohio, and wherever he found a clearing, he planted his seeds; then felling trees around the plantation to protect the young growth, he went on to new fields, and as years passed on he returned to exact payment from those who had settled on the lands. He was a benevolent, inoffensive man, distributing religious books among the people with whom he put up at night. His name was Jonathan Chapman. He was born in Boston in 1775. Very little was known about him. On account of his strange employment he received the name of Johnny Apple-seed.

In the door-yard of Delos Hotchkiss, at Cheshire, Conn., stands an apple-tree which is supposed to be the oldest, largest, and most fruitful in New England. It is the last survivor of the orchard which was set out by the first settlers of that neighborhood, and popular belief fixes its age at 180 years. The tree is sixty feet high, and the tips of its outermost branches are one hundred and four feet apart. Mr. Hotchkiss affirms that he has picked 125 bushels of sound apples from it in a single year.

Use. — The apple must be regarded as foremost among the fruits of the temperate zone. No other fruit is so agreeable to all palates, and so generally used. There are so many ways in which it serves man, and they are so generally known, that it seems superfluous to attempt to name them. A moderately sized apple, of any variety, either sweet or sub-acid, is a very popular fruit for dessert. For pies, puddings, dumplings, and sauces it should be sour or tart. The farmers of Pennsylvania make "*apple butter*" by boiling sliced apples to a pulp in new cider. In the same manner apples are cooked in sweet wine in France, and the preparation is called *raisiné*. Apple juice, when fermented, is cider, and forms a common table drink among farmers, as wine does in the wine-making districts of France, Germany, Italy, and Spain. Verjuice is the fermented juice of the *crab apple*. Cider when exposed to the air soon becomes sour or hard, from the formation in it of lactic acid.

Apples are preserved by drying them in the sun. In late years large quantities have been dried by steam heat. Apples form an important food, and large quantities are exported to tropical and subtropical countries.

The wood of the apple-tree is close-grained, hard, and it takes a polish. It is valuable for turners and cabinet-makers, and is largely used in the manufacture of shoemakers' lasts.

2. **P. communis**, L. (Pear.) Stem from 20 to 40 feet in height, and from 8 inches to 20 in diameter, branching; the branching is upright, forming a pyramidal head. Leaves ovate, lanceolate, acute, sometimes acuminate, somewhat crenate, serrate, glabrous. Flowers in corymbs, white and fragrant, appearing in May. Fruit pyriform, ripening from July to October. Carpels 2-seeded.

The number of varieties of this tree is very great. There is no single fruit upon which more care and expense has been lavished than upon the pear. The nurserymen in the United States catalogue about 3,000 varieties, each one of which is represented to possess excellencies to recommend it to cultivators; but the pear fanciers of France and Belgium publish lists of far greater numbers. It is related of a single nurseryman in Belgium that he had growing at one time 80,000 seedlings for the purpose of developing new varieties.

The varieties have reference to the character of the fruit alone.

Geography. — The geographical zone of the pear is from 35° to 55° north latitude. It is native to China, Syria, Persia, central and northern Europe, and Great Britain. It was brought by colonists to northeast America. It thrives wherever the apple flourishes, but arrives at its maximum excellence in size and flavor in Belgium and northern France.

Etymology. — The specific name of this tree signifies ordinary or common, from the Latin word *communis*.

The varieties are named usually to indicate some quality of the fruit, or in honor of some person, or they take the name of the places whence they originate.

History. — The pear is a native of Europe. It is spoken of by Homer, who says it was one of the trees in the garden of Laertes, the father of Ulysses.

Pliny also mentions several varieties growing about Rome. There is good authority for believing that the Romans cultivated about thirty-six varieties; and it is believed that they took their choice varieties to England after the middle of the first century, during the administration of Agricola. It could not have been largely cultivated, and was no doubt confined to the gardens of the wealthy for a long time; for, when spoken of, reference is made to its delicacy. A record is preserved that in the days of Henry VIII. twopence was awarded an old woman for presenting pears to the king. During the sixteenth century much attention was paid to its cultivation. Near the end of the sixteenth century Gerard published his herbal, in which he says that the number and sorts of pears and apples would require a book to hold their description.

The best pears have originated in France and Belgium, so that nurserymen have come to regard Belgium as the Eden of this fruit.

There are some remarkable pear-trees whose size and age may be mentioned in this connection. Several on the continent of Europe are known to be 400 years old, but the most wonderful pear-tree is one in Hertfordshire, England, from which were made in one year fifteen hogsheads of perry. In 1805 it covered more than half an acre of land; its branches had bent to the ground, taken root, and thrown up shoots. By favoring this habit, a grove of trees all connected has been produced. A very remarkable pear-tree is now growing in Indiana, about ten miles north of Vincennes. It was planted by Mrs. Ockletree about the year 1805. The circumference of this tree a foot above the ground is twelve feet, or about four feet in diameter. In 1834 it bore 184 bushels of pears, and in 1840 the yield was 140 bushels. The fruit is of large size, of good flavor, and ripens in early autumn. It bore its first crop in 1820, and has borne a crop every year since.

The old Stuyvesant pear-tree, which was planted by Peter Stuyvesant, one of the Dutch governors of the New Netherlands, on his farm in the city of New York about the year 1645, was blown down in 1867, making it about two hundred and twenty years old. It stood on the edge of the sidewalk in Third Avenue on the corner of Thirteenth Street.

When the pear was first used for fruit or brought under cultivation is not known; it has been stated that there is evidence that it was in use in the days of the lake-dwellers of Switzerland, and it is represented in the pictures of Pompeii.

Use. — The apple is without doubt the most important fruit of the temperate zone, on account of the great variety of forms in which it may be prepared as an article of food, and the ease with which it may be preserved; but for delicacy of flavor, the pear takes the precedence. It is more highly valued as a dessert than any other fruit, and is found upon the tables of the wealthy throughout its season. The favorite mode of use is without cooking or any other preparation.

It moreover constitutes one of the most delicate preserves, either as a candied or brandied fruit. It cooks well as a preserve or a baked food; but on account of its lack of a pungent acid it is not suitable for pies, puddings, or sauces.

ROSACEÆ.

The juice of the pear when fermented is called perry, and is used as a table beverage, as wine is used.

3. **P. Cydonia**, L. (Quince.) Stem branching very near the ground, from 3 to 6 inches in diameter; branches very crooked and irregular, sparingly armed with spines. Leaves oblong, ovate, obtuse at the base, and acute at the apex; entire, smooth above, woolly beneath. Flowers solitary, large, on short pedicels, at the ends of the branchlets; calyx lobes expanding into small tomentose leaves. Fruit globular or pear-shaped, golden-yellow when ripe; clothed with a soft down in an unripe state. Seeds numerous. Flowers appear in May. Fruit ripens in October.

Like other fruits grown from the seed, the quince sports, and we have varieties. The nurserymen in the United States catalogue about 20 varieties, each of which has some quality to recommend it to growers.

Geography. — The geographical zone of the quince is between 35° and 60° north latitude, extending from northern Persia both east and west. It was brought to northeast America by European colonists. It reaches great perfection in Portugal.

Etymology. — The name *Cydonia* is derived from *Cydon*, in the island of Crete. The common name is supposed to be a corruption of one of the various names by which it is known, most likely from the Middle English name *coine*.

PYRUS CYDONIA (Quince).

History. — The quince was known to the ancient Greeks and Romans, and was highly esteemed by them. It is a native of northern Persia, and was found in the island of Crete. It has spread westward, through middle and northern Europe.

Use. — The quince is a well-known fruit of the apple family. It is not edible in a raw state, but is valued as a flavorer of other fruits, especially of apples. It is highly esteemed for preserving, and for marmalades and jellies. It is also preserved by drying. In England it is used for wine-making. The seedlings are largely employed by nurserymen for grafting pears upon. The seeds yield large amounts of mucilage used in preparations for hair-dressing, and as a lotion for sore eyes.

4. **P. Japonica** (Pyrus Japonica, or Japan Quince) is a native of Japan. The fruit is not edible; but the flowers are very showy, and the tree is a favorite shrub for ornamental purposes and for hedging.

Order XXIII. SAXIFRAGACEÆ.

Flowers perfect, regular, occasionally irregular, variously arranged. Sepals 4-5, more or less cohering, adherent to ovary. Petals 4-5, inserted on the rim of the calyx. Stamens equal to the number of petals alternating with them, or 2 to 10 times as many. Ovary usually more or less inferior. Fruit mostly a 2-celled capsule or berry; seeds small. Leaves alternate or opposite, sometimes whorled. Herbs or shrubs. Number of genera, 73; of species, 540.

Ribes rubrum (Common Red Currant).

RIBES, L. (Currant. Gooseberry.) Calyx tube adherent to the ovary, 5-parted. Petals 5. Stamens 5, alternating with the petals. Ovary 1-celled, with 2 opposite parietal placentæ. Styles 2 in number. Fruit a succulent berry crowned by a persistent calyx. A shrub.

1. **R. rubrum**, L. (Common Red Currant.) Stems numerous, slender, sparingly branched, 2 to 4 feet high. Leaves obtusely 3-5-lobed, smooth above, pubescent beneath, 2 to 3 inches long, about as wide as long; unequally toothed, incised; petioles as long as the leaves. Flowers in pendent racemes, not axillary; bracts ovate; petals small, greenish-yellow. Berries globular, from two to three tenths of an inch in diameter, red when ripe. Flowers in April; fruit ripens in June and July.

The currant sports freely, and many varieties are under cultivation. About 70 varieties are catalogued by American nurserymen, differing from the species in size and quality of the fruit.

Among the favorite varieties are: Cherry, Versailles, Red Dutch, Red Grape, White Dutch. One or all of these may be found in every well-furnished garden.

2. **R. floridum**, Heretier. (Black Currant.) Like R. rubrum as to habit of the stem. Leaf blunt or subcordate at base, sharply 3-5-lobed, sprinkled with resinous dots, and doubly serrate. Flowers abundant and showy, white. Fruit ovoid, and black when ripe. Sometimes cultivated for the flowers.

SAXIFRAGACEÆ.

3. **R. nigrum**, L. (Black Currant.) Stem as in R. rubrum, but higher. Leaves dotted with resinous spots, 3-5-lobed. Flowers whitish-green. Stamens sometimes more than 5; for every additional stamen over 5, there is one less petal. Stigma bifid. Fruit globose, black. Flowers in May; fruits in June and July. Introduced from Europe.

Geography. — The zone of the currant is from Lapland to southern Europe, extending quite across the continent of the Old World and the northern parts of the United States and southern and middle Canada. It delights in cool, damp grounds.

Etymology. — The word *ribes* is supposed to be the Arabian name for the plant. De Candolle, however, believes it to come from the Danish word *ribs*, by which the plant is known. The specific names *rubrum* and *nigrum*, red and black, from the Latin, are due to the color of the fruit of these species. The wild black currant of North America, R. floridum, was thus named on account of the showy flowers of that species. The common name, *currant*, is supposed to be a corruption of the word Corinth, a name applied to the small seedless grapes of southern Greece, which are dried and taken to Corinth for export under the name of currants. (See Grape.)

History. — When or where the currant of our gardens was first cultivated is not known, but its usefulness and popularity point to its early cultivation, so that we may date its origin at a very early period in the history of agriculture, as we are justified in the inference that as soon as man began to reside in fixed habitations he commenced to gather such shrubs and trees around his dwellings as he found producing edible fruit.

The Dutch have been very successful in producing from seedlings varieties that are now under cultivation both in Europe and America. The currant is found in a wild state in north and middle Europe from eastern Siberia to Great Britain, and in North America quite across the continent, and north to the Mackenzie River. In northeastern Russia and Siberia the currant is employed for wine-making.

Use. — The currant is used sparingly for a dessert fruit, but largely for jellies, and to some extent for wine-making. It is preserved in sugar, and canned. It is used in a green state for tarts. The dried currant of commerce is a small grape.

No other small fruit is more generally cultivated. It is not only grown in the gardens of the rich, but is also to be found in the planted grounds of the most humble cottager.

Marts. — The markets, on account of the perishable character of the fruit, are local, except for the dried and preserved fruit.

4. **R. Grossularia**, L. (Gooseberry.) Stem 2 to 3 feet high, numerous, slender, 2 or 3 prickles under each bud. Leaves 3-5-lobed, villous. Flowers greenish; pedicels 1-2-flowered; calyx bell- or pear-shaped; segments reflexed, shorter than the tube; petals rounded at the apex, bearded in the throat; style beset with long down. Fruit usually dark-red when ripe, globular or ellipsoid, and in the cultivated varieties from an inch to an inch and a half in diameter. Flowers in early summer, fruits in August. Indigenous in Europe. Its varieties are the cultivated gooseberry.

There are about 100 varieties under cultivation in the United States. One of the most popular, though by no means the largest, is "Smith's improved." The American varieties are inferior to those of the British Isles.

5. **R. cynosbate**, L. (Wild Gooseberry of Canada.) This species grows without cultivation in the northern United States and Canada. The stem is spar-

ingly beset with sharp prickles, in pairs, just below the leaf. Leaves cordate, lobed, cut-toothed, and pubescent; fruit armed with prickles. Flowers in May; fruits in August. This has not been improved by cultivation, though for many years it has been an occupant of our gardens in the Northern States and Canada.

6. **R. oxyacanthoides**, L. (R. hirtellum, Mx.) (American Gooseberry.) Stem as in the last; rarely prickly; spines, when present, short and solitary. Leaves rounded, cordate, 3–5-lobed, cleft half-way to the middle, toothed, pubescent underneath. Flowers drooping, green, on short 1–2-flowered pedicels; calyx tube smooth, bell-shaped; segments much longer than the petals; stamens protruding; style hairy, 2-cleft. Berry purple and smooth, small. This species is cultivated in gardens, but does not improve by cultivation. Flowers in May; ripens in August. Northern United States and Canada, and west and north.

7. **R. rotundifolium**, Mx. Stems 3 to 4 feet high, numerous, slender; bark whitish, frequently without spines, subaxillary one solitary. Leaves roundish, smooth, crenate-dentate, slashed, with 3–5 lobes, truncate at base, shining above; petioles ciliate, 2 inches long. Flowers yellowish-white; calyx cylindrical, smooth, segments linear, reflexed; petals spatulate; stamens protruding; style smooth. Fruit smooth, purple, very pleasant to the taste. Flowers in May; fruit ripens in August. Northern United States and Canada to North Carolina. Edges of open woods.

The last three are frequently found in gardens, but they have not been improved either in quality or size.

RIBES GROSSULARIA (Gooseberry).

Geography. — The varieties of the gooseberry under cultivation are the offspring of plants found in a natural or wild state in England, France, and Germany.

It also has its representatives in the Alps, the Himalaya mountains, and throughout the northern United States and Canada.

It is indigenous in northeastern Russia and Siberia, and along the Valdai hills, and the cold bogs of the lowlands of central Russia.

Etymology. — The specific names are all Latin words. *Grossularia* signifies "thick," referring to the size of the fruit. *Hirtellum* has reference to the roughness of the plant. *Rotundifolium* means "round-leaved." *Cynosbate* signifies "briery," referring to the spines on the stem. *Gooseberry*, the common name is said to have arisen from the fact that a sauce made of gooseberries was eaten with roast goose. Another derivation is from *groise*, a berry (old French), corrupted into gooseberry.

History. — The gooseberry, like the currant, has no history which points to its first introduction into the garden. The fruit arrives at its greatest perfection as to quality in the Scottish highlands; but as to size, the operatives in the factories of Lancashire, in England, raise in their little gardens the largest gooseberries known. Specimens have been exhibited measuring 2 inches in diameter. The large varieties do not arrive at perfection in the United States. The climate of England, and the damp, cool atmosphere, seem exactly suited to their full development, while the hot suns of the northern United States seem to induce a mould that prevents perfection in the fruit while the plant flourishes. The great berries grown in England do not possess the high flavor of the smaller berry grown in Scotland.

Use. — The gooseberry is well known, and almost as common as the currant in our gardens.

It is used in a green state for pies, puddings, and tarts. When quite ripe, the fruit is used as a table dessert. It is also canned, preserved in sugar, and, when nearly ripe, bottled in water. It is set in a vessel of cold water, brought to a boiling heat, then corked and kept in a cool cellar, with the neck of the bottle down. In this way it will keep for an indefinite period. It is also a favorite fruit for making jam.

Marts. — The markets, like those of the currant, are local.

Order XXIV. **COMBRETACEÆ**.

Flowers perfect, or occasionally imperfect by arrest, in axillary or terminal spikes, racemes, or heads; a bract to each flower, also 2 lateral opposite bractlets. Calyx superior, 4–5-lobed, valvate in the bud. Corolla sometimes absent; when present, the petals equal in number the parts of the calyx. Stamens are inserted alternately with the petals on the calyx, and are either equal in number to the parts of the corolla, or double. Ovary inferior, with 2–5 pendulous ovules, 1-celled. Fruit a drupe, or berry; seed solitary, frequently winged; testa thin, membranous. Leaves alternate or opposite, simple, penninerved, entire or toothed, leathery. Trees or shrubs, sometimes climbing. Number of genera, 7.

TERMINALIA, L. Flowers polygamous. Calyx tube cylindrical, adherent to the ovary, contracted above; limb bell-shaped, 1–5-toothed, deciduous; corolla wanting. Stamens 10, inserted on the calyx; filaments awl-shaped; anthers 2-valved, egg-shaped or globular, opening lengthwise. Ovary inferior, 1-valved; ovules 2–3. Style awl-shaped; stigma sharp. Large trees.

1. **T. chebula**, Retz. (Myrobalans.) Trunk 40 to 70 feet in height, regularly branched, in verticils, forming a symmetrical head. Leaves ovate, on short petioles, alternate, entire or slightly toothed, collected at the ends of the branchlets, spotted. Flowers in spikes or racemes. Fruit a drupe, about the size of a prune.

Species. — There are about 80 species of Terminalia; those besides the T. chebula whose products enter into commerce are —

2. **T. bellerica**, Roxb.
3. **T. citrina**, Roxb.
4. **T. catappa**, L.
5. **T. angustifolia**, Wight.

Geography. — The homes of all the species which yield the myrobalans of commerce are in tropical India, along the southern fringes of the Ghaut mountains, and in Burmah.

Etymology. — *Terminalia*, the generic name, is from the circumstance that the leaves are usually at the ends of the branches, and is derived from the Latin *terminalis*, belonging to the end. The specific, *chebula*, is Arabic, but the signification is not known.

TERMINALIA CHEBULA
(Myrobalans).

Myrobalans is the old Latin name for the fruit, through the Greek μύρον, sweet juice, and βάλανος, a drupe-like fruit.

History. — The products of terminalia were unknown to the early botanists. Their medical qualities were revealed in the writings of Arabian naturalists, but especially by those of Prince Mesues, a learned physician who lived about the middle of the twelfth century. The fruits were first introduced into Europe by the way of Arabia and the Red Sea. They are brought to market in a preserved state, and the bark and pits are shipped, either entire or in a pulverized state, for tanning and dyeing.

Use. — The medicinal properties are purgative, tonic, and astringent. In India and China it is highly prized, and supposed to possess curative properties for every ill. The wood is hard takes a fine polish, and is used for cabinet work.

The fruit, bark, and leaves are all charged with tannin. They also yield a dye which, with alum, produces a beautiful yellow, and with iron, a fine black. The leaves and bark of the T. catappa furnish a pigment from which the celebrated India ink is made.

T. angustifolia is charged with a fragrant juice of a creamy consistency. This, when dried, is used in the temples for incense, and for tanning and dyeing.

The tree itself is sacred, and has a mythological origin and history.

ORDER XXV. **MYRTACEÆ.** (SUBORDER MYRTEÆ.)

Flowers perfect, regular, axillary, solitary, or in spikes, cymes, corymbs, or panicles. Calyx superior, limb 4–6-parted, persistent, or falling, valvate in the bud, occasionally entire, falling away with the expansion of the flower. Petals inserted in the throat of calyx. Stamens inserted on the calyx throat, mostly numerous, frequently double or treble the parts of corolla, or indefinite; filaments threadlike, free or in bundles. Ovary with 2 or more cells; seeds numerous. Leaves opposite, rarely whorled, entire, exstipulate. Berry or capsule

2- or more- celled, 1- to many-seeded. Small trees and shrubs; seldom herbs. No. of genera, 76.

MYRTUS, Tourn. (Myrtle.) Calyx 4-6-parted, tube attached to the ovary. Petals 4-6, inserted together with the many stamens in the throat of the calyx; sometimes absent. Filaments long, free, or combined in groups. Anthers opening lengthwise, dehiscence inwards. Style solitary; seeds attached to a central column.

1. **M. communis**, L. (Common Myrtle.) Stem 6 to 8 feet high, branched. Leaves opposite, with punctured spots, ovate, lanceolate, variable in breadth,

Myrtus communis (Common Myrtle).

evergreen. Flowers axillary, solitary, white. Fruit a berry, 2-3-celled. Evergreen shrub.

Varieties. — There are many species, and of the communis there are 5 varieties known to the florists.

Var. **Romana**, broad-leaved. Leaves leathery.
Var. **Tarentina**. Leaves like those of the box.
Var. **Bœotica**. Leaves like those of the orange.
Var. **Belgica**. Broad-leaved, Dutch.
Var. **mucronata**. Leaves like those of the rosemary.

There are other varieties of this species, but these 5 are the most important.

Geography. — The geographical home of the myrtle is tropical and subtropical, but it grows well in regions of light frost, gradually becoming acclimated. By some it is supposed to be native in southeastern Italy, and it is growing now in all the countries around the Mediterranean sea.

Etymology. — *Myrtus*, the Latin name, is through the Greek μύρτος, a myrtle-tree, derived from the Greek μύρον, perfume, due to the pleasant odor discharged from the bruised leaves. *Communis* is the Latin for common, or usual. *Myrtle*, the popular name, is Dutch, and is a corruption of *myrtus*, the generic name. In America the periwinkle, which belongs to a different order of plants, is popularly called *myrtle*.

History. — The home of the myrtle is western Asia, Asia Minor, and other countries of the Levant. It has been known from the earliest historic periods, and is said to have been growing upon the site of the city of Rome when it was founded; it was common in Egypt before the beginning of the present century.

Pickering makes its home near the Persian Gulf, whence it has been carried to Egypt and other countries of the Mediterranean. It is spoken of by the earliest historians. Anecdotes are rife illustrating its use and value, but we have not room for them.

Use. — The medicinal properties of the myrtle are mostly stimulant and astringent.

It was formerly a favorite flavorer of wine and food; the flavoring substance resides in the young twigs, the leaves, and the berries. The leaves are said to make a very tolerable tea. It was held in great esteem by the ancient Greeks; and a place was set apart in all their markets for its sale. It was used by both the Greeks and Romans for wreaths to decorate victors in the Olympian and other games. The Jews held it in great veneration as an emblem of peace, and among them it constituted a part of the bride's decoration. It is frequently mentioned in the Scriptures. The Mahometans hold that it is one of the pure things that Adam carried with him out of Paradise. It was an emblem of authority as well as of honor, and worn by the magistrates of Athens when in the exercise of their duties.

The fruit and leaves are both used for tanning goat-skins.

The plant is a beautiful object, a favorite in planted grounds, and on that account has an important commercial value.

EUGENIA, L. Calyx 4-lobed, rarely 5. Petals 4 or 5, free or united. Stamens numerous, inserted in the throat of the calyx, and on the receptacle, in several rows; filaments free, threadlike; anthers 2-celled. Ovary 2-3- or more- celled, ova numerous. Style simple; stigma terminal. Berry crowned with the persistent limb of the calyx. Leaves opposite, entire, dotted with pellucid spots, without stipules. Flowers axillary or terminal, in solitary cymes, or panicles, 2-bracted, white, or purple. Fruit black, red, or purple. Trees.

1. **E. caryophyllata**, Thunb. (Cloves.) Trunk 20 to 40 feet high, branching regularly into a hemispherical or conical head of great beauty. Bark yellowish-gray. Leaves opposite, numerous, evergreen, oval, acute at each end, entire, smooth, thick, dotted with pellucid spots, dark-green and shining above, paler beneath, midrib and lateral veins prominent, petioles short, blade 3 to 5 inches long. Flowers axillary or terminal, in loose, small cymes; bracts small and falling off; calyx half an inch long, fleshy, round below, upper part divided into 4 triangular, spreading teeth. Petals 4, tightly imbricated in the bud, forming a smooth, spherical head, fringed by the teeth of the calyx, falling off early. Stamens many, inserted on a raised disk; filaments as long as the petals, spreading; anthers small, roundish, opening lengthwise. Ovary inclosed in the calyx, small, 2-celled; ovules many; style simple, shorter than the stamens, slender, tapering. Fruit in shape like an olive, but not so

MYRTACEÆ. 135

large; seed solitary; all the ovules but one become abortive; outer covering membranous.

Geography. — The zone of the clove tree is narrow. Its home is the Molucca Islands. It has been planted in Brazil, the West Indies, and extended to distant islands of the Indian and Pacific Oceans; but out of the latitude of the Moluccas, its spicy character is very inferior.

Etymology. — *Eugenia*, the generic name, is for Prince Eugene of Savoy, a patron of Botany. *Caryophyllata*, from the Greek κάρυον, a nut, φύλλον, a leaf, due to the appearance of the flower-bud, a nut in a leaf, or among leaves. *Clove*, from the Spanish *clavo*, a nail, on account of the fancied resemblance to a nail.

History. — The clove has been in use among the western nations of the Old World for more than two thousand years, and was taken to Europe overland by the Persians and Arabs; but its native country was not revealed until after 1511, when the Portuguese came into possession of the Molucca Islands. As it is not known in a wild state, the exact locality of its nativity is not clearly determined.

The enterprise and boldness that the discovery of America gave to navigators and merchants led them to the Indian Ocean, and discovery of the islands of the coast of Asia thus opened the great storehouse of the spices of India to the commerce of the world, and the homes of the clove, cinnamon, allspice, and pepper became known to the wondering nations.

In the 17th century the Dutch came into the possession of the Spice Islands and established a monopoly of the spice trade. They raised prices to exorbitant figures, and confined the cultivation of the clove to the Island of Amboyna. During the French war in 1810 the English for a short time held possession of these islands. They transplanted the trees to other islands, and broke the monopoly.

EUGENIA CARYOPHYLLATA
(Clove).

Mode of Harvesting. — Just before the flower-buds develop they are picked or shaken off and dried over a fire or in the sun, then packed in bags made of the leaf of the cocoa-nut, and thus sent to market. A tree yields about five pounds for a crop, and bears two crops in a year.

Use. — The tree is used for ornamental purposes in subtropical countries. The wood of the clove-tree is hard, takes a fine polish, and is used by the cabinet-maker in fine and ornamental articles of furniture. The clove, in medicine, is a stimulant, aromatic, and irritant, and largely employed to cover up the taste of disagreeable drugs. The odor resides in the essential oil, of which the clove yields a very large percentage. Its principal use is as a spice for flavoring cake, sauces, and confectionery. The oil and tincture are both used in the manufacture of cordials and bitters.

2. **E. pimenta**, DC. (Allspice.) Trunk 25 to 30 feet high, much branched; branches long and horizontal, forming a hemispherical head, in form and size like an apple-tree. Bark light-gray. Leaves elliptical, lanceolate, opposite, evergreen, obtusely pointed, conspicuously veined, deep-green, shining above. Flowers small, inconspicuous, in terminal, 3-forked panicles. Fruit a globular

berry, crowned with the persistent calyx, smooth, black or purple, and shining when ripe.

Geography. — The geographical zone of the allspice is tropical and subtropical, and its distribution is very limited. Jamaica supplies the markets of the world.

Etymology. — *Pimenta*, the specific name, is said to be derived from the Portuguese *pimenta*, which signifies "a color," from the Latin *pigmentum*. This name is probably derived from the fact that a decoction of the fruit, bark, or leaves, treated with sulphate of iron, produces an inky black, and the bark and leaves are highly charged with tannin. *Allspice*, the popular name, is said to be due to the circumstance that the taste of this spice was thought to resemble that of cloves, nutmeg, and cinnamon combined, hence was said to possess the properties of all the spices.

EUGENIA PIMENTA
(Allspice).

History. — When this spice first became known to civilized man is not recorded, but it was no doubt taken to Europe soon after the discovery of the West Indies.

Its home is the island of Jamaica, and it is abundant in the mountains on the northern side of the island. It also grows in Yucatan, but the fruit is not exported from any locality but Jamaica. Attempts have been made to introduce it to cultivation in Cuba and in Brazil, but all efforts have failed to improve the quality of the fruit or the size of the tree.

The leaves and bark, as well as the fruit, are aromatic.

Cultivation. — The tree grows best without cultivation, or, at least, is not improved thereby. As the groves are exhausted, new ones are obtained by removing all trees from a suitable spot in the forest near an old or exhausted grove, and very soon a thicket of pimenta trees appears from seeds which have been sown, carried by the wind or birds to the clearing. The young trees are allowed to reach the age of two or three years, when they are thinned out by removing the weaker, after which the grove (or *walk*, as it is called) needs no attention till harvest, which commences as soon as the berries are full grown but not mature. The trees are full grown in about seven years from the time the grove is begun.

The mode of harvesting is to break off the ends of the branches which are laden with fruit, and drop them to the ground (the tree is greatly benefited by removing the fruit before it matures), where women and children pick off the berries and place them in bags, in which they are carried to a place to cure either by the rays of the sun or by artificial heat, when they are packed in bags for market.

The harvest occurs in July and August.

Use. — The tree is sparingly used for ornamental purposes. The leaves are used for tanning leather. The fruit forms one of our most popular spices, used for flavoring sauces, cakes, bread, and for spicing wines, pickles, and cordials.

Its medicinal properties, as to the fruit and the oil, are identical with those of cloves.

MRYTACEÆ.

BERTHOLLETIA, Humb. and Bonpl. Calyx 4-parted. Corolla made up of 4 fleshy petals. Stamens united at the base in 5 concentric circles; filaments thread-like, short. Stigma cruciform, sessile. Ovary inferior, 4–5-celled. Inflorescence in terminal panicles. Fruit large, globular, woody. Nuts numerous, obovoid, triangular. Leaves alternate. Large tree.

B. excelsa, Humb. and Bonpl. (Brazil Nut.) (Cannon-ball Tree.) Trunk 3 to 4 feet in diameter, rising to the height of 150 feet, branching into a symmetrical head. Leaves 2 to 3 feet in length, broad, glabrous, prominently veined underneath, leathery. Fruit subglobular. Shell or husk woody, 6 inches in diameter, 4-celled, each cell containing 3 or 4 nuts an inch and a half long, three quarters of an inch in diameter, the testa hard, horny, and rough, kernel creamy white, oily, and possessing a delicate flavor. There is only one species of this magnificent tree.

Geography. — The geographical distribution of the Bertholletia is limited to the tropical regions of South America, extending to the Isthmus of Panama. Large tracts along the Amazon and the lower reaches of its tributaries are covered by this gigantic tree.

Etymology. — *Bertholletia*, the generic name, was given to this plant by De Candolle in honor of Berthollet, a celebrated chemist. *Excelsa* is from the Latin *excelsus*, grand, or lofty, due to the gigantic size and character of the tree. *Brazil Nut* is named from Brazil, its home; *Cannon-ball Tree*, from the shape of the fruit.

History. — At the time of harvest the natives ascend the rivers and enter the vast groves to gather the crop. For the same reason the vegetable-eating animals assemble to secure their share of the delicious fruit. When the great seed-vessels, weighing several pounds, fall from the height of 60 to 100 feet and burst open as they strike the ground, scattering the seed in the midst of the assembled men and their monkey cousins, the imagination must

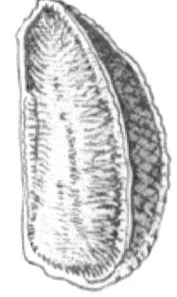

BERTHOLLETIA EXCELSA (Brazil Nut).

be drawn upon to picture the scene. Men, women, and children, monkeys of all the Brazilian varieties, rodents, and other nut-eating brutes all rush to secure the prize, and mingle in the scramble. The Indians club and pelt the monkeys, while they in turn seize the unbroken balls, flee to the branches of the trees, and hurl them at the heads of the Indians, thus presenting a feast, frolic, and fight combined.

The seed-vessels are forced open either by dashing them upon the ground or striking them with mallets made for the purpose. The nuts are collected into bags and baskets, carried down the rivers, and shipped to Europe and the United States from the seaports near the mouth of the Amazon, principally from Para.

Use. — The nut affords an important food to a large number of the inhabitants of Brazil. It is an esteemed dessert, and though very indigestible, is highly prized by children. The oil obtained from it is an excellent table and salad oil; it is also much used in the compounding of hair-dressings and for illuminating and lubricating purposes.

Statistics. — From Para alone it is estimated that upwards of two and a half million fruits, yielding about fifty million nuts, are exported annually, in addition to the large quantities which leave other Brazilian harbors.

Order XXVI. LYTHRACEÆ.

Flowers perfect, symmetrical, perigynous. Calyx inclosing, but not adhering to the ovary. Petals sometimes wanting; when present, free. Stamens equal to petals in number, or twice as many or more, inserted on the calyx-tube; anthers versatile. Ovary 2 to 6, rarely 1-celled; style 1. Seeds numerous, usually on axile placentæ. No albumen. Fruit a pod, more or less inclosed by the calyx. Trees or shrubs, branches frequently 4-angled. Leaves mostly opposite, entire, exstipulate.

No. of genera, 30; of species, 250.

PUNICA, L. Calyx large, broadly tubular, an inch and a half long, thick and leathery, attached to the ovary below; lobes 5-7, thick, triangular, acute, smooth, shining, scarlet, valvate. Petals 5-7, inserted on the calyx, alternating with its lobes, spreading, imbricated, crumpled, roundish, with a short, broad claw, crimson. Stamens numerous, free, inserted on the calyx-tube beneath the petals, crowded, erect. Anthers turned inwards, yellow, opening lengthwise. Ovary thick and leathery, with two tiers of cells, upper tier 5-8 in number; ovules many. Style tapering; stigma simple, head-shaped. Fruit hard, sub-globular, abruptly contracted at the top into a short neck, crowned by the thick calyx, size of an orange, reddish-yellow. Seeds numerous, angular, covered with a pinkish, juicy coating.

P. granatum, L. (Pomegranate.) Arborescent, 14 to 15 feet high. Bark brownish-gray, furrowed. Branches straight, strong, sub-angular, armed near the ends with spines; young shoots and buds red. Leaves opposite or fascicled, short-stalked, and without stipules. Flowers large, solitary, or two to three together in the axils of the leaves, near the ends of the branchlets. A beautiful object for planted grounds.

PUNICA GRANATUM (Pomegranate).

Varieties.—This is the only species, but as it is grown from seed it sports, consequently varieties have been produced. The two most prominent are:
Var. alba, white-flowered.
Var. plena, double-flowered.

Geography.—The zone of the pomegranate is within the region of no frost, and along its outer fringes, in the north temperate zone, all around the globe.

Etymology.—*Punica*, the generic name, is from the Latin *punicus*, red, due to the color of the fruit. *Punicus* also means Carthaginian, signifying "deceitful" (the Carthaginians had the reputation of being unfaithful), applied to the fruit, which is beautiful in appearance, but not delightful to the taste. The name may have been applied to the plant because it was largely planted near

Carthage, hence called Pomum Punicum, or Carthage Apple. *Granatum* is the Latin for grained or seeded, due to the abundance of seed in the fruit. *Pomegranate*, the common name, is made up of *pomum*, apple, and *granatum*, seeded.

History.—The home of this beautiful little tree is Persia and the adjacent countries, whence it has spread throughout Syria, Asia Minor, the Levant, southern Europe, Africa, China, and Japan. It has also been brought by European colonists to southern North America.

It was known to the ancients and is spoken of frequently in the Bible. The Hebrews on their journeyings through the desert of Arabia, complained to Moses, saying. "It is no place of seed or of figs or of vines or of pomegranates." And Moses himself describes the promised land as a country of "wheat, barley and vines, fig-trees and pomegranates." Solomon speaks of an "orchard of pomegranates with pleasant fruits."

By these quotations we are led to the inference that the pomegranate was an important food-plant at that time.

Use.—The plant is cultivated largely in the regions of no frost as an ornamental tree, and in colder climates in conservatories, throughout Europe and the United States.

The fruit is used for a dessert, being prepared by cutting it into halves, removing the seeds, filling their places with sugar, and sprinkling the whole with rose-water.

The bark is highly charged with tannin, and produces a beautiful yellow dye, with which the yellow Levant morocco is colored. The bark of the root is used as a vermifuge, and was formerly considered a specific for tape worm.

Statistics.—Outside of the countries of Asia and Africa, where the pomegranate is grown, it is of very small commercial importance.

Order XXVII. **CUCURBITACEÆ.**

Flowers monœcious, or diœcious, seldom perfect, solitary, sometimes fascicled, or racemed, usually white or yellow. Calyx bell-shaped, 5-toothed or lobed, imbricate in the bud. Corolla with petals united, wheel- or bell-shaped, 5-lobed. Stamens 5, mostly 3, one of the anthers 1-celled, the others 2-celled. Ovary inferior, 1- or many-celled. Fruit a many-seeded berry. Leaves alternate, petioled, usually cordate. Stem succulent, climbing or clambering over undershrubs, etc. Mostly herbs.

Genera, 68.

CUCUMIS, L. Calyx tubular, bell-shaped, 5-parted or toothed, teeth awl-shaped, about as long as the tube. Petals 5, slightly attached to the calyx. Stamens in three groups. Stigmas 3, nearly sessile, stout, and 2-lobed. Fruit globular, sometimes flattened at the poles, and again lengthened into a prolate spheroid or short cylinder. Seeds numerous, white or yellowish, oblanceolate, acute at the base, and flattened.

1. **C. sativus**, L. (Cucumber.) Stem trailing, rough, hairy, 5 to 12 feet long, branched; tendrils simple. Leaves cordate, 3 to 6 inches long, angularly

lobed, terminal lobe largest. Fruit cylindrical, 5 to 10 inches long, and from 2 to 4 inches in diameter. When young, the surface is besprinkled with tubercles, armed with rigid, sharp bristles, which fall off at a later state. Green, turning yellow when ripe. Seeds very numerous, yellowish-white, three eighths of an inch long and less than two eighths wide, oblanceolate, flattened; about twelve hundred weigh an ounce, and they retain their vitality about ten years, if kept from the air.

Varieties. — The cucumber sports freely, and many varieties are under cultivation. There are about thirty choice kinds recommended by seedsmen. The Cluster, Early French, White Spine, and Early Russian are among the most desirable for the market garden.

Amateurs favor other varieties, but the above four are the most popular.

Geography. — The zone of the cucumber is very broad. It grows well in rich soil wherever there are three or four months without frost, but requires warm nights and hot days to be prolific.

CUCUMIS SATIVUS (Cucumber).

Etymology. — *Cucumis* is Latin, and signifies a vessel, alluding to the rind of the fruit, which when the pulp is removed forms a cup which may be used for drinking. It is said to be derived from the Celtic word *cuce*, a hollow vessel, or from the Latin *cucuma*, a cooking-vessel. *Sativus*, the specific name, is Latin, and signifies sown, or cultivated. *Cucumber*, the common name, is a corruption of the word *cucumis*, the generic name.

History. — The home of the cucumber is the northwest of India and the region north of Afghanistan, and it was no doubt taken into the Levant and southern Asia at a very early period in history. It was under cultivation in Hindustan three thousand years before the Christian era, and was known to the ancient Greeks.

It is by no means certain that the plant referred to under the name cucumber in Scripture was the *Cucumis sativus*. Nothing has appeared on the Egyptian monuments to prove that the Israelites became acquainted with it during the period of their bondage, but it is possible that it reached them in Syria from the East.

It worked its way into southern Europe and Africa by commerce and travel, was brought to America in the days of Columbus, and has become one of our most important garden crops about our great cities.

Use. — It is largely used raw when in an unripe state, as a salad, with a salt and vinegar dressing, and as a pickle, in America, Europe, and especially in southern Russia among the peasantry, by whom it is stored in casks under heavy weight, and allowed to heat and reach the vinous fermentation, when it is eaten with coarse bread, serving the purpose of butter or oil. In the Southern States, in North America, it is sliced, fried in oil or butter, and served up as egg plant is.

2. **C. melo, L.** (Muskmelon. Cantaloupe.) Stem rough, hairy, 5 to 10 feet long, trailing. Leaves heart-shaped, or somewhat kidney-shaped, with rounded

lobes, rough, hairy, 3 to 5 inches long. Flowers axillary, on short stalks, yellow. Fruit globose, from 3 to 12 inches in diameter, generally ridged and furrowed, sometimes much flattened at the poles, while in some varieties it is much elongated, forming a short cylinder, or oval. Seeds yellowish-white, oblanceolate, flattened; about a thousand to an ounce; when kept in a uniform temperature they retain their germinating properties about ten years. Flowers in June. Fruits in August.

There are many species, and as it sports freely, very many varieties are under cultivation. The leading varieties in America are: —

The *Beechwood*, an early variety, flesh sugary.
The *Black-Rock*, large-fruited, very sweet.
The *Citron*, rich, juicy, and sugary.
The *Large-ribbed*, very large, oval in form, flesh sweet.
The *Nutmeg*, delicious in flavor, and popular.

Geography. — The muskmelon grows to perfection in rich, sandy soil, in all the countries of the Levant, on the shores of the Mediterranean, in India, China, Japan, and in fact in all tropical and subtropical countries throughout the world. It is a very important crop on the southern plains of New Jersey and throughout the Middle and Southern States.

Etymology. — *Melo*, the specific name, is from the Greek μῆλον, an apple, hence an apple-shaped fruit. Latin *melo*, a melon. *Melon*, the popular name, is a corruption of the same word, or rather an Anglicizing of the Greek word. *Muskmelon* is due to the peculiar aroma shed by some of the varieties, which has fancifully been compared to the odor of musk. *Cantaloupe* arose from the circumstance that one of the varieties was cultivated or originated at a country-seat of the Pope, called Cantalouppi.

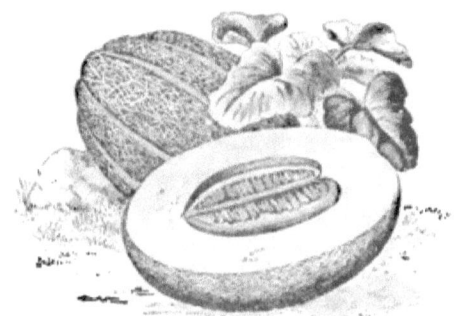

CUCUMIS MELO (Muskmelon).

History. — The muskmelon or Cucumis melo, is indigenous to British India and Baluchistan. It has also been found wild in western Africa, in Guinea, and along the banks of the Niger.

It was cultivated by the ancient Egyptians, but how early there is no means of knowing. It was also known to the Greeks and Romans at an early period in history, brought either from the East by commerce or travel or introduced from Africa.

Use. — The muskmelon is the richest and most juicy fruit of all the pepo family. In the Atlantic States it is the favorite first course at breakfast, and is highly esteemed as a dessert. The rinds are preserved in ginger syrup as a sweetmeat, and also brandied.

The following anecdote is told of Frederick the Great, who was extravagantly fond of a small Egyptian melon, which he caused to be cultivated in his grounds. He one day called his physician to treat him for an attack of indigestion. The doctor, aware of the king's fondness for the melon, inferred that his indisposition was due to its excessive use. He therefore advised the king to abstain from its use; to which the monarch replied: "I will only eat five for my breakfast." At the same time calling his gardener, he ordered him to send the doctor a dozen for his breakfast.

CITRULLUS, Schrad. (Melon.) Divisions of the calyx 5, narrow, lanceolate. Petals 5, united at the base and attached to the bottom of the calyx. Stamens, in three groups, connected. Style 3-parted. Stigmas convex, heart or kidney-shaped. Fruit globular or in the form of a prolate spheroid, from 6 inches to 2 feet in length, and from 6 to 15 inches in diameter. Rind leathery, greenish-brown, mottled or striped, with alternate green and yellow lines from an inch to an inch and a half wide, filled with a fleshy, juicy placenta, or core, of an orange-red color, sweet and edible. An herbaceous vine.

C. vulgaris, Schrad. (Watermelon.) Stem 8 to 15 feet long, angular, branched, rough, hairy, slender, trailing. Leaves 3 to 6 inches in length, lobed, and the lobes pinnately divided, glaucous beneath, petioles 2 to 3 inches long. Flowers axillary, on hairy pedicels, about an inch and a half in length, corolla yellow. Flowers in June to July. Fruits August to October.

CITRULLUS VULGARIS (Watermelon).

Varieties. — There are numerous varieties. Among the most popular are—

The *Black Spanish*, somewhat globular, deeply ribbed lengthwise, skin dark or blackish-green. Quality excellent, and grows well in New Jersey, and as far north as southern New York, especially on Long Island.

The *Bradford*, or Carolina watermelon, a favorite in the Southern States; one and a half to two feet long, striped or mottled; pulp tender and delicious.

The *Mountain Sweet*, a hardy and greatly esteemed variety; grows well in New Jersey, Delaware, southern Pennsylvania, and Long Island. Pulp dark-red, and delicious.

Odell's Large, of enormous size, round, gray; seeds large, grayish-black. A single melon of this variety has been known to weigh sixty pounds.

Citron Watermelon, 6 to 10 inches in diameter, color pale-green, marbled with darker shades; not edible raw, but highly prized for preserving.

Geography. — It arrives at perfection only in tropical and subtropical countries, but succeeds well in all southern Europe and southern and middle North America, up to the 41st degree of latitude in North America.

CUCURBITACEÆ.

Etymology.—*Citrullus* is derived from the color of the pulp, which is sometimes of an orange red, hence *orange* color, or citrus color. *Vulgaris*, common, is from the Latin. *Watermelon* refers to the watery pulp. Melon is an adoption of the Greek word μῆλον, an apple, because of the supposed apple-shape of the fruit.

History.— It is not known where or when the watermelon was first brought under cultivation. Its home is in Africa, in the torrid zone. Livingstone states that large districts are covered with it in a wild state, and it has not been seen growing without cultivation out of Africa. It was cultivated by the ancient Egyptians. It was grown in Asia at an early date, and was known in all the Mediterranean countries in the beginning of the Christian era. It was brought to the New World by European settlers, and is common in all the warm countries of both North and South America, where Europeans have settled.

Use.— The watermelon is a very popular dessert fruit, and the rinds are preserved in sugar and also brandied. In Egypt it is not only prized by the wealthy as a dessert, but it constitutes a very important article of food for the poorer classes, who eat it with their bread, and in fact largely subsist upon it during its season, which is long.

CUCURBITA MAXIMA (Gourd Squash).

CUCURBITA. L. (Pumpkin. Gourd. Squash.) Calyx egg-shaped, corolla bell-shaped, petals united half-way up, yellow. Flowers monœcious, axillary, on angular stalks. Stamens with anthers cohering. Fruit-stalk deeply grooved. Fruit globose, flattened or prolonged at the poles; seeds yellowish white, obovate, with slightly convex sides. Leaves heart or kidney shaped, stem trailing with branching 2–3-cleft tendrils. Annuals.

1. **C. melopepo,** L. (Flat Squash.) Stem 5 to 20 feet in length, branching, tendrils branched, or partially developed into leaves. Leaves heart-shaped, 5-angled, 6 to 8 inches long on a foot-stalk as long as the blade. Flowers yellow, large. Fruit wheel-shaped, flattened and dished about the stem, and convex on the opposite side; circumference lobed or escalloped. Flowers and fruits July to Oct. Fruit cooked in a green state, with or without salt meats.

2. **C. verrucosa,** L. (Warty Squash. Long-necked Squash.) Stem 10 to 15 feet long, tendrils branched. Leaves from 6 to 12 inches long, and nearly as wide; leaf-stalk same length as the blade. Flowers large and yellow. Fruit obovoid, or club-shaped, neck frequently crooked or curved, roughened with warty tubercles; when ripe the shell becomes hard or bony. Like the above,

the fruit is prepared for the table in an unripe state. Flowers and fruits from July to Oct.

3. **C. maxima**, DC. (Winter Squash. Gourd Squash.) Stem 8 to 20 feet long, trailing. Leaves large, with rounded lobes. Corolla yellow, segments curved, or rolled outwards; flower-stalks smooth. Fruit ovoid or pear-shaped, neck sometimes crooked or curved. Variable in size, frequently reaching the length of 3 feet, and specimens have been known to weigh 70 pounds. The neck is usually solid, the end farthest from the stem enlarged, and contains the seeds. This squash has the characteristics of a pumpkin, and is used much in the same way that the cheese pumpkin is.

Var. **corona** (Crowned Squash). Fruit expanded near the stem into a broad, circular, turban-like process, much larger in diameter than the extended part. This expanded part is solid, and the cell containing the seed is in the contracted end.

Varieties. — These are the principal marrow squashes. There are many forms, but the above are the favorites with gardeners and amateurs.

The *Custard Squash*, a large variety, is grown for stock.

Puritan Squash, grown largely in New England, is very constant, very hardy and productive, and raised both for the table and for stock; skin white, marked with green mottled stripes.

The *Valparaiso Squashes*, of which we frequently have specimens brought from California, some of them weighing more than a hundred pounds, have not been fully described.

4. **C. ovifera**, Gray. (Orange Gourd Squash.) The fruit of this species is small and egg-shaped, and by cultivation it is supposed to have given rise to the following forms: —

Autumnal Marrow. Stem 10 to 15 feet in length, and stout. Fruit ovoid or spindle-shaped, furrowed and ridged, the blossom end tipped with a short nipple; skin very creamy, yellow; flesh sweet and delicate. Ripens early in August. Keeps well.

Hubbard Squash is in shape and quality very much like the *Autumnal Marrow*, color a bluish-green, flesh orange-color and delicate; smoother than the last, 8 to 10 inches long, and 6 to 8 in diameter.

Sweet Potato Squash resembles the above two in shape and character, a foot long, 7 or 8 inches in diameter, skin ashy green, smooth and polished; flesh salmon-yellow; thick-fleshed and fine-grained.

Var. **Medullosa**.

Vegetable Marrow. Its stem 12 to 15 feet in length, leaves deeply 5-lobed; fruit 8 to 10 inches long, elliptical in shape, ribbed and furrowed lengthwise; flesh white and delicate. Keeps well through the winter.

Etymology. — *Melopepo* is from the two Greek words μῆλον, an apple, and πέπων, a melon, an apple melon. The fruit in a natural state is of the size and shape of an apple. *Maxima* is Latin for *great*, and is due to the size of this species. *Corona* is Latin for *crown*, given on account of the *turban* or crown-like process at the stem end of this variety. *Verrucosa* is from the Latin *verrucosus*, warty, on account of the warts that abound on the skin of the fruit of this variety. *Ovifera*, from the Latin *ovum*, an egg, is due to the oval shape of the fruit of this species.

History. — Dr. Gray believes that the *C. ovifera* is the ancestor of all the American squashes. It is claimed, however, by some authorities that Europe or western Asia is the home of the *C. maxima*. But Pickering claims that the C. maxima is an American plant, and says it has been carried thence by colonists to the Pacific islands, to southern Asia, and to Europe and Africa. If this be so, it leaves us in the dark as to what the C. maxima known to the ancients was.

5. **C. pepo,** L. (Pumpkin.) Stem prostrate, 5 to 20 feet long, rough, hairy, sparingly branched, with branched tendrils. Leaves large, 9 to 13 inches long, and 5 to 10 wide, heart- or kidney-shaped, 5-lobed. Flowers yellow and axillary. Fruit cheese-shaped or club-shaped, or sub-globular, on deeply grooved peduncles, flesh yellow, sweet, solid, but not hard; cavity of the fruit filled with a stringy pulp and seeds. Flowers in July; fruit ripens in October.

Varieties. — There are many varieties of the Pumpkin under cultivation, the most popular of which are the following: —

Cheese Pumpkin, which is flattened at the poles, and from 10 to 20 inches in diameter, and 4 to 10 inches from pole to pole, deeply ribbed, dished about the stem, skin reddish-orange color, leathery; flesh yellow, sweet, and delicate. The cheese pump-

CUCURBITA PEPO (Pumpkin).

kin holds the highest place among the varieties of this plant, on account of its hardy character, its size, productiveness, and the delicacy of its flesh. It has been claimed that it is a variety of the C. maxima, brought to America by European colonists; but history favors the belief that it is an American plant. It was extensively cultivated throughout the Middle States at the time of the Revolutionary War, and was carried to New England by the soldiers returning home from service in New Jersey, southern Pennsylvania, and adjacent states further south, where it is still found growing, with great constancy as to form, size, and qualities, though the cultivation has been in many cases careless and slovenly.

Canada Pumpkin is in the form of a flattened globe, deeply ribbed, 10 to 15 inches in diameter, and 8 to 10 inches at the poles. Skin yellow and hard; flesh yellow. Much cultivated for cattle, and also for table use. It grows better in a higher latitude than the cheese pumpkin.

Common Field Pumpkin, or *Leather Back.* Globose, ends flattened, rather longer than broad, 10 to 14 inches long, and 8 to 12 in diameter. Grown for stock, and sparingly for the table. Ribbed, yellow; skin hard, flesh yellow. Leaves deeply lobed.

Sugar Pumpkin. Grows in the form of a flattened sphere, about 9 inches in diameter, and 6 at the poles. The smallest of the varieties under cultivation; a prolific bearer, and of excellent quality. Grooved skin, bright orange-yellow; flesh yellow, sweet, delicate, and fine grained. Stem long, ridged, and grooved.

There are other varieties, but the above are the favorites, and most important to gardeners and agriculturists.

Geography. — The Gourd family, of which the pumpkin and squash are members, delights in a warm climate, but fruits well as far north and south of the equator as the middle of the temperate zones.

Etymology. — *Cucurbita* is the Latin for gourd, a hollow vessel or a cup, and must allude to the circumstance that these plants are hollow, or become so when allowed to ripen on the vine. Some derive this from the Latin *curvitas*, crookedness, alluding to the form of some of the club-shaped gourds, whose necks are curved. *Pepo* is from the Greek πέπων, a melon. *Pumpkin* is a corruption of the French word *pompon*, a melon. The popular names all explain themselves.

History. — The home of the pumpkin is believed to be America. It has been found growing wild in Mexico, and was under cultivation by the aborigines in Florida, Mexico, and the West Indies, when these regions were first visited by Europeans. Dr. Gray believed that all the species except C. maxima are American. The species and varieties of this genus have been so confused that this is not certain.

Use. — The cheese pumpkin and the sugar pumpkin are esteemed for making the celebrated New England pumpkin pies. They also, like the other varieties, are grown for feeding cattle. They are valuable for milch cows because they not only promote the flow of milk but improve its quality. In Europe the pumpkin pie is prepared by making a circular orifice in the top, the center of which is the stem. Through this hole the seeds and pulp are removed, and the cavity filled with sliced apples, spices and sugar. The whole is then baked, and served.

Order XXVIII. UMBELLIFERÆ.

Flowers small, 5-merous, superior, in simple or compound umbels. Calyx lobes minute, tube adnate to ovary. Ovary 2-celled, each with a pendulous ovule. Fruit, 2 dry indehiscent akenes, separating from a carpophore; each akene with 5 primary and often 4 secondary ribs. Number of genera, 152.

APIUM, Hoffm. (Celery.) Calyx without teeth, base of style flat. Petals white, entire, with a small apex bent in. Fruit, egg- or globe-shaped. Carpels nearly straight, with 5 thread-like ribs; channels with single oil-tubes, except the outer ones, which sometimes have more. Leaves pinnately or ternately divided; divisions wedge-shaped; umbels opposite the leaves. Biennial herb.

A. graveolens, L. Stem 2 to 3 feet high, branching, channelled. Leaves from the root, on long, stout stalks, green; stem leaves on short stalks. Flowers terminal and axillary, those in the axils on very short foot-stalks; rays unequal; petals greenish-white. Fruit subglobular. Flowers in July. Fruit in September.

The celery sports freely, and many varieties have arisen, for the names of which the student is referred to the seedsmen's catalogues. There are about 20 choice varieties under cultivation by the market gardeners and amateurs.

Geography. — Its geographical distribution is very wide. It is indigenous to Great Britain, all the coast of western Europe, the shores of the Mediterranean, and it is found in the Peloponnesus, on the foothills of the Caucasus, and in Palestine. It is also native to South America, and along the western coast as far north as southern California. Watson, in his "Flora of California," speaks of it as very rare, but says it is found in the salt marshes down the coast.

Etymology. — *Apium* is traced to the Celtic word *apon*, water, due to the habitat of the plant, which is in wet places. *Graveolens*, the specific name, is from the Latin *gravis*, heavy, and *oleo*, smell, whence "heavy smell," or "strong smell," on account of the peculiar odor of the plant. *Celery*, the common name, is a corruption of the Greek word σέλινον, parsley.

History. — We do not know where or when the celery was first used as a table vegetable. It was known to the ancient Greeks and Romans, and is mentioned by Theophrastus, Dioscorides, and Pliny. It was also used by the ancient Egyptians.

Cultivation. — The seed is sown like a cabbage-seed; and when the plants are from three to six inches high, they are pricked out in beds. In August they are set in well fertilized trenches, about 8 to 12 inches apart in the row, and allowed to grow till October, at which time a number of long-stalked leaves are developed; these are then held together to prevent the earth from getting among them, and banked, and thus left to bleach till frost appears, when they are ready for the table.

Use. — Celery is the most delicate and highly esteemed of all salads in use. When properly blanched, the leaf-stalks are a delicate creamy white, and the flavor is greatly admired. It is either served in this form and eaten with salt or with prepared dressing; occasionally it is cooked and eaten with a vinegar dressing. The Turnip-rooted variety is cooked and eaten with salad dressings. It is used also for flavoring soups and gravies. Its medicinal properties are said to be diuretic and tonic, producing biliary secretion, and it is recommended for rheumatism. The Egyptians used it to prevent and to cure sea-sickness.

PIMPINELLA. L. (Anise.) Calyx limb indistinct, teeth wanting; petals white, obcordate, unequal, notched, flowers usually perfect, but sometimes the stamens and pistils are on different flowers. Styles long and slender. Bracts of involucre few, small or wanting. Leaves decompound. Fruit egg-shaped, ribbed, with convex intervals. Perennial herb.

P. anisum, L. Stem 2 feet high; branches slender; lower leaves roundish-heart-shaped, cut into three lobes by deep incisions; leaves on the middle and upper parts of the stem pinnate, parts wedge-shaped; umbels large and loose; stalks of the umbellets unequal in length; flowers yellowish-white, appearing in July.

There are many species, but the anisum is the one under cultivation for the production of the oil of anise.

Geography. — The anise seeds of commerce are produced in Egypt, Syria, in the island of Malta, and in Spain, and in late years the plant has been largely cultivated in Southern Germany. It has worked its way east of Syria to Hindustan, and to Japan. The best seed is brought from Egypt. The plant is supposed to be indigenous to Egypt, Asia Minor, and the Greek Islands.

Etymology. — *Pimpinella* is a corruption of *bipinnate*, due to the divisions of its leaf. *Anisum* is from the Greek word ἀνόμοιος, unequal, the ancient name referring to the inequality in the length of stalks of umbellets, or perhaps to the unequal petals. *Anise*, the popular name, is a corruption of *anisum*.

History. — The anise was known to the ancients, and is mentioned by Dioscorides, and described by Pliny, who says the best is brought from Crete and from Egypt. It is one of the plants Charlemagne ordered his gardener to cultivate in the royal gardens. It is one of the products carried into England early in the fourteenth century, and upon which an impost duty was levied. It was brought to North America about the middle of the seventeenth century by European colonists, and is sometimes planted in gardens in the Middle States, but seldom fruits. The anise named in the New Testament is held by some to be another plant.

PIMPINELLA ANISUM (Anise).

Use. — The essential oil of anise is obtained by distillation from the seed. It is used in medicine as a stimulant and carminative, as a stomachic and antispasmodic, and to cover the disagreeable taste of other medicines in compounding remedies such as paregoric, cough mixtures, and cordials.

Rats and mice are very fond of it, and it is used by vermin-destroyers to perfume their bait. The sprigs are sometimes used to garnish dishes at the table, and as a condiment for meats. In Germany and Middle Europe it is used to flavor bread, cakes, and cheese.

NOTE. — The fruit of the Illicium anisatum (Star Anise), a small evergreen tree of the order Magnoliaceæ, when distilled yields an oil identical in odor, chemical analysis, and medicinal properties with the oil of pimpinella, for which it is sometimes substituted.

FŒNICULUM, Hoffm. Calyx limb indistinct. Petals roundish-obovate, entire, truncate, involute. Cremocarps oblong or ellipsoid, ovoid, not flattened. Columella 2-parted, the branches attached to mericarps; mericarps with 5 prominent obtusely keeled ridges, side ones a little broader and marginal, a single oil-vessel in each space. Flowers, small, deep yellow, not radiant. Umbels large and compound, without involucres. Leaves decompound, with thread-like segments. A biennial herb.

1. **F. vulgare**, Gaert. (Fennel.) Stem erect, terete, thick, striate, smooth, bright green, large pith, through the center of which extends a small tube.

UMBELLIFERÆ.

Rootstock thickened. Leaves on short, flattened footstalks embracing the stem, triangular in outline, three or four times pinnate, segments or divisions thread-like and bristly. Flowers small, on short pedicels, bright yellow, in large, regular, 10-30-rayed umbels, without involucres; petals entire, involute. Cremocarps olive-colored, oblong-oval, barely flattened, one fifth of an inch in length, prominently ridged; whole plant deep green. Fruit aromatic, stimulant, stomachic. Root and leaves aromatic, medicinal, nutritive, and stimulating.

Biennial herb. July.

2. **F. dulce** is a species found in Italy and used for food.

Geography. — The geographical range of Fœniculum is the middle and southern parts of the north temperate zone.

Etymology. — *Fœniculum* is from the Latin *fœnum*, hay, due to the odor of fennel, which is that of new-mown hay. The specific name, *vulgare*, means "common," and *dulce*, "sweet."

History. — Indigenous in the countries of the Levant, carried by Europeans to Hindustan, and brought to the Atlantic States by colonists. Cultivated in France, Germany, Great Britain, and all southern Europe.

FŒNICULUM VULGARE (Fennel).

Use. — The medicinal qualities of fennel are carminative, and it is frequently administered to disguise the disagreeable taste of other medicines. It is used in Germany to flavor bread and cakes. The leaves and the root of F. dulce are used in southern Europe as a table vegetable, both as a salad and cooked. It is a favorite vegetable with fish. In former days it had the reputation of curing all sorts of poisons, restoring sight, and imparting strength to the body. The Roman gladiators mingled the seeds with their food, and wore the leaves as crowns of victory.

An essential oil is obtained from the seed by distillation, called Fennel Oil.

FERULA, L. Calyx entire, or obscurely toothed; petals broad-acuminate, frequently short and turned in. Disk small, stylopodium flattened. Fruit orbicular or ovate, margined; seeds with 3 lines along the back, intervals and commissure grooved or channelled. Common involucre falling off, involucels many-leaved. Radical leaves decompound. Flowers yellow, in globose umbellets.

F. narthex, Boiss. (Asafetida.) Stem 6 to 8 feet in height, cylindrical, smooth, solid, and furrowed. Leaves, at the root, 2 feet long, bipinnate, stem leaves numerous, alternate, lower ones bipinnate, on sheathing petioles, sheaths increasing in size towards the middle of the stem and decreasing from the middle upward.

Flowers polygamous, staminate flowers much smaller than the others, crowded into dense globular umbellets; involucre wanting. Calyx slightly striate. Petals oblong-oval acute and entire, pale yellow, unequal in the staminate flowers. Filaments as long as the petals; styles long and falling off. Fruit half an inch long, oval, smooth, yellowish. The root which yields the gum is in the form of a carrot, and 3 to 4 inches in diameter. The plant dies after flowering, but sometimes does not flower until the third or fourth year.

Geography.— The geographical distribution of the plants whose roots furnish asafœtida is not large; it is confined to middle and western Asia. It has been taken to England and to Africa, but it is not known to be successful in the production of the gum outside of Asia. It has been found in the Himalaya mountains at an elevation of 7,000 feet.

Etymology.— *Ferula*, the generic name, is from the Latin *ferio*, strike, alluding to the use made of the stem as a rod for scourging, employed by schoolmasters. *Narthex* is from the Greek word, νάρθηξ, a box or magazine to contain medicine, alluding to the medicine stored up in the plant. According

FERULA NARTHEX (Asafœtida).

ing to the fabulous history of the doings of the gods, Prometheus used a stalk of this plant, which had been hollowed out, to conceal the sacred fire which he brought down from heaven, hence a magazine for fire; it is also said to have been employed to make boxes or magazines for other purposes. *Asafœtida*, the common name, is from the Arabic *aza*, gum, and the Latin word *fœtidus*, fetid, or bad-smelling.

History.— The home of the Ferula is Persia, Afganistan, the Punjaub, and northern Hindustan. When its products were first used as a medicine, or as a condiment for food, is not known. The Greeks and Romans made use of it; Dioscorides and Pliny both speak of it. The plant is described in the books of the Buddhists as one of the ornaments of the Himalayan forests.

Preparation.— The gum is the dried milky juice which issues from the wounded root of F. narthex and other species of the genus. There seems to be good authority for the belief that most of the asafœtida is from the F. narthex.

To obtain it, the leaves and stalk are twisted off. The earth is removed from the upper part of the root, which is covered to protect it from the sun, and left for forty days, when it is exposed and the crown of the root sliced off. Two days later, the juice which is deposited upon this wound is removed, and a thin slice taken off, making a new wound. This operation goes on from the middle of May to the end of July, when the root is exhausted; a root yields from half an ounce to two pounds. The sap of the first scrapings is thin, and to make it more easy to handle it is mixed with earth; therefore the gum reaches market containing much earthy impurity. The gum, when hardened, has a brownish appearance, and in cold weather is brittle.

It is taken to Bombay from the ports of the Persian Gulf, whence it reaches the markets of the world.

Use.— Asafœtida contains a resin, a gum, and an essential oil. The odor resides in the oil, and is like that of onions, or garlics, accompanied by a separate fetid odor which is very offensive when the drug is warmed.

UMBELLIFERÆ. 151

Its medicinal qualities are stomachic and antispasmodic; it is administered in hysteria, hypochondria, and nervous disorders. The tissues of the human system absorb it readily, and it affects the kidneys with great activity.

The essential oil is distilled in Germany, and is used in medicine on the continent of Europe.

Asafœtida is used in India as a condiment for food; it was also employed by the Romans for the same purpose.

The peculiar and inimitable flavor of the celebrated Worcestershire sauce is supposed to be due to the presence of asafœtida.

PEUCEDANUM, L. Calyx 5-toothed, or obsolete; petals broad-lanceolate, point long and turned in. Fruit flat, oval, with a broad margin. Carpels obscurely 5-ribbed, secondary ribs wanting; valleys furnished with single oil-vessels. Flowers yellow, involucre many or few-leaved or absent; involucels the same. Root conical, large, and fleshy. Biennial.

P. pastinaca, L. (**Pastinaca sativa**, L.) (Parsnip.) Stem grooved, tapering, hollow, 2 to 4 feet high, branching. Root-leaves of the first year, orbicular, cordate, and crenate. Stem or upper leaves of the second year, compound; leaflets 2 to 3 inches long, cut, toothed, ovate, and obtuse, in 3 to 4 pairs, with a terminal one which is 3-lobed. Flowers in June. Fruits in August to October.

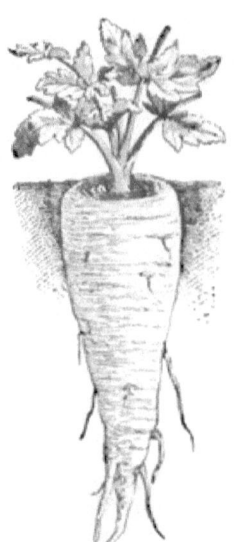

PEUCEDANUM PASTINACA (Parsnip).

NOTE.—There are several varieties: a great favorite is the *Guernsey*, which is an improved form of the common, or P. pastinaca. It is said that in the island of Guernsey, where deep cultivation is practiced, the root reaches the length of four feet. The *Student* has come into favor, and is said to excel all others in its sweet, mild, and pleasant flavor.

Geography.—The parsnip is found all over Europe south of 60 degrees of latitude, southern Greece, western Asia, Hindustan, and Siam. It sows its seeds easily, and escapes from gardens and cultivated grounds, and therefore is found growing outside of cultivation wherever it has been cropped.

Etymology.—*Peucedanum* is derived from the Greek πεύκη, a pine tree, and δανός, burnt, a burnt pine tree, due to the peculiar odor of the parsnip. *Pastinaca* is by some authors derived from the Latin word *pastinum*, a dibble, an instrument for digging into the ground, in reference to the deep piercing of the ground by the root of this plant. By others it is derived from the Latin word *pastus*, food, fodder, pasture, on account of the edible character of the root, and its use for feeding stock. *Sativa* is Latin for "sown" or "planted." The common name, *parsnip*, is supposed to be a corruption of the Latin word *pastinare*, to dig up, hence something dug up.

History.—The home of the parsnip is Europe. It was cultivated in Britain in early times, and eaten with salt fish during Lent; we have no means of knowing when or where it was first used for food, but it is known that it has been in use a long time. It is now cultivated largely throughout Europe for the table and for feed for cattle and horses. It is also found growing in a state of nature. It was brought to eastern North America by British colonists, and carried to other countries by emigrants from other parts of Europe.

Use.—It is an important table vegetable, eaten with meats, as potatoes are. It is prepared by simply boiling, or, after being boiled, it is sliced and fried in lard or butter, or mashed, made into balls, and fried; it is also stewed with pork, bacon, or other meats. It is extensively raised for stock. Horses, cattle, and sheep are said to fatten on it with great facility, and it is especially valuable for milch cows.

CORIANDRUM, Hoffm. Calyx-teeth conspicuous. 5 in number. Petals obcordate, turned in at the point, outer ones radiate and 2-parted. Fruit globose, smooth. Carpels cohering; 5 primary ribs depressed, the 4 secondary ones more prominent; seeds concave on their faces. Involucre 1-leaved or wanting. Involucels 3-leaved, unilateral. Annual.

C sativum, L. (Coriander.) Stem 1 to 2 feet high, slender, striate, and branched at top. Leaves bipinnate, with deeply cut, wedge-shaped segments below, segments of the upper leaves linear. Ovary inferior and globular, with 2 short diverging styles; stigmas flat or obtuse. Stamens 5, filaments slender, anthers roundish and yellow. Petals 5, white or purplish, obcordate, and turned in at the top, outer ones 2-parted. Calyx 5-toothed; teeth sharp and unequal. Umbels terminal, rather small, rays 5 to 8, bracts about 3 in number. Flowers in July.

There are only two species of coriandrum. The C. sativum, however, furnishes all the seeds of commerce.

Geography.—The coriander grows well in subtropical regions, and flourishes high up in the north temperate zone. It is found east of the Black Sea, in the cultivated fields of Tartary, in Hindustan, and Burmah, and is in cultivation in middle, southern, and western Europe. It was brought by European colonists to North America.

Etymology.—*Coriandrum* is from the Greek κόρις, a bug, due to the disagreeable odor of the bruised leaves. *Sativum* is Latin for " sown " or " planted." *Coriander*, the popular name, is a corruption of the botanic name.

History.—The home of this plant is said to be southwestern Tartary, but it is now spread over western Europe, is found in all the countries of the Mediterranean, and has made its way to the gardens of North America, whence it frequently escapes to the fields and roadsides in the northern and middle United States. Theophrastus, who wrote about three centuries before the Christian era, mentions it, and Pliny speaks of it as growing both in Italy and Africa in the middle of the first century.

Use.—The seeds and the oil of the coriander are used for flavoring desert sauces, confectionery, cordials, and English gin. In Germany and the countries of northern Europe they are employed as a condiment in both bread and cake. The ground seeds are used in the mixture known as curry powder, and in other culinary mixtures. The flavor depends upon an essential oil which is obtained from the seeds by distillation.

Its medicinal properties are stimulant, carminative, sedative, and pectoral, and it is frequently administered to modify the griping effects of active purgatives. It was formerly prescribed for gout, St. Anthony's fire, and that class of difficulties. The Mahometan practitioners prepare from the seed an eye-wash which they believe preserves the sight in small-pox.

CUMINUM, L. Calyx-teeth bristle-like, persistent, the outer ones longer. Petals nearly equal, deeply 2-lobed, white or rose colored; style short, erect. Umbels stalked, somewhat irregular, with few rays; general involucre composed of a few long, spreading and deflexed, narrow, stiff, 3-parted or entire bracts; the umbellets with 2 to 4 small bracts. Flowers few in number. Fruit aromatic, bitter.

C. **cyminum**, L. C. **sativum**. (Cumin.) Stem 10 to 15 inches high, branched, cylindrical, solid, striate, smooth; branches spreading. Leaves nearly sessile above, longer stalked below; stalks flattened and clasping, blade ternately divided into long, entire, acute segments, smooth and pale green. The oil-vessels small. Cultivated annual. The only species of the genus.

CUMINUM CYMINUM (Cumin).

Geography. — The geographical zone of this plant is northern Africa, middle and southern Europe, and extends eastward through Syria, Hindustan, Bombay, and Burmah.

Etymology. — *Cuminum*, the generic name, is from the Arabic *Gamoun*, the ancient name of the plant, of obscure signification. *Cyminum*, the specific name, is a variation of the same word. *Cumin* is an abridgment of the generic name.

History. — The exact home of this plant is not known; it is found under cultivation in southern and western Asia, throughout the countries of the Levant, and in northern Africa. It is sparingly cultivated through middle Europe, and fruits as far north as southern Sweden. It is no doubt the plant spoken of in Scripture in the 23d chapter of Matthew, as a minor crop on which tithe was paid. It was cultivated in Asia Minor in the early part of the first century, and is mentioned by Dioscorides. The bruised seeds emit a heavy, disagreeable odor.

Use. — The seeds are used in Germany as a condiment and for flavoring, and the Dutch use them to flavor gin. Their medical properties are carminative, stomachic, and astringent, and they furnish a favorite medicine among the Hindus for dyspepsia and chronic diarrhœa. They are used in external applications for dispersing swellings and allaying pain and irritation. At the present day their use is nearly confined to veterinary practice.

The medicinal qualities are due to an essential oil obtained from the seeds by distillation, known as the oil of cumin.

DAUCUS. Tourn. (Carrot.) Calyx 5-toothed; petals notched at the end, with point turned in, the two outer larger and deeply cleft. Leaves 3-pinnate. Fruit oblong; carpels with 5 primary bristly ribs and 4 secondary, the latter more prominent, winged, and divided each into a row of prickles having a single oil-gland beneath. Flower envelope pinnate. Bracts of the involucels entire or 3-cleft. Fruit oblong-ovate, bristly. Biennial herb.

D. carota, L. **Var. sativa.** (Garden Carrot.) Stem rough, 2 to 3 feet high, clothed with rough hairs, terete, branched from below the middle

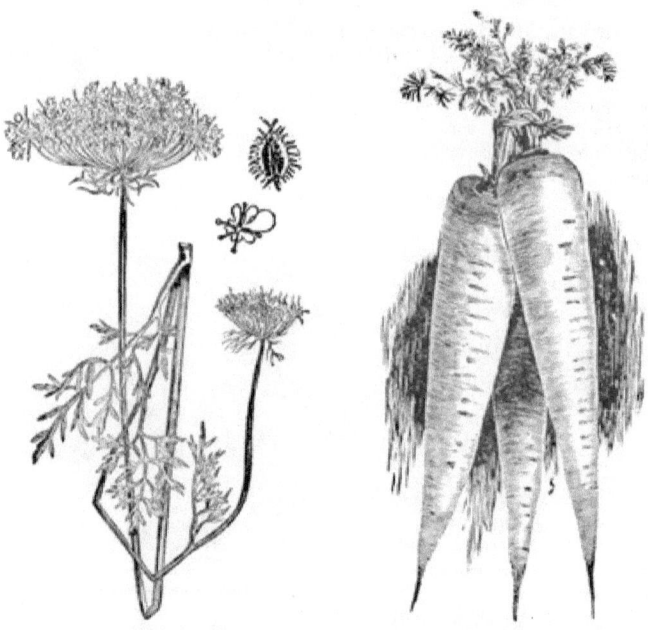

Daucus (Wild Carrot).   Daucus carota (Carrot).

upwards. Leaves 3-pinnate, deep-green; segments linear, pointed. Flowers white or yellowish-white. Root fleshy, fusiform or conical, either white-orange or reddish-yellow.

There are many varieties known to the market gardeners, differing from each other only in the size, shape, and color of the root, which is a valuable culinary vegetable.

Geography. — The carrot is found under cultivation in the British isles, all over the continent of Europe south of 60 degrees, especially in France and Germany, in northern Africa, southwestern Asia, China, Japan, and many of the Pacific islands; in fact, it has found its way into all parts of the world where European settlers have established themselves. It was brought to North America by English colonists, where it has run wild and become a pest.

UMBELLIFERÆ. 155

Etymology. — Daucus is from the Greek word, δαῦκος, a carrot. *Carrot* is said to come from the Celtic word *kar*, red. *Sativa*, the specific name, is Latin, meaning "sown" or "cultivated."

History. — When or where the carrot was first introduced into culinary use is not known. It was known to the Greeks and Romans at least three hundred years before the beginning of the Christian era.

Use. — The carrot is an important culinary vegetable, used to flavor soups, sauces, etc., and is eaten with meats; boiled and reduced to a pulp it is used for pies and custards. It is a very important feed for horses and cattle, and especially for milch cows.

CARUM, L. Flowers hermaphrodite or polygamous. Sepals wanting, or very small. Petals white, 5, unequal, dilated, emarginate, sub-two-lobed or entire, point short, sometimes long and turned in. Fruit oval or oblong-ovoid, compressed, and without wings; carpels 5-ribbed, lateral ribs marginal; umbels perfect. Leaves pinnate. Biennial.

1. **C. petroselinum**, Willd. (Parsley.) Stem angular or striate, 2 to 4 feet high, branched. Leaves smooth and glabrous, decompound, parts incised segments of the lower ones wedge-shaped, terminal ones trifid. Flowers in terminal and axillary umbels. Involucre, a single linear leaflet, occasionally made up of two or more bracts. Petals roundish, incurved, greenish. Fruit ovate. Carpels 5-ribbed. July.

CARUM PETROSELINUM (Parsley).

In common with all plants that are propagated from seed, parsley sports freely, hence cultivation develops varieties. Dwarf Curled, Mitchell's Matchless, Myatt's Triple Curled, Hamburg Large-Rooted and Naples or Celery Parsley, are the most prominent.

Geography. — Parsley is found in the middle and southern edge of the north temperate zone, and grows well in moderately fertile soil.

Etymology. — *Carum*, the generic name of the parsley, is from Caria, in Asia Minor, where it was first brought to the notice of man. *Petroselinum*, the specific name, is from the Greek πέτρα a rock, and σέλινον, parsley, hence rock parsley, due to the place where it grows. — among the rocks. *Parsley*, the common name, is a corruption of *Petroselinum*.

History. — Parsley is found wild in the Mediterranean countries of Europe and in Asia Minor. It has been seen in Japan under cultivation, and is common in the gardens throughout middle Europe and the British Isles; it was brought to northeastern America by English colonists, and has spread over the whole country, but is rarely seen outside of cultivation. It was used by the Greeks and inhabitants of the Levant to decorate the bridesmaids at the marriage feast, to make wreaths, and to adorn graves. A superstition is also attached to it which no doubt arose from its connection with the dead.

Use. — Parsley is used as a flavoring herb in soups, sauces, and in meat and fish stews. To garnish meats, fish, and salads, there is perhaps no flavoring herb more widely used.

The medicinal properties of parsley, as well as its flavoring quality, are due to an active principle known to chemists as *apiol*. This substance is said to have the same effect upon the human system as *quinine*, and was formerly used in intermittent fevers. Infusions of the roots are administered as a cure for fevers and affections of the liver.

2. **C. Carui**, L. (Caraway.) Stem smooth, channeled, branching, 3 feet in height. Leaves smooth, deep-green, bipinnate, cut; segments narrow, linear, pointed. Flowers numerous, in terminal umbels; involucre composed of narrow leaflets, sometimes wanting; petals 5, nearly equal, white or pinkish; filaments slender, rather longer than the petals; anthers small, roundish; ovary inferior, bearing short capillary styles with simple stigmas. Seeds two, bent, one quarter of an inch long, brown; striæ 5, the interspaces furrowed; seed ripens at the end of the second season. Root fusiform and edible.

Geography. — The geographical range of the caraway is between 42° and 60° of the north temperate zone, and it is said by A. de Candolle to be indigenous in a belt from Lapland to Siberia. It grows in Great Britain and all parts of the continent south of 60°. It is also found in northern Africa, Hindustan, and Burmah.

Etymology. — *Carui* is derived from *carum*, whose etymology is given under parsley.

History. — Mention was made of the caraway in an account of Morocco in the twelfth century. In the fourteenth century a custom of eating caraway seeds with apples had been established in England. To this custom Shakespeare refers in Henry IV., Part II., Act V., Sc. 3, where Shallow addresses Silence: "Nay, you shall see mine orchard, where, in an arbor, we will eat a last year's pippin of my own graffing, with a dish of caraways, and so forth." The custom of eating baked apples with caraway seeds is still kept up at one of the colleges of the University of Cambridge, England, and at the ceremonial feasts of some of the London livery companies.

Pliny makes Caria the home of the caraway; if so, it must have spread rapidly, for it is growing in most parts of Europe without cultivation. It is common in the gardens of northeast America, and is frequently found outside of cultivation.

Use. — The root of the caraway plant is eaten as a table vegetable in the north of Europe. The seeds are employed to flavor bread, cake, confectionery, pastry, and cheese; and in Russia, Sweden, Germany, and in parts of the United States, to flavor alcoholic cordials.

An essential oil (oil of caraway) is procured from the seed by distillation, large quantities of which are made in Leipzig. As a medicine, it is aromatic and stimulant, and much used to disguise the unpleasant taste of other drugs. Manufacturers of fancy soap use it in large quantities.

Order XXIX. RUBIACEÆ.

Flowers perfect, rarely unisexual, sometimes defective, but usually regular. Calyx superior, tubular, 2–6-toothed, or wanting. Corolla superior; petals united; limb 4–6-toothed or lobed; segments valvate in the bud. Stamens, 4–6 inserted on the tube of the corolla. Ovary inferior. Style simple, bifid or multifid; stigmas at top or sides; ovules, 1 or more in a cell. Fruit a capsule, berry, or drupe. Seeds in various positions. Leaves opposite and stipulate or whorled, simple, entire. Trees, shrubs, and herbs.

Number of genera, 337.

CINCHONA, L. Calyx cup-shaped, 5-toothed; corolla tubular, limb 5-parted; stamens 5-epipetalous; anthers 2-celled; pistil divided at top; capsule 2-celled, many-seeded, opening at the base. Fruit winged. Trees.

CINCHONA OFFICINALIS (Peruvian Bark).

1. **C. officinalis**, L. (Peruvian Bark; Jesuit's Bark.) Trunk 40 to 50 feet high, and 12 to 18 inches in diameter; branches stout. Leaves opposite, elliptical, entire, and nerved, smooth on the upper side, hairy beneath; petioles short. Flowers panicled; calyx campanulate, margin 5-toothed; corolla tubular, spreading at the throat, and divided into 5 segments; edges serrate; stamens 5; corolla downy on outside.

2. **C. calisaya**, Wedd. (Calisaya or Yellow Cinchona Bark.) Trunk 50 to 100 feet in height, and 5 feet in circumference. Leaves oblong and obtuse, varying in size and shape, 3 to 6 inches long. Flowers in pyramidal panicles, pink. Fruit in ovate capsules, 2-celled; seeds winged.

3. **C. micrantha**, Wedd. Trunk 30 to 40 feet in height, and from 10 to 15 inches in diameter. Leaves from 4 to 12 inches long, and 2 to 6 inches wide, oblong, smooth, and shining above, pitted beneath at the axils of the veins. Flowers small, in loose, leafless panicles.

4. **C. succirubra**, Pavon. (Red Cinchona Bark.) Trunk from 60 to 80 feet high. Leaves broad, oval, 12 inches long, glabrous above, pubescent beneath. Flowers, in large terminal panicles, rose-colored; seed-vessels an inch in length. Sap and wood red. Western slope of the Andes, near the equator, three to five thousand feet above the sea.

Species. — There are thirty-six well marked species of cinchona, and numerous varieties, all natives of the Andes. Those described here furnish the barks of the shops of the United States and Great Britain; but the trees that furnish the materials for the production of quinine are the C. Pitayensis, C. cordifolia, C. lancifolia, and other species which grow in the United States of Colombia. The barks of these species are admitted into market only when they yield two per cent, or over, of alkaloids.

Geography. — The geographical home of the cinchona is in the tropical Andes, extending from 10° north to 20° south latitude, a region about two thousand miles long, mostly on the eastern slopes, in Bolivia, Peru, Ecuador, New Granada, and Venezuela, from 5,000 to 11,000 feet above the level of the sea; the best barks are produced where the temperature ranges from 54° to 68° Fahrenheit. The minimum height for the best barks is not below 5,000 feet. It grows well in similar heights and temperatures in India, where it has been introduced and is under successful cultivation. It is also cultivated in Ceylon and in Jamaica.

Etymology. — The generic name *cinchona* was given to this plant by Linnæus, to honor the Countess of Chinchon, who while residing in Lima was cured of a fever by the use of the bark. The specific names are derived as follows: *Officinalis* signifies "useful," or "of the shops," and is derived from the Latin word *officina*, a shop. *Micrantha* is from the Greek word μικρός, small, alluding to the size of the flower of this species. *Calisaya*, which produces the "yellow bark" of commerce, has a history which is somewhat obscure. Markham gives the following derivations: 1st, from *calla*, a remedy, and *salla*, rocky, meaning a medicine growing among the rocks. 2d, from *ccali*, strong, and *sayay*, become, meaning a medicine that will strengthen the patient. 3d. In Caravaya is a family of caciques, by the name of Calisaya, one of whom distinguished himself in the revolt of 1780-1781, and it is suggested that this species was named to honor him. *Succi rubra* is from *succus*, juice, and *rubra*, red; hence, red-barked cinchona.

History. — It is stated upon good authority that the aborigines were not acquainted with its medicinal properties before the country was visited by Europeans. Humboldt states that it is not upon the list of native remedies.

There is a story as follows: A savage was taken ill with a fever in the forest, near a pool of water, into which a number of cinchona trees had fallen, whereby the water had been made bitter. He was offered some of this water, as no other could be procured, and drinking, was speedily cured. Thus the curative qualities of the bark were revealed.

In 1638 Ana de Osoria, wife of the fourth count of Chinchon, viceroy of Peru, lay dangerously ill with a tertian fever at Lima. When accounts of her sickness reached Don Francisco Lopez de Canizares Corregidor of Loxa, he sent the bark to her physician, Don Juan de Vega, who administered it to his patient, and thereby effected a speedy cure.

The countess was so grateful for her recovery that she determined, on her return to Europe, two years later, to take with her a quantity of the powdered bark, to be administered to the sufferers from chills and fever upon her husband's estate. From this circumstance it was called Countess Powder, and for a long time retained that name. De Vega, on returning to Spain, carried with him large quantities, which he sold in Seville at 100 reals a pound. Linnæus to honor the countess, named the tree cinchona, which was intended to be "Chinchon," and the error in spelling has never been corrected.

Another account relates that it was made known to the civilized world through a monk, who, lying at the point of death with a fever, received a decoction of the bark from the hands of a native medicine man, and was cured. It is not improbable that both these accounts are correct, and that the circumstances occurred as stated. The tree is a native of the tropical Andes, on a chain of mountains in Peru.

The history of the cinchona would be defective without some account of its introduction into India. In 1839, Dr Royle, the English East India botanist, drew the attention of the home government to the importance of providing a febrifuge for medical practice among the natives, and advocated the introduction of the cinchona into India.

In 1859 the British government sent an expedition to South America to procure seeds and plants of all the species possessing commercial value. The party encountered great difficulty and endured great hardships, but secured seeds and living plants, which were taken to England and sent to India, where suitable localities in the mountains were selected, and a successful plantation was commenced. Previous to this, the Dutch East India Company had sent an agent to South America who had procured seeds, but when the trees came to maturity, they proved to be worthless. Those taken to India and Ceylon were very productive, and they far exceed in value the trees in a native state.

In 1878 a German company established a plantation in Bolivia, where there are now about 10,000,000 trees under cultivation. Though only a little more than a quarter of a century has elapsed since the first plantations were commenced, the barks which supply the markets of the world are nearly all from cultivated trees.

Chemistry. — The cinchona barks yield to the chemist a number of alkaloids, the principal of which are the following : —

Quinia, C_{20}, H_{24}, N_2, O_2.
Cinchona, C_{20}, H_{24}, N_2, O.

The lowest per cent of alkaloids in the barks of commerce is 2, and the highest $13\frac{1}{4}$. This high per cent is obtained from the C. officinalis, var. lanceolata, under cultivation; 9 per cent of the $13\frac{1}{4}$ is Quinia.

Quinia and Cinchonia were discovered in 1820 by the chemists Pelletier and Caventon, who secured the prize of 10,000 francs offered by the French Academy of Science.

Preparation. — The mode of collecting the bark is to cut down the tree, and then strip the bark, after which it is dried in the sun and sewn up in green ox-hides, and exported in large bundles or packages.

This wasteful mode of collecting the bark is not practiced upon the trees planted in India, but alternate strips are removed ; the wounds are then bound up, and when properly healed the other spaces are stripped, by which means the tree is indefinitely preserved. Trees under cultivation yield barks far richer in alkaloids, and the successive new layers of bark are more and more highly charged with the valuable products.

Use. — The substances yielded by the cinchona barks are powerfully tonic, antiseptic, and antiperiodic, and the bark itself as a whole is highly astringent Quinine, the sulphate of quinia, contains the properties of the bark in the most concentrated form.

No substance in the *materia medica* is of such importance in the healing art where malarial and intermittent fevers prevail. Though it is a specific as a febrifuge, it is administered in all complaints that attack the system at intervals, as neuralgia, rheumatism, etc.

It has been found that a solution of quinine in 20,000 parts of water will destroy bacteria. It is believed that malarial fevers are due to the direct introduction into the blood of living organisms. Quinine is supposed either to destroy these organisms or to render the condition of the blood unfavorable to their development.

COFFEA, L. (Coffee.) Calyx, tubular, 5-toothed; corolla funnel-shaped, separated at the crown into 5 reflexed lanceolate divisions; stamens 5 in number; anthers oblong; style, with 2 stigmas; ovary 2-celled. Shrubs.

1. **C. Arabica,** Alpinus. Stem 10 to 15 feet in height, and 2 to 4 inches in diameter, diffusely branched; branches slender and drooping; bark greenish-brown. Leaves elliptical-lanceolate, entire, crenate or wavy, 3 to 5 inches long, on short foot-stalks, opposite and evergreen. Flowers white, in axillary small, nearly sessile, clusters. Fruit, a dark-red berry, in form of a cherry, with a glutinous, tasteless pulp inclosing 2 plano-convex seeds about three eighths of an inch in length, two eighths wide, a groove extending along the longer axis of the plane side.

2. **C. occidentalis** is no doubt a variety of C. Arabica, from which have arisen most of the varieties which are known in South America.

Like all plants grown from the seed, coffee sports freely, hence we have many species, among which the following are well marked and constant.

Growing in Brazil:—

3. **C. Australis.**
4. **C. nodosa.**
5. **C. biflora.**
6. **C. paguiodes.**
7. **C. jasminoides.**
8. **C. parvifolia.**
9. **C. magnolifolia.**
10. **C. sessilis.**
11. **C. meridionalis.**
12. **C. stipulacea.**
13. **C. minor.**
14. **C. truncata.**

Growing in the East Indies are four species:—

15. **C. semiexserta.**
16. **C. Travancorensis.**
17. **C. tetrandra.**
18. **C. Wightiana.**

Three species are cultivated in Mexico:—

19. **C. Mexicana.**
20. **C. obovata.**
21. **C. rosea.**

In New Granada there is one species:—

22. **C. spicata.**

The following species grow in Peru:—

23. **C. nitida.**
24. **C. racemosa.**
25. **C. subsessilis.**
26. **C. umbellata.**
27. **C. verticillata.**
28. **C. longifolia.**
29. **C. ciliata.**
30. **C. acuminata.**

In Java there are two prevailing species under cultivation:—

31. **C. densiflora.**
32. **C. Indica.**

In the Molucca isles one single species prevails:—

33. **C. pedunculata.**

In the Sandwich Islands two species are cultivated:—

34. **C. Chamissonis.**
35. **C. kaduana.**

In Arabia and Abyssinia the **C. Arabica** prevails, and is very constant. On the western coast of Africa two species are cultivated: **C. laurina,** and **C. Liberica**; while on the eastern coast the **C. Mozambicana** and the **C. Zanguebarica** are grown.

Geography. — Coffee will grow and ripen its fruit in all regions of no frost and is grown in all tropical and subtropical countries. The market is largely supplied from South America and the East India islands. The principal coffee-growing regions are Brazil, Guatemala, Cuba, British West Indies, St. Domingo, Java, Padang, Sumatra, Maccassar, Ceylon, British India, and Manilla.

Etymology. — Coffee is said to have derived its name from the Turkish *qaveh*, a decoction of berries. The common name is a corruption of the botanic name. The specific name of the principal species comes from Arabia, where it was first used.

History. — Persia, the home of delicious fruits, seems to have given birth to coffee. Thence sometime in the fifteenth century it was carried by Magalleddin Mufti of Aden into Arabia Felix, where it was first used, not as a beverage, but for medicinal purposes.

Coffee was not known in commerce till about the middle of the sixteenth century, when it became an article of trade in the markets of Constantinople.

The government of Syria forbade its use, ostensibly because of its intoxicating qualities. After its introduction into Constantinople, the Mohammedan priests complained that the mosques were neglected, while the coffee-houses were thronged. The government interfered and forbade its sale, and a strict police espionage was instituted; but as it was found impossible to suppress its sale, the state levied an excise tax on it, and thus reaped a large income from it.

Coffee is consumed in Turkey in large quantities. There was a time when it was regarded as so necessary to the people that it became one of the legal causes for divorce when a man refused to furnish his wife with coffee.

Coffea (Coffee).

Though the coffee tree in cultivation is supposed to have been brought to Arabia from Persia, yet there is good reason to believe that it is indigenous in Arabia Felix, and in Africa, on the opposite shores of the Red Sea. Rauwolfins took it into Europe in 1573; but its introduction is traced also to the Dutch, who procured berries at Mocha, which were planted at Batavia. In 1690, a plant was sent to Amsterdam, which was planted and bore fruit (under glass). The seeds of this fruit were then planted, and many young trees were produced therefrom. These trees were sent to the gardens in the Dutch possessions in the East Indies, and some of the plants were presented to Louis XIV, by the Dutch authorities; these were placed under the charge of Jussieu, by whom young plants were sent to the French West Indies, whence the coffee-tree has spread, not only throughout the islands, but to the continent of South America.

All the coffee grown in the new world is said to have sprung from a single plant which a French naval officer carried to Martinique in 1720, depriving himself of water when parching with thirst in order to nourish his coffee-plant. From this tree, it is said, all the American tropical colonies obtained their seed, which has multiplied to such an extent that Brazil, Mexico, and the West Indies produce as much coffee as Java and Ceylon.

It is not known when coffee came into use in western Europe as a beverage. The Venetians, who traded with the East, no doubt first used it.

It was not known in Italy in 1615. In 1645, some men returning home from Constantinople to Marseilles took with them a supply of coffee, with suitable vessels for preparing it for the table; it was thus introduced into France. About twenty years later a house was opened at Marseilles for the sale of coffee.

In 1671 the first coffee-house was established at Paris. Other places were soon opened for its sale, but upon a very humble scale, and fashionable people did not resort to them. Some Frenchmen, shrewdly guessing the reason for a want of genteel patronage, fitted up a coffee-house in a liberal and elegant style, to which well-bred people were attracted. About the same time a successful coffee-house was opened at London.

Nieber, in his account of coffee, maintains that it was grown upon the hills of Yemen in Arabia, introduced from Abyssinia by the Arabs, long before it was used by Europeans.

Chemistry. — Coffee yields to chemical analysis the same substances that are found in tea, though in different quantities, hence the effects upon the nerves and the circulation are similar to those produced by tea.

In 100 parts of tea in a dry state there is one part of thein; in the same quantity of coffee there is only one half as much. Of nitrogenous substances, there are in tea 25 parts, in coffee, 13; but of essential oil, tea has about $\frac{3}{4}$ of 1 per cent, while coffee has only $\frac{3}{100}$ of 1 per cent.

Tea has 12 parts of tannic acid, coffee has $5\frac{1}{2}$. Potash, phosphoric acid, and oxide of iron are found in both in nearly equal quantities, amounting to about five per cent.

Preparation. — Coffee is prepared by first browning it over a gentle heat, called burning or roasting; it is then crushed or ground; hot water is applied to it and kept at the boiling-point for a short time; some substance is then mixed with it to precipitate the grounds or powdered coffee held in suspension; after which, the liquor is poured into cups, and milk and sugar are added to suit the taste.

Use. — In most families it is used for breakfast, and for dinner. It is so well known that further description is not necessary.

Statistics. — No other warm dietetic beverage is so largely used as coffee. It is the daily drink of more than 100,000,000 people. We have no means of knowing the actual consumption in Turkey and Africa, but the tables of import show the consumption in other countries. In the several coffee-drinking countries the consumption is as follows: —

In the United States of North America, 400,000,000 of pounds are annually consumed, which is equal to 8 pounds for every man, woman, and child. The amount used in Holland is equal to 21 pounds for each person. In Belgium and Denmark the consumption is equal to 13 pounds for a person; in Norway, 10 pounds; in Switzerland, 7; and in Sweden, 6. These are the great coffee-drinking peoples of Europe and America. In the kingdom of Great Britain, a greater amount of tea is used than in any other nation, amounting to about 4 pounds for each individual; the amount of coffee consumed is only one pound to each person.

CEPHAËLIS. Swartz. Calyx bell-shaped, toothed; corolla tubular, inflated at throat, 5-parted; stamens 5; stigmas 2-parted. Flowers crowded into a head, inclosed in a 5-leaved envelope. Berry 2-seeded. Shrub.

RUBIACEÆ.

C. Ipecacuanha. Richard (Ipecac.) Stem pubescent at top, 18 to 24 inches high; root 4 to 6 inches long, about the size of a goose-quill. Leaves, about 6 in number, opposite, petioled, oblong-obovate, acute, entire, 4 to 6 inches long, 1 to 2 wide, rough above, downy and veined beneath; stipules clasping, membranous at base, split above into numerous bristle-like divisions, falling. Flowers, 8 to 10, small, white, each with a green bract, forming a little head on an axillary foot-stalk, and inclosed by a 1-leaved involucre, cut into 4 or 6 segments. Fruit, an ovate, purple berry, becoming black; seeds small, plano-convex, 2 in number.

Geography. — The cephaëlis is a tropical and subtropical plant, and flourishes in rich, damp woods.

Etymology. — Cephaëlis is from κεφαλή, Greek for "head," alluding to the form of inflorescence. Ipecacuanha is from the Brazilian *ipecaayuen*, road-side, sick-making plant.

CEPHAËLIS IPECACUANHA (Ipecac).

History. — This drug is said to have been introduced into the materia medica by John Helvetius, a Dutch physician practicing in Paris. He first used it as a secret remedy in dysentery, and was induced by Louis XIV. to reveal his secret, for which the sum of 25,000 francs was awarded him. The home of the Cephaëlis is the damp, rich woods of the valley of the Amazon. It is found in Bolivia and Colombia, has been introduced into the West Indies and Hindustan, and is under successful cultivation in India.

The American ipecacuanha is the root of the Euphorbia Ipecacuanha; it has a local reputation as an emetic, and is occasionally used as a substitute for the South American drug.

Chemistry. — The active principle of ipecac is emetine, of which it contains less than 1 per cent. Pelletier discovered, or rather isolated, it in 1817, and found it to be an alkaloid.

Use. — The medical properties are astringent, diaphoretic, expectorant, and emetic. The active principle is largely in the bark of the root, though the woody part of the root also possesses it.

It forms, in combination with opium, the well-known Dover's powders. It is an important medicine in dysentery, and is an ingredient in most cough medicines.

RUBIA. Tourn. (Madder.) Calyx-tube egg-shaped, 5-toothed; corolla rotate, 5-parted; stamens 5, short, 2 styles, united at the base. Fruit in twos, berry-like, smooth and subglobular. Perennial, herbaceous, does not flower until the third year.

1. **R. tinctorum**, L. Stem weak, 4-angled, angles armed with prickles turned backwards, or downwards, trailing or climbing. Leaves in whorls of

6, lanceolate, margins and midribs aculeate; flower-stalk axillary or terminal, trifid. Flowers, brownish-yellow. Root consists of many long prickly shoots half an inch in diameter, and 3 to 4 feet long, descending deep into the ground, all united near the surface in a sort of head.

2. **R. Chiliensis** is used as a dye in South America.

3. **R. cordifolia**, a native of Persia, is largely used in Hindustan, both as a dye and as an article of medicine. Other species are used for dyes in the countries where they grow.

Geography. — Its geographical range is the middle and southern parts of the north temperate zone. It is indigenous to western Asia and the Mediterranean countries of eastern Europe. It is cultivated successfully in Hindustan, China, Japan, and Northern Africa, Turkey, Greece, Spain, France, Germany, and Holland, and in the middle and central United States of North America. It has been cultivated for market in Ohio and Delaware. It is shipped to England from the ports of India and from the eastern Mediterranean, and thence to America.

Etymology. — *Rubia* is derived from the Latin word *ruber*, red, from the color of its root. *Tinctorum* is from the Latin word *tinctor*, a colorer. *Chiliensis* is derived from *Chile*, the home of this species. *Cordifolia*, from the Latin, refers to the heart-shaped leaves. *Madder* is derived from the Sanscrit *madhura*, sweet, or tender, alluding to the character of the root.

History. — The madder was known to the ancients, and it is believed that some of the cloths in which the mummies were rolled were colored with madder. It was one of the most important dyes known to the Greeks and Romans.

Preparation. — The long, slender roots are dug when the plant is three years old, and when dry are about the thickness of a goose-quill, and of a deep red color. The method of preparation is to grind the root and wash it with water, which takes out the inert matter, among which is sugar. Straining through woollen cloth leaves the bruised root upon the cloth; the root is then dried and reground. The liquor is preserved in vats and allowed to ferment, after which it is distilled, and yields a quart of alcohol for every 100 pounds of root.

Use. — Madder is used for a dye, especially in the printing of muslins. The fine Turkey red is produced by madder. By the use of chemicals, every shade of red, purple, lilac, and rose-color can be obtained from the madder-root. The color is suspended both by alcohol and by water. As a medicine, it excites the secretory organs, and especially the kidneys. When fed to cattle, it enters into the milk and other fluids of the body, and even colors the bones.

The Alizarine of the shops is artificial, and is a derivative of anthracine, a coal-oil product. Since its introduction, the cultivation of madder has almost ceased; over a million acres of madder land have gone out of cultivation in France alone.

RUBIA TINCTORUM (Madder).

Order XXX. COMPOSITÆ.

Flowers in close heads, polygamous, monœcious. The central flowers in a head are called the *disk*; the marginal flowers, if of different shape from the disk, form the *ray*. Flower heads each on a common receptacle, inclosed by an involucre of scale-like bracts; the whole head resembling a flower, and the involucre like its calyx; each of the proper flowers (termed *florets*) having the calyx adhering to the ovary, its limb represented by a hairy pappus or scales, or wanting. Calyx tube adhering to the ovary, its limb usually made up of hairy bristles or scales, occasionally wanting. Corolla either tubular or strap-shaped, generally 5-toothed or lobed; stamens 5, inserted on the corolla; anthers united in a tube around the 2-cleft style. Fruit an akene, one-seeded. Leaves alternate or opposite, frequently divided or cut, without stipules. The florets, or little flowers, are aggregated upon a receptacle, the tubular florets forming a circular disk, while the strap-shaped ones form a circular ring outside the disk ; in some cases all are strap-shaped. The under side of the receptacle is clothed with or included in a foliaceous aggregation of bract-like scales, which take the place of a common calyx, and are called the *involucre*. Nearly all herbs. A very large order, containing 766 genera, about one tenth of all flowering plants.

INULA. L. Heads many-flowered, with an imbricated involucre. Ray flowers numerous, pistillate ; disk flowers perfect. Receptacle naked ; pappus simple, scabrous ; anthers with two bristles at the base.

1. **Helenium,** L. (Elecampane.) Stem 5 feet high, stout, coarse, furrowed, downy, and branching above. Leaves clasping above and petioled at the root, ovate, rough, downy underneath, very large, 2 feet long and 1 foot wide ; serrate, crowded with a network of veins, midrib large. Flower-heads large, solitary, and terminal ; rays linear, yellow ends, 2-3-toothed. Flowers in August.

Geography.— It grows freely throughout the middle of the temperate zone, both in Europe and North America as well as Asia, in rich, damp soil.

INULA HELENIUM (Elecampane).

Etymology.— *Helenium*, the specific name, comes from the Greek name of the plant, ἑλένιον, given in honor of Helen of Troy. *Inula*, the generic name, is the Latinized form of the same. *Elecampane* is derived from the Greek ἑλένιον and Latin *campus*, a field.

History.— The Elecampane was eaten by the ancients in the countries of the Levant, and was used by the Egyptians for medicine. Dioscorides describes it. Thunberg saw it in Japan, near Jeddo, both under cultivation and without. He heard no Japanese name, hence inferred that it was introduced there by Europeans. It *is* indigenous to middle Asia, and was carried west throughout middle Europe by travellers, and was brought to northeast America by European colonists.

Use. — It is used in the *materia medica* as a tonic in weak digestion and nervous complaints, throat diseases, and as an expectorant.

ANTHEMIS, L. (Camomile.) Flower-envelope, in form of a hemisphere; scales equal; rays numerous, pistillate; akenes terete, angular, or striate, crowned with a border. Receptacle convex. A perennial herb.

1. **A. nobilis, L.** (Roman Camomile.) Stems numerous, 6 to 12 inches long, spreading, and decumbent. Leaves 1 to 2 inches long, sessile, and velvety; divisions of the leaves linear. Flower-heads terminal, on long, axillary pedicels; rays white; disk yellow; scales of receptacle broad and obtuse; whole plant has a pleasant aromatic odor and bitter taste, especially the flowers.

Var. **flore pleno** has double flowers.

2. **A. arvensis, L.** (Common camomile. Field camomile.) Differs from A. nobilis very little, except that the flowers are smaller, and the bitter taste is less agreeable. It is a troublesome weed in corn and potato fields and on roadsides.

Geography. — The anthemis grows well throughout the middle and southern portions of the north temperate zone, in Europe, America, and Africa. The home of the camomile is western Europe. It is cultivated in Italy, France, Spain, and Germany, but the best is produced in England, and was brought to northeast America by European colonists. By cultivation, the yellow-disk florets change into ray florets, and become white; they are then called "double." But their medicinal character is thereby damaged.

ANTHEMIS NOBILIS
(Roman camomile).

Etymology and History. — *Anthemis* is from the Greek ἄνθος, a flower, due to the profusion of flowers it produces. *Nobilis*, the specific name, Latin for "noble" or "grand," is due to the large size and showy character of the flower of this species. *Arvensis* is from the Latin *arvum*, a plowed field, where the plant loves to grow. *Camomile* is from the Greek χαμαί, on the earth, and μῆλον, an apple, due to the apple-like smell of the flower. There is a popular belief that the worse the usage it receives the better it grows. Shakespeare, in his Henry IV., says: "For though the camomile the more it is trodden on the faster it grows, yet youth the more it is wasted the sooner it wears."

Use. — The medicinal properties of camomile are stimulant and tonic, and it is a favorite domestic remedy for stomach disorders and loss of appetite. The effects, besides being stimulant and tonic, are also carminative and anodyne. A strong infusion, administered warm, is emetic. The flowers are used in domestic practice for fomentations and poultices. The extract of the flowers, as well as the essential oil, is used in the manufacture of bitters. The flowers are chewed and the saliva swallowed for stomach disorders; and it is recorded that the ancient Egyptians made an ointment by bruising the flowers with oil, which they used for skin diseases. The constituents of the flowers are a fixed oil, contained in the seeds; an essential oil, upon which the odor depends; and a substance obtained by extraction, to which the bitterness is due. The essential oil is obtained by distillation.

CHRYSANTHEMUM, Tourn. Heads many-flowered, with numerous pistillate rays; disk flowers fertile. Involucre, a flattened hemisphere; scales short, appressed, thin, and imbricated. Receptacle flat or convex, naked. Rays usually elongated; corollas of the disk flowers flattened or 2-winged, below 4- to 5-toothed. Akenes short, ribbed or angled, truncate at the tip, often destitute of pappus. Herbaceous perennials, or annuals.

There are many species belonging to this genus. The celebrated Persian insect-powder is the pulverized flowers of C. roseum, C. carneum, and C. Wilmoti. The Dalmatian insect-powder is the product of C. cinerariifolium, Tres., var. rotundifolium. In the south of Europe the C. corymbosum also furnishes an insect powder.

C. carneum, M. B. (Chrysanthemum.) Stem 18 inches high. Leaf smooth, bipinnate; segments of prismal acute. Flower-heads one and a half inches broad; involucre imbricated; margin brown and scarious; receptacle convex, naked. Ray-flowers 20 to 30, ligulate, nerved, and 3-toothed; disk-flowers numerous, tubular, 5-toothed. Akenes dark brown, angular, wingless, crowned with a short membranaceous pappus. Ray florets pale pink. Anthers projecting. Perennial.

CHRYSANTHEMUM CARNEUM (Chrysanthemum).

Geography.— The C. carneum, C. roseum, and C. Wilmoti are found native in the mountainous regions of northern Persia, and the country east of the Black and Caspian Seas. The C. cinerariifolium is indigenous to Dalmatia, Montenegro, and Herzegovina, and has been introduced into southwestern Europe and California. The C. corymbosum is native in southern Europe.

Etymology.— *Chrysanthemum* is from the Greek χρυσός, gold, and ἄνθος, a flower, due to the yellow color of some of the species. *Carneum*, the specific name, is Latin for "fleshy," possibly derived from the thick fleshy leaf of this species. *Roseum* refers to the color of the flower. The name *cinerariifolium*, from the Latin word *cinis*, ashes, and *folium*, leaf, ash-leaved, is due to an ashy down, with which the leaves are clothed. *Wilmoti* is for Wilmot. *Rotundifolium*, from *rotundus*, round, and *folium*, a leaf, round-leaved.

History.— The use of these flowers as an insecticide was known to the ancients in the countries of Asia, but their introduction as articles of commerce in western Europe and America is quite recent.

It is advertised under the name of Persian Powder, but very extravagant claims are made for the Dalmatian Powder.

Use.— The insect-powder is scattered about the flower, or blown into cracks and crevices where the insects hide. It is especially useful in the destruction of croton bugs, roaches, fleas, the house-fly, mosquitos, spiders, ants, etc., and all forms of insect life that infest dwellings. Lice upon poultry or cattle are exterminated by it. Scorpions and centipedes also succumb before its potent presence.

The best powder is produced from the flower when the pollen is just mature; in fact it has been supposed that it is the pollen alone that is effective, but it is claimed that it is the odor that effects the destruction of the insect, and there seems to be reason for this belief; if that be true, it cannot be the pollen that causes the death of the insect, for the odor does not reside in the pollen alone. Again, it has been asserted that the pollen possesses an independent odor from the plant, which is destructive to insect life.

The powder is produced by grinding the flowers, which have been harvested just as the pollen is ripe. The powders of commerce are said to be adulterated with the pulverized flowers of anthemis.

It is destructive to the caterpillar family, as well as the coleoptera tribe, and this has been used as an argument against the odor theory. Professor Riley, in 1878, showed that it destroys the cotton-worm. The reason for its destructive character to insect life, while harmless to higher forms of existence, is not understood. Men, quadrupeds, and birds breathe it with impunity.

TANACETUM, L. Heads many-flowered and corymbosely cymose, staminate flowers occupying the central part of the head. Pistillate flowers, with a tubular 3–5-toothed corolla, sometimes imperfect, or partly ligulate, arranged around the outer edge of the head; the little seed-vessels, ribbed or angled, with 3–5 ridges, flat on top.

TANACETUM VULGARE (Tansy).

T. vulgare, L. (Tansy.) Stem erect, strong, angular, leafy, and branched above; smooth and purplish, 2 to 3 feet high. Leaves numerous, alternate, clasping, bipinnate; segments oblong, cut, and serrate; the lower leaves bipinnate; the little leaflets trifid, spreading at the base along the petiole, deep green, roughish, though not hairy, deep green, paler beneath. Flowers yellow, in a terminal flat corymb; involucre hemispherical; scales imbricated, numerous, linear, lanceolate, acute. Ray flowers few and inconspicuous, limb toothed. Disk flowers many, perfect, tubular, 5-cleft. Stamens 5; anthers united; all included within the corolla tube. Ovary oblong; style setaceous; stigma forked. Fruit small, obovate, angular, crowned with a 5-sided membranous pappus containing a single seed. August.

Geography —The geographical range of the tansy is not very wide. Indigenous to the Crimea and adjacent parts of western Asia, it has spread through middle and western Europe and northern Africa, where it is found in gardens and by the roadsides near dwellings. It was brought to North America by European settlers early in the colonization of New England, whence it has spread throughout the Atlantic States, escaping from gardens, and has become naturalized.

Etymology. — *Tanacetum* is said to be altered from *athanasia*, which is derived from the two Greek words, α, without, and θάνατος, death, in allusion to the durable character of the flowers. *Vulgare,* the specific name, is from the Latin adjective *vulgaris,* and signifies "common." *Tansy* is a corruption of *Tanacetum.*

History. — The Egyptians had a legend that their deity Isis discovered the properties of tansy. The plant was known to the ancients, but when and where it was introduced into medical practice is not known.

Use. — The medical properties of tansy are stimulant, carminative, sudorific, and anthelmintic. It is used principally at the present day in domestic practice. The pulverized leaves, mixed with sirup, are said to be a specific for ascaris. A tincture is used for stomach bitters, and in some rural districts the bruised leaves and flowers are used to flavor gin, taken for stomach troubles. It is also administered for ague, in the form of tea.

The medical properties depend upon an essential oil, called oil of tansy, obtained by distilling the whole plant

CARTHAMUS, L. (Safflower, Saffron.) Heads discoid, flower envelope imbricated, outer bracts leaf-like Florets all tubular and perfect; filaments smooth; without pappus; receptacle with bristly bracts or paleæ. Akene 4-angled. Annual.

C. tinctorius, L. (Safflower.) Stem smooth, 3 to 4 feet high, much-branched near the top. Leaves ovate, lanceolate, sessile, and subamplexicaul, teeth armed with sharp spines. Flowers orange-colored; heads large, terminal; florets long and slender.

Geography. — The Carthamus is indigenous to all eastern Asia and the Levant, and has been introduced into Egypt and western Europe. It thrives well in France and southern Germany, and was brought by European colonists into the eastern United States of North America, where it is cultivated for ornament.

Etymology and History. — *Carthamus* is derived from the Arabic word *quortom,* paint.

The specific name, *tinctorius,* is from the Latin word *tinctura,* a dyeing. *Safflower* is supposed to be a contraction of saffron-flower, but its origin is not clear. The flowers were brought into western Asia and southeastern Europe, overland, as early as 115 B. C.

CARTHAMUS TINCTORIUS (Safflower).

Preparation. — To obtain the dyeing principle — carthamine — the young florets are picked and washed to free them from a soluble yellow coloring matter which they contain. They are then dried in kilns and powdered, and placed in an alkaline solution in which pieces of clean white cotton are immersed. The alkaline solution having been neutralized with weak acetic acid, the cotton is removed and washed in another alkaline solution. The second solution is again neutralized with acid, and carthamine in a pure condition is precipitated. Dried **carthamine** has a rich metallic green color.

Use. — The coloring matter of the safflower is used in cosmetics for delicate red tints. Its principal value is as a dye. The Chinese, by the use of mordants, alkalies, and acids, produce from this plant the delicate rose, scarlet, purple, and violet colors that make their silks so valuable. The Spaniards employ the flowers to color their soups. The Poles mix them with their bread and cakes. The seed is a valuable food for parrots and other caged birds; domestic fowls eat it greedily, and fatten rapidly when fed upon it. The yellow coloring-matter is an extract, but the red is known to the chemist under the name of carthamine.

As a medicine the safflower is purgative when taken in large doses. The seeds yield an oil which is prescribed as a remedy for rheumatism and paralysis.

ORDER XXXI. **CAMPANULACEÆ.** (BELL FLOWER.)

Flowers superior, 5-merous, symmetrical; perianth and stamens adhering to the ovary; anthers distinct or united; ovary usually 2-3-celled; seeds numerous. Herbs or shrubs, with milky juice. Leaves usually alternate, exstipulate.

No. of genera, 53.

LOBELIA, L. Calyx 5-parted; tube short, egg-shaped; corolla irregular, 2-lipped, upper lip 2-lobed, lower lip 3-cleft; stigma 2-lobed; seed-vessels 2-celled, many-seeded, opening above. Leaves alternate. Biennial.

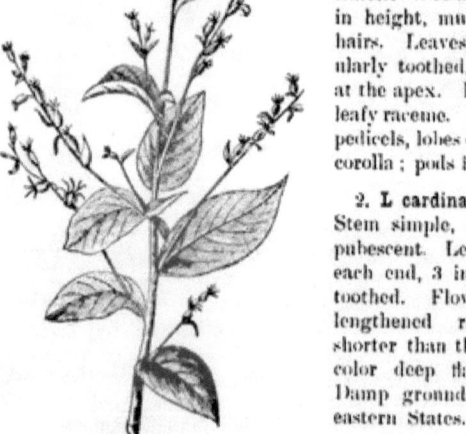

LOBELIA INFLATA (Emetic Weed).

1. **L. inflata**, L. (Indian Tobacco. Emetic Weed.) Stem 10 to 20 inches in height, much-branched, clothed with hairs. Leaves ovate or lanceolate, irregularly toothed, sessile, lower ones blunt at the apex. Inflorescence a paniculated leafy raceme. Flowers pale-blue, on short pedicels, lobes of the calyx as long as the corolla; pods inflated. Flowers in July.

2. **L. cardinalis**, L. (Cardinal Flower.) Stem simple, erect, 1 to 2 feet high, pubescent. Leaves lanceolate, pointed at each end, 3 inches long, pubescent and toothed. Flowers scarlet, in a terminal lengthened raceme, 1-sided, pedicels shorter than the bracts; stamens exsert, color deep flaming red, very showy. Damp grounds throughout the northeastern States. Easily cultivated.

3. **L. syphilitica**, L. (Blue Cardinal Flower. Great Lobelia.) Like the last, except in the color of the flower, which is bright blue, rarely white. Wet grounds by the roadsides, common in the northeastern states. Bears cultivation.

This genus is very large, containing about 400 species, about 20 of which are indigenous to the United States. Both the L. cardinalis and L. syphilitica possess the narcotic poisonous properties of L. inflata, but in a milder degree.

Geography. — The geographical distribution of this genus is very wide, and it has representatives in all parts of the world; but the L. inflata is confined to North America, ranging from North Carolina to Canada, and west to Kentucky.

Etymology. — The name *Lobelia* was given to this plant in honor of Matthias de Lobel, a native of Lisle, botanist and physician to James I. The specific name *inflata* is Latin, and due to the circumstance that the pods are inflated. *Emetic weed* derives its name from the powerful emetic qualities which the plant possesses. *Indian tobacco* owes its name to the fact that this plant is used by the North American Indians, and that its effects are similar to those of tobacco. It is expectorant and diaphoretic in small doses, but in full medicinal doses, nauseating and emetic. *Cardinalis* and *cardinal* are names due to the large, showy, intensely red flowers of this species. *Syphilitica* derives its name from the fact that this plant is used as a remedy in syphilitic diseases, while the common name, *blue cardinal*, is due to the fact that this species, though otherwise similar to the L. cardinalis, has bright blue flowers.

History. — The only history the Lobelia inflata can boast of is due to the controversy carried on some years ago by the physicians of the old school and the Thomsonian empirics, the latter claiming marvelous curative properties for it, and proclaiming it a useful and harmless medicine, while the regular physicians denounced it as a dangerous poison, to be avoided or used with great caution.

Chemistry. — The exact chemical character is not known. It yields to analysis an alkaloid liquid *lobeliana*, and an acid called *lobelic acid;* these substances reside in all parts of the plant.

Use. — Lobelia inflata has gained renown as an empiric remedy. It enters into almost every preparation of the Thomsonian physicians; they place great reliance upon its virtues as a tonic, emetic, and bilious excitant. The root of the L. syphilitica is extensively used by the North American Indians.

Order XXXII. VACCINIACEÆ.

Flowers 4-5-merous, regular. Calyx adnate to the ovary; petals united; 8-10 stamens; anthers opening at the apex; ovary several-celled. Fruit, a berry or drupe. Shrubs, with alternate exstipulate leaves.

No. of genera, 26.

GAYLUSSACIA, H. B. K. (Huckleberry.) Corolla a short, egg-shaped tube, with a 5-cleft edge; limb reflexed; stamens 10; anthers without awns; cells tapering upwards, forming a sub-tubular process opening at top; style longer than stamens; stigma flat. Fruit globular, flattened at top with 4-5 cells; seeds many. Flower solitary, pedicellate, racemose, drooping, pinkish. Fruit black or bluish.

1. **G. dumosa,** Torr. and Gray. (Dwarf Whortleberry, or Huckleberry.) Clothed with fine hairs, and glandular. Leaves oblong-ovate, mucronate, both sides green, shining when old; racemes long, bracts oval, and as long as the pedicels, persistent; corolla campanulate. Fruit black, tasteless. Var. Hirtella is distinguished by having the young branchlets, racemes, and leaves

clothed with hairs. This variety is found along the coast of New Jersey, and south.

2. **G. frondosa**, Torr. and Gray. (Blueberry. Tangleberry. High Blueberry.) Smooth branches, slender and spreading. Leaves ovate, blunt, twice as long as wide, pale beneath. Racemes slender, loose; bracts sublinear; corolla globular, bell-shaped. Branches slender, with grayish bark. Flowers small, nearly globular, reddish-white. Fruit large, clothed with a glaucous bloom.

3. **G. resinosa**, Torr. and Gray. (Black Huckleberry.) Branched, rigid, somewhat hairy when young. Leaves petioled, oblong, egg-shaped, entire, sprinkled with resinous dots, 2 inches long, sometimes acute, shining beneath; racemes short, 1-sided, bracteate; corolla conically egg-shaped or cylindrical, narrowed at the mouth. Flowers reddish; corymbose in dense clusters, small and drooping, greenish- or yellowish-purple, longer than the stamens; style exserted. Fruit black, globular, sweet, and edible. Ripe in July and August.

GAYLUSSACIA RESINOSA (Huckleberry).

Geography. — G. dumosa is found common along the coast of North America, from Newfoundland to Florida. The G. frondosa is common in New England, Pennsylvania, and Kentucky, and south to Louisiana and Florida. G. resinosa is found in damp woods, from Newfoundland to Georgia.

Etymology. — The name *Gaylussacia* was given to the genus in honor of M. Gay Lussac, the eminent French chemist. *Dumosa*, the Latin for "bush," is applied to this plant to denote its character in that respect. *Frondosa* is Latin for "leafy," and was given to the species on account of the length of the leaf. The specific name *resinosa* was applied on account of the presence of resinous dots, or globules, on the leaves. *Whortleberry* is derived from the Anglo-Saxon *wyrtil*, a small shrub. *Huckleberry* is a corruption of *hurtleberry*, derived from *whortleberry*. The derivation of *tangleberry* is obscure. *Blueberry* is named from the color of the berries.

History and Use. — G. resinosa is the huckleberry of the markets. It was the favorite berry used by the natives of North America in their celebrated *attitash*, consisting of huckleberries of several sorts dried and beaten to powder. Another favorite dish, called *santaash*, consisted of the *attitash* mixed with corn meal, and was always prepared for their festivals.

OXYCOCCUS. Pers. (Cranberry.) Calyx adhering to the ovary, 4-cleft. Corolla 4-parted: segments narrow and turned back; stamens 8, convergent; anthers tubular, 2-parted, opening by oblique pores. Fruit a globular, 4-celled, many-seeded berry. Shrubs, with slender, creeping, assurgent stems and branches.

1. **O. macrocarpus**, Pers. (**Vaccinium macrocarpum**, Ait.) (Large-fruited Cranberry.) Stem from 1 to 5 feet long, prostrate, throwing up assurgent flower-

ing and fruit-bearing branches. Leaves elliptical, margins rolled over, upper side dark green, glaucous underneath; flower large; anthers more than twice the length of the filament. Fruit varying from campanulate to orbicular, dark when ripe. The cranberry sports freely as to the fruit; the leaf is very constant, except as to size.

There are three well-marked and quite constant varieties known among cultivators, as follows: Bell-shaped, or Pear-shaped; Bugle-shaped; Cherry-shaped. This last form is sometimes flattened at the poles, and is then called Cheese-shaped. Cultivators who have made careful observations think they have detected from 20 to 40 varieties. The following five forms are very constant: —

1. Fruit pear-shaped, from six tenths of an inch to an inch in length, and four to seven tenths in diameter; dark when ripe.

2. Varying in shape from fusiform to cylindrical, sometimes smaller in the middle than at the ends, and every way larger than No. 1; dark when ripe. This is what the growers call bugle-shaped.

3. Nearly globular, from a quarter to three quarters of an inch in diameter; very dark when ripe.

4. About the size and shape of No. 3, but creamy white when ripe.

5. Very much smaller than Nos. 3 and 4. Globular, three to four tenths of an inch in diameter; very prolific, and very dark when ripe.

2. **O. palustris**, Pers. (**Vaccinium oxycoccus**.) (Small Cranberry.) Differs from O. macrocarpus in bearing very much smaller leaves and fruit. The fruit also yields a sharper acid, indicated by its name, *sour-berried*. It is collected where it grows spontaneously, but is not cultivated.

OXYCOCCUS MACROCARPUS (Cranberry).

Geography. — The geographical range of the cranberry is very wide; it reaches from 38° to 60° north latitude, and covers a belt trending east and west from Siberia to the British Isles, and in North America from the Atlantic coast to the Mississippi.

Etymology. — *Oxycoccus*, the generic name, is from the Greek ὀξύς, sour, and κόκκος, a berry, — sour berry. The specific name, *macrocarpus*, is from the Greek μακρός, long, and καρπός, fruit; hence we have *long fruit*. This name was given by Aiton to distinguish it from the small-fruited species O. palustris. The genus was formerly known by the name of *vaccinium*, the meaning of which is obscure. The common name *cranberry*, which is a corruption of *crane berry*, is said to have been given because the assurgent branches bend over in a curve resembling the neck of the crane.

History. — The history of this plant furnishes little of interest. It has for some years been under cultivation, but its importance as a food-plant or an article of commerce is of recent date. It is largely cultivated in the middle Atlantic States, Massachusetts, Rhode Island, New York, New Jersey, Delaware, and Maryland. More attention has been given to it in New Jersey than elsewhere.

Cultivation. — The mode of cultivation in Ocean, Monmouth, and Burlington counties, in New Jersey, in the Beaver Dam Company's plantations, and those of the Hon. Ephraim P. Empson at Collier's Mills, which are among the most extensive, and are fair specimens of the others, is as follows: The location is selected along a small stream, whose valley is of some width and the adjacent banks of which are high enough to allow flooding by constructing a low dam. The ground is then cleared of the trees and shrubbery, roots and all. These are piled in heaps, and as soon as dry they are burned, and the ashes are spread. The whole is then covered with sand, into which the plants are set in rows, so that the following year they may be kept free from weeds with the hoe. When the plants begin to fruit they are flooded for several months, beginning in November, and in May the water is let off. In September picking commences, which is paid for by the crate, or bushel, the price varying from forty to sixty cents a crate. The owners of the bogs, as the plantations are called, erect cabins on or near the grounds to accommodate the pickers, who come from far and near; old men and women, girls and boys, flock to the cranberry harvest. The quantity picked in a day varies from one crate to five; the women and young girls are the most dexterous, and frequently earn as much as two and a half dollars a day when the fruit is abundant.

Use.—The cranberry has a sharp, acid, and astringent taste, in a raw state. Cooking destroys the astringency, but does not neutralize the acid. It is a favorite sauce with poultry and game, and is largely used for jellies and preserves. The ripe fruit can be kept for a long time in vessels of water tightly sealed. It may be kept for any length of time immersed in molasses, and kept in a uniform and cool temperature. Large quantities are shipped to France, where the berries are used in the manufacture of colors.

Marts.—The great market for cranberries is New York City. The prices have ranged during the last ten years from two to five dollars a crate or bushel.

ORDER XXXIII. **SAPOTACEÆ.**

Flowers perfect, regular, axillary; calyx 4–8-parted; corolla with united petals, hypogynous, 4–8-lobed, imbricated in the bud; stamens on the corolla, fertile ones equalling number of, and opposite to, the corolla-lobes; ovary several-celled; style cylindrical; stigma acute or capitellate; ovules solitary, in the cells. Fruit a berry, with one to many cells. Seeds with a bony testa, embryo large. Leaves alternate, entire, and coriaceous; stipules wanting or falling early. Trees or shrubs, with milky juice and stellate leaves.

No. of genera, 26; species, 325. Tropical or subtropical.

DICHOPSIS. Thu. Calyx 6-parted in two series, outer row valvate; corolla 6-lobed, usually acute; stamens 12, attached to the base of the corolla, every alternate one shorter; anthers lanceolate; ovarium villous, 6-celled; style awl-shaped. Large trees. Leaves leathery, clothed underneath with rusty-yellow, short, woolly, pubescence. Flowers axillary, stalked.

D. gutta, Bentley and T. (Gutta Percha.) Trunk 60 to 70 feet high, 2 to 3 feet in diameter; bark rough; twigs tomentose. Leaves alternate; stipules deciduous; petioles long, stout, thickened at the base; blade obovate, oblong, short, acuminate, tapering at the base, entire; margin revolute, glabrous

SAPOTACEÆ.

above, densely tomentose underneath, and leathery, with parallel veins nearly at right angles with the prominent midrib. Flowers small, on short, recurved, silky pedicels, clustered in the axils of the leaves. Calyx bell-shaped; segments 6, in 2 imbricated rows, persistent. Corolla tube scarcely longer than the calyx, with 6 segments; stamens 12, inserted on the throat of the corolla; filaments in one row, equal, slender, extending beyond the segments of the corolla; anthers ovate acute, 2-celled, opening lengthwise outwardly; ovary globose, slightly pubescent, 6-celled, with an ovule in each cell; style simple, slender, longer than the stamens; stigma terminal, blunt. Fruit one and a half inches long, ovoid, pointed, and rusty-pubescent. Seed not described.

Geography.— The tree that furnishes the gutta percha of commerce is tropical, found native in the East Indies, Ceylon, and the Malay Islands.

Etymology.— *Dichopsis* is from the Greek δίχα, diversely, and ὄψις, aspect. *Gutta*, the specific name, is from the Malay, and signifies "sticky juice;" and *percha* is Malay for "tree;" hence "sticky-juiced tree."

History.— Gutta percha was first introduced to the notice of Europeans by voyagers who had visited the coasts of Malacca and the Malay Islands. The sailors obtained it from the natives, in the form of bowls, cups, etc., as drinking-vessels, knife-handles, and other useful articles. Dr. Montgomerie carried gutta percha to England in 1843, and showed the method of forming it into domestic utensils and surgical instruments; it has now attained a point of wonderful importance in domestic economy.

Some notice of the tree was taken to Europe in 1656, nearly two hundred years prior, but the world seemed not yet ready for it.

DICHOPSIS GUTTA (Gutta Percha).

Preparation.— It was at first obtained by felling the tree, stripping off the bark, and removing the cambium layer, which is charged with sap containing the substance. Now a less wasteful method is practiced, which consists in tapping by boring into the sap-wood and inserting a tube. The sap thus caught soon coagulates, and can be kneaded into cakes for market, at which time it has the appearance of leather and the odor of cheese. Though not elastic, it is made plastic by heat, when it takes any form to suit the workman's fancy.

Use.— The timber of this magnificent tree is not valuable for building purposes, as it is soft and weak. The gum it produces, however, has become of great economic importance. It is of intermediate consistence between wood and leather, softens by heat, and is immersed in hot water for that purpose; while in a soft state it takes delicate impressions, which remain sharp when cool.

It is formed into knife-handles, whips, surgical instruments, splints, combs, soles of shoes, and covers for books; but the most important use to which it has been applied is to insulate telegraphic cables. Being impervious to water and a bad conductor of electricity, it is the best of all materials for that purpose.

NOTE. — **Achras sapota**, Wight, a tree of this order, bears a pear-like fruit, which, when a little over-ripe, is sweet and edible. Native of Panama.

Bassia latifolia, Roxb., another tree of this order, is one of the *Butter Trees*. The flowers are used for food; a wine is also made from them. By distilling, a spirit comes over, and by expression, an oil is obtained from the seeds. Native of Bengal.

The **Bassia Parkii** is the butter tree of Africa, which supplies material for candles and soap.

The **Mimusops elata**, the Cow Tree of Para, in northeastern parts of Brazil, — is another of this order, the sap of which resembles rich cream. The fruit, of about the size of an ordinary apple, is edible.

ORDER XXXIV. **EBENACEÆ**.

Flowers seldom perfect, usually diœcious, in crowded cymes or scattered along the ends of the branchlets; calyx 3–6-parted; corolla on the receptacle; petals united, urn-shaped, leathery, smooth within and pubescent outside; limb 3–6-parted, imbricated in the bud; stamens inserted on the bottom of the corolla, twice as many as the number of lobes in the corolla, occasionally equal or 4 times as many; filaments free or in pairs; ovary sessile, 3 to many-celled. Fruit a berry, globular or ovoid, succulent, few-seeded. Seeds inverted; testa membranous. Leaves alternate, leathery, entire, and without stipules. Trees and shrubs, with hard wood.

No. of genera, 6.

DIOSPYROS, L. Flowers diœcious and polygamous; calyx 4–6-lobed; corolla bell-shaped, 4–6-parted, rolled together in the bud; stamens 4–8, or numerous; filaments short. Fruit globular, an inch to an inch and a half in diameter, surrounded at the base by a fleshy, persistent calyx, 4–8-celled, with 8 to 12 seeds. Tree.

1. **D. Virginiana**, L. (Persimmon.) Trunk 30 to 50 feet in height, 10 to 18 inches in diameter, forming a symmetrical head. Leaves elliptical, bluntly acuminate, entire, dark green, paler underneath, smooth, and 3 to 5 inches long; petioles and veins somewhat hairy, glaucous underneath; calyx 4-parted; stamens 8. Flowers greenish yellow. Fruit globular, an inch to an inch and a half in diameter. Sap-wood yellowish white, light and soft; heart-wood dark, heavier, and harder.

2. **D. ebenum**, Retz. (Ebony.) Trunk 50 to 80 feet high, 2 to 3 feet in diameter, branching into a beautiful head. Leaves elliptical, pointed, and mucronate. Fruit large, 1 to 2 inches in diameter, edible. Wood black, very close-grained, sinks in water, and takes a fine polish.

There are many species of the Diospyros, all yielding a hard, dark wood. The wood of the D. ebenum, however, is the wood known as *ebony*, or *iron-wood*.

3. **D. melanoxylon**, Roxb., produces the black ebony found native in southern Asia. Wood very hard, heavy, sinks in water, and takes a fine polish.

Geography. — The Diospyros Virginiana is a native of North America, throughout the eastern part of the north temperate zone up to 40° of north latitude. D. ebenum is a native of Ceylon, and is not indigenous outside of the tropics; it is found, however, at an elevation of 5,000 feet, which suggests that it would grow in the edge of the temperate zones, in regions of no frost. The D. ebenum is also found native in Mauritius.

Etymology. — *Diospyros* is derived from the Greek words Διός, Jove, and πυρός, fruit, — the fruit of Jove, or the fruit of the gods, or heavenly fruit. *Ebenum*, from the Hebrew *eben*, a stone, is in allusion to its hardness. *Virginiana* is derived from *Virginia*, the country where it was first found by botanists. *Melanoxylon* is derived from the Greek μέλας, black, and ξύλον, wood, hence black wood. *Ebony* (the common name of D. ebenum) is a corruption of the scientific name *Ebenum*. *Persimmon* is the name given to this species by the American Indians; its meaning is unknown.

DIOSPYROS VIRGINIANA (Persimmon).

Preparation. — The best ebony is produced by the D. ebenum, and is taken from Ceylon to England in logs from 18 to 24 inches in diameter, and 12 feet long; large quantities of excellent ebony are also sent from Mauritius (D. reticulata) in poles or logs from 10 to 20 feet long and 12 to 15 inches in diameter. It is seasoned under water. When felled it is immersed in water, where it remains for six to eighteen months; when removed from the water the ends are hooped with iron rings, to prevent checking and splitting. An ebony is also obtained in the Island of St. Helena from the Dombeya melanoxylon, of the order Byttneriaceæ.

Use. — Ebony is highly prized by turners as a material for their purpose; largely used for wind instruments, as flutes, fifes, piccolos, etc.; it is used for veneering, inlaying, and for piano keys, nuts for violins; also the finger boards and tail pieces, with the screws for tuning, are made of ebony.

ORDER XXXV. **OLEACEÆ.**

Flowers perfect, occasionally diœcious and without petals, in a raceme or trichotomous panicle; calyx, with sepals united, 4-lobed or toothed, sometimes wanting; corolla hypogynous; petals 4, united at the base in pairs, or all united; stamens 2, inserted on the corolla, and alternating with its lobes; ovary free, 2-celled. Fruit a drupe, a 2-celled berry or a samara. Leaves opposite, petioled, simple, or odd-pinnate. Trees and shrubs.

No. of genera, 18; species, 280.

OLEA, Tourn. (Olive.) Calyx short, 4-toothed, persistent; corolla tube short; limb 4-parted, and spreading; stamens 2, inserted in the base of the tube, longer than the corolla tube; ovary with 4 suspended ovules, 2 or 3 of which prove abortive. Fruit a fleshy drupe, and oily. An evergreen shrub or tree, with opposite leaves.

O. Europæa, L. (Olive.) Trunk 20 to 30 feet high, much-branched and spreading, forming a symmetrical head; branches angular. Leaves lanceolate, entire, deep green above, light hoary beneath, and evergreen Flowers axillary, in short, compact racemes, small and white Fruit greenish, or whitish-violet, sometimes nearly black, size of a pigeon's egg, oval, sometimes globular or obovate, and very abundant.

There are several species, but the O. Europæa is the one usually cultivated. Of this species there are five important varieties: —

Var. **longifolia**. Leaves linear, lanceolate, flat and silky beneath.
Var. **latifolia**. Leaves broad, oblong, flat and hoary beneath.
Var. **ferruginea**. Leaves narrow, acute at each end, rusty beneath.
Var. **obliqua**. Leaves oblong, bent obliquely, pale beneath.
Var. **buxifolia** Leaves oblong-ovate, and branches divaricate.
Var. **sylvestris**, found outside of cultivation, is characterized by smaller fruit. It is common in the Mediterranean countries, the Canary and Madeira Isles. There are other varieties, but those mentioned are the most approved.

Geography. — The geographical range of the olive-tree is not very broad; it grows best just on the edge of the region of no frost, and on the seacoast. Its area of growth is especially the countries around the Mediterranean sea. Asia Minor, Greece, Syria, Spain, Italy, northern Africa, and the islands of the Mediterranean are the great olive-growing regions. The ports of export are Trani, Barletta, Bari, Mola di Bari, Molfetta, Otranto, Taranto, and especially Gallipoli.

Etymology. — *Olea*, the generic name, is from the Greek ἐλαία, through the Celtic or Gothic word *olew*, oil, due to the oil-sacs deposited near and just under the skin of the fruit. *Europæa*, the specific name, is due to the circumstance that this species is the one under cultivation throughout Europe. *Olive*, the common name, is plainly a corruption of the generic name, or a contraction of the French *Olivier*.

History. — The home of the olive has been traced to Syria, whence it easily spread through the countries of the Levant and southern Europe. Though it has no Sanscrit name, it is referred to as one of the plants upon Mount Ararat at the time of the Deluge. The wood is found in the stone coffins of Egyp tians. But we have no means of determining the exact date of its introduction into Europe. It propagates itself freely, and is growing without cultivation in all the countries around the Mediterranean. The olive tree is of slow growth, but where its natural development is allowed for ages, the trunk often attains a considerable diameter. De Candolle records one 23 feet in girth whose age was supposed to be seven centuries. Some Italian olives are credited with an antiquity reaching back to the days of republican Rome; but the age of such ancient trees is always doubtful during growth, and their identity with old descriptions is still more difficult to establish.

Cultivation. — Its mode of culture, or rather the method of making a plantation, is singular and interesting. The province of Susa, in Morocco, produces great abundance of olive oil, which has the reputation of being of such excellent quality as to rival the celebrated Florence oils. In Jackson's account of the empire of Morocco there is a description of an extensive olive plantation. The order and arrangement of the trees struck him as being very curious, and on inquiring the cause of the arrangement he was told by an official high in authority that during the Saddia dynasty, a king, on the march with his army to the Soudan, encamped on the spot, and that the pegs to which his horses were picketed were cut from an adjacent olive grove, and each one became a tree. This explanation he regarded as fabulous, but goes on to relate that he had occasion to plant some fruit-trees in a garden. The person employed to make the plantation procured some olive-branches, cut them up into pieces a foot long, sharpened one end with a knife, and proceeded to drive them into the ground with a stone. Supposing the fellow was imposing upon him, he ordered him away; but on being assured that it was the usual method, he allowed him to proceed, and each peg grew into a thrifty olive-tree.

OLEA EUROPÆA (Olive).

Engrafting the better varieties upon wild stocks greatly increases the production of fruit.

Preparation of olive oil. — The fruit is crushed in a mill, the pulp then placed in woollen bags and subjected to pressure and the application of hot water. The oil is skimmed off the water and placed in tubs, barrels, bottles, crocks (a sort of earthen jar or pot), and other vessels. In the remote districts, where it is made in small quantities, it is taken to market in bottles made of goatskins. On that part of the Italian peninsula skirting the mouth of the Adriatic, the entire country is an olive orchard. In the oil season, hundreds of mules and donkeys crowd the highways going into Gallipoli, the seaport, laden with oil, where may be seen at the same time fifteen to twenty ships taking in their cargoes of oil and olives. When the oil is first brought in from the mills it is emptied into a large vat, or cistern, at which time it is dark and turbid. After remaining for some months in this mass, it settles, becomes clarified, and takes on a beautiful amber color, when it is drawn off into barrels and other vessels for shipping to the United States, England, France, Genoa, and elsewhere. The best oil comes from Florence, and was formerly shipped exclusively in glass flasks, called Florentine flasks.

Use.—The olive is eaten when in an under-ripe state. The taste is bitter and slightly astringent, hence a taste for it must be acquired. The fruit is put up in either glass or wood, and preserved in salt. To prepare it for the table it is soaked in water, and afterwards placed in vinegar or oil. The principal use of the olive is for the manufacture of oil. The oil obtained from the olive is esteemed for its soft, delicate flavor, and is largely used for culinary purposes and for table use, for salad dressings, etc. It is also an excellent lubricator, and valuable for illuminating purposes. It solidifies at 10° to 15° Fahrenheit. Its specific gravity is .9176; it is frequently adulterated with cotton-seed oil, the specific gravity of which is .9300.

The wood of the olive tree is hard, of a fine, close grain, and takes an excellent polish; it is used largely by turners for the manufacture of small articles. The root is esteemed on account of its gnarls and curls.

Order XXXVI. **LOGANIACEÆ**.

Flowers perfect, regular, in axillary or terminal cymes; calyx, with united sepals, valvate in the bud, or 4 to 5 free imbricated sepals; corolla hypogynous, with sepals united, wheel-shaped, or bell-shaped; limb 5-10-cleft, sometimes 4-lobed; stamens on the corolla equal to and alternate with its lobes; filaments thread-like or awl-shaped; ovary superior, 2-4-celled; style thread-like; stigma shield-shaped; ovules numerous. Fruit a capsule or berry. Trees and shrubs. Leaves simple, opposite or whorled.

No. of genera, 30; species, 350; mostly tropical.

STRYCHNOS. L. Calyx somewhat bell-shaped, or wheel-shaped, with 4 lobes, whose edges just meet; stamens 4 or 5, on the corolla; filaments short, attached to the backs of the short anthers; ovary 2-celled; style thread-like. Fruit a berry, globular, covering hard, and without valves. Leaves opposite. Flowers in cymes, axillary or terminal, small and white.

1. **S. nux vomica**, L. (Nux Vomica.) Stem from 20 to 35 feet high; bark smooth, gray; much-branched, the branchlets swollen or knotted at the nodes. Leaves 5-nerved, with 2 ribs each side of the midrib, reaching from the base to the apex, ovate, pointed; calyx tubular, 5-toothed; corolla tubular, greenish-white, lengthened; limb 5-parted, parts lanceolate; stamens 5; anthers erect; pistil longer than the stamens; stigma globose. Fruit as large as a middling-sized orange, with a hard, bitter, smooth, yellow peel, inclosing fleshy pulp, in which are embedded a number of flat, circular seeds, concave on one side and convex on the other, an inch in diameter, and a quarter of an inch thick, covered with a gray, velvety down, hard and horny, containing a gummy, resinous matter, soluble in alcohol. Wood hard, bitter, and very durable.

Chemistry.—The seeds yield to the chemist two substances,—strychnine and brucine.

Strychnine, $C_{21} H_{22} N_2 O_2$.
Brucine, $C_{22} H_{21} N_2 O_4$.

LOGANIACEÆ.

Brucine differs little in composition from strychnine, but is not so active in its poisonous qualities. These two substances in combination form one of the most active poisons known. Strychnine is intensely bitter; one grain gives to 110 gallons of water a perceptible bitter taste. It requires 2,000 parts of boiling water to dissolve it; but alcohol suspends it more readily, and it is very soluble in chloroform.

2. **S. Ignatii**, Berg. The seeds of this species are said to be far richer in the yield of strychnine than the nux vomica. The tree is native in the Philippine Islands. The seeds are known in commerce as the beans of St. Ignatius. They were brought to the notice of Ray, the English botanist, in 1699, by a Jesuit, who obtained them through missionaries.

3. **S. colubrina** (Snakewood), L., yields strychnine from the wood of the root.

4. **S. tieuté**, Lech, yields it from the bark of the root.

5. **S. potatorum**, L. (Clearing Nut.) Like the nux vomica, but a larger tree; fruit similar in form, but does not possess the same poisonous qualities. Found in the mountains of East Indies. The pulp is eaten by the natives, and the seeds are used to purify water. One of the seeds is rubbed smartly upon the sides of a water vessel, which is then filled with water, that in a very short time becomes clear and pure; the effect of the seed is to precipitate not only suspended vegetable matter, but impurities of every sort. This effect is attributed to some albuminoid property of the seed.

When the fruit is ripe it is attacked by birds and climbing animals: they eat the rind, and throw down the pulp and seeds. The seeds are collected, washed, and sold to country merchants for a quarter of a cent a pound.

6. **S. toxifera**, Bth., found in the silvas of the Amazon and Oronoco, furnishes the celebrated poison, *curari*, with which the natives prepare their arrows for battle. The poison resides in a resin found in the bark of the tree, and is separated by maceration in water. The substance is harmless when taken into the stomach, but fatal when introduced into the blood. The preparation for the poisoned arrows is a mixture of the product of several different species, but the mode of preparation is a secret.

Strychnos nux vomica (Nux Vomica).

Geography. — These trees are tropical and subtropical, natives of India and the islands south of Asia.

Etymology. — The name *strychnos* is from the Greek word, στρύχνος, which signifies to "strew" or "throw down," in fancied allusion to the stupefying effects it produces upon the animal system. *Nux vomica* signifies *foul nut*, — from the Latin *nux*, a nut, and *vomica*, a plague or loathsome disease, alluding to its poisonous qualities. *Ignatii* is named after *St. Ignatius*. *Colubrina*, Latin, is a general name for innocuous serpents, hence also *snake wood*. Chettik is the Javanese name for the S. tieuté, which is a native of Java, and tieuté is probably derived from chettik. *Potatorum*, Latin, *drinking*, and *clearing nut*, are names due to the fact that it is used in the East Indies for clearing muddy water. *Toxifera*, from Greek τοξικόν, poison, and φέρειν, bear, derives its name from the fact that the Indians use it to poison their arrows.

History. — This poison was discovered by the chemists Pelletier and Caventon in the seeds of Strychnos Ignatii and S. nux vomica.

Use. — The medical properties are stimulant, tonic, and narcotic, — and it is used as a remedy in rheumatic paralysis and lead poisoning. In large doses it attacks the brain and spinal cord, producing dizziness, contraction of the heart, and muscular spasms. Thirty grains of the powdered nut have proved fatal, and three grains of the extract. Half a grain taken by mistake caused the death of Dr. Warner. It is said that swine and goats are not injured by it. It enters into the medical preparations of homœopathic practitioners for stomach disorders, and is by them regarded as a specific in dyspepsia. It is largely used in the United States to destroy vermin, and especially animals and birds injurious to agriculture.

It has been charged that large quantities of strychnine are used in the preparation of whiskey; this is a mistake, as its intensely bitter properties would render the liquor unpalatable and unsalable. The wood is hard, durable, and takes a good polish; some of the species yield a snake-wood.

The spinal cord is the seat of strychnine poisoning, and the effects are intermittent tetanic convulsions. In some cases the respiratory muscles become rigid, and death ensues from suffocation. Large doses of opium are said to neutralize the effects of strychnine.

Order XXXVII. BORRAGINACEÆ.

Flowers perfect, usually regular, axillary or terminal, solitary, or mostly in 1-sided scorpioid cymes; calyx persistent; sepals united, 4-5-parted; corolla regular, with scales under the middle of lobes, hypogynous, with united petals, deciduous, bell-shaped or wheel-shaped; throat naked, or clothed with hairs or scales; limb 5-lobed, imbricate in the bud; stamens 5, on the throat of corolla, alternate with its divisions. Fruit, 4 distinct, nut-like akenes, sometimes united in pairs. Mostly rough, hairy. Herbs.

No. of genera, 68; species, 1,200; cosmopolitan.

SYMPHYTUM, Tourn. Calyx 5-parted; corolla tubular, bell-shaped; mouth closed by 5 awl-shaped scales, forming a cone. Fruit smooth and ovoid. A perennial herb.

S. officinale, L. (Comfrey.) Stem stout, winged, 4 feet high, branching towards the top, hairy. Leaves large, coarse, petioled, lower ones broad, lanceolate, upper ones lanceolate. Flowers in racemes, and terminal; sepals lanceolate; border of corolla divided into 5 recurved teeth; yellow, white, pink, or red.

Var. **Bohemicum**, Sch. has bright red flowers.

Geography. — The geographical range of this plant is not great, but it grows well about the middle of the temperate zone, and is found throughout middle Europe and the older parts of the United States of America.

Etymology. — *Symphytum* is from the Greek σύν, together, and φυτόν, a plant, in allusion to the gummy character of the mucilage contained in the root of this plant. *Comfrey*, from the Latin *confirmare, strengthen,* owes its name to its healing properties.

History. — This plant is indigenous to the Peloponnesus and Greek islands, whence it has worked its way westward to the British Isles; it was introduced by European colonists into northeast America, where it has become naturalized about old dwellings and around ruins, having escaped from gardens. It loves damp, rich soil.

Use. — The root abounds in a gummy, glue-like mucilage; a decoction of it is used to bind up wounds. It is also used for throat and lung troubles, on account of the soothing properties of its mucilage. It is grateful in irritable stomach complaints. It likewise serves as a remedy for bleeding at the lungs; and the bruised heated root is sometimes applied to wounds, in the form of a poultice.

SYMPHYTUM OFFICINALE (Comfrey).

Order XXXVIII. CONVOLVULACEÆ.

Flowers perfect, regular; peduncles axillary or terminal, simple or dichotomous, usually bibracteate; calyx 5-sepaled, usually free and persistent; corolla hypogynous; petals united and funnel-shaped, twisted in the bud; stamens 5, inserted at the bottom of the corolla-tube, alternating with its lobes; filaments swollen below, thread-like above; style simple, or nearly so; seeds few, 2 in each of the 2-3 cells of the ovary. Fruit capsular; carpels connate. Herbaceous, woody, or sub-woody plants, climbing or trailing.

No. of genera, 32; species, 800; cosmopolitan; mostly in warm sands.

IPOMŒA, L. Calyx 5-parted; sepals green; corolla salver or funnel-shaped, spreading; number of stamens 5 in the throat; style simple, terminated by a head-shaped stigma, which is sometimes 2-lobed; seed-vessel 2- or 3-valved, 2- or spuriously 4- or 3-celled; seeds 4-6.

1. **I. batatas,** Lam. (Sweet Potato.) Ovary spuriously 4-celled; stem from 2 to 10 feet long, creeping and rooting at every node, from an eighth to a quarter of an inch in diameter. Leaves very variable, usually triangular or 3-lobed; general outline heart-shaped, the sinus at the base broad, 5-veined, smooth; blade 2 to 5 inches long, on long petioles. Flowers on long peduncles, 2 to 5 in a cluster, purple; root gives rise to long, spindle-shaped tubers. An herbaceous perennial.

Geography. — The sweet potato is largely cultivated in southern United States, and comes to perfection as far north as the Carolinas. North of North Carolina it was not formerly supposed to be perfect; but for the last quarter of a century it has been successfully cultivated as a market-crop in eastern Virginia, Maryland, Delaware, and southern New Jersey; in fact the sweets, as they are called in New York market, from south Jersey are as popular as the Carolinas. It is grown in southern Spain and Italy. The British Isles are too damp for it.

Etymology. — *Ipomœa*, the generic name, is derived by London from the Greek Ἰψ, a worm, and ὅμοιος, like, — like a worm. *Batatas* comes from the Spanish *batata*, the native name of the sweet potato. *Potato* is a corruption of *batatas*, and *sweet* refers to the taste of this species.

History. — Its home is held by some authorities to be America; by others, Asia; and there seem to be good reasons for believing it to be indigenous both to Asia and America. It was introduced into southern Europe by the Spaniards soon after the discovery of America. It now forms an important article of food throughout tropical and subtropical countries, and needs a high temperature to develop the peculiar delicate sweet taste.

It is not known when this tuber was introduced into the kitchen.

Cultivation. — The mode of cultivating the sweet potato is to place the tubers in a hot-bed, where they sprout. The sprouts, when six to ten inches long, are taken off and transplanted, in the same manner as cabbage-plants are treated. They grow to greatest perfection as to quality in loose sand. A shovelful of well-rotted barnyard manure is dropped, and over it with a hoe is formed a conical hill, in the top of which the plant is transplanted; in about two months the tubers begin to form about the base of the plant, which by that time has become a prostrate vine, six to ten feet in length, rooting at every node. A part of the labor of cultivating is the destruction of these rootlets, by frequently lifting the vine from the ground, which violence breaks them.

IPOMŒA BATATAS (Sweet Potato).

Use. — The sweet potato in tropical and subtropical countries is an article of food of vast importance. The rudest modes of cooking are roasting and boiling, but it is also largely used for pies, custards, and other delicacies; it is also minced while raw, roasted with Maracaibo coffee, then ground and sold for coffee. It is in common use in the southern and middle States as a vegetable at breakfast and dinner.

2. **I. purga**, Hayne. (Bind weed Jalap.) Stems twining, 12 feet long, many from the same globular, tuberous, fleshy root; branched. Leaves alternate, on stout foot-stalks, which are 4 to 5 inches long; base cordate, lobes pointed, narrowed at the apex, entire, smooth both sides, paler beneath, with conspicuous veins. Flowers in cymes, axillary, few-flowered; peduncles long and twisted; pedicels bracted; calyx, short, smooth, 5-parted; corolla large, tubular, with flattened, spreading limb, contracted just where the limb begins to flatten, dull pink; stamens inserted in the tube near the base; filaments flattened, three longer than the other two, all extending beyond the mouth of

SOLANACEÆ. 185

corolla-tube, **anthers** small; **ovary tapering** into the slender style, which **is a** little longer than the **stamens, 2-celled**, 2 ovules in each cell. The **root** is somewhat the size and **shape of a medium-sized** Swedish turnip, tuberous in character, giving off **stems from** all points near the crown; fleshy and soft when growing, very hard when dry. Perennial.

Geography. — Its home is in the high lands **of Mexico**, near the 20th parallel, a rainy district; it grows at an elevation **of four** to six thousand feet above **the sea level. It** is now cultivated in British India, in corresponding latitudes and altitudes to its American home. The species and its locality were in doubt till 1829, when Dr. Coxe of Philadelphia obtained **living** plants from Mexico, and settled the question.

Etymology. — *Purga*, the **specific name, is** the Spanish name **for the plant,** and indicates **its medicinal property.** *Jalap*, the popular name, is derived **from** the name **of the city Xalapa in Mexico, near** which **the plant was first found.**

History. — It was carried to Europe by the Spaniards for its medicinal properties early in the seventeenth century.

Preparation. — The medicinal properties reside in a resin found in the root. The root when in perfect condition, yields about 20 per cent of the resin.

The roots are washed, and the larger ones cut into slices and suspended in nets over fires till dry, when they are very hard; they are then ready for the market.

Use. — The medicinal properties of jalap are especially cathartic; when administered in small doses it is alterative, and, in still smaller doses, tonic. In ordinary doses it is a safe but violent cathartic; ginger mixed with it modifies its activity. It was formerly administered with calomel. Its tincture constitutes a part of the black draught, and it is regarded as a very valuable cathartic **in brain troubles.** Though violent, it does not irritate and inflame **the** intestinal canal, and is hence a safe **medicine.**

Order XXXIX. SOLANACEÆ.

Flowers perfect, generally regular, axillary or terminal, solitary, fascicled, or subcorymbose; calyx, with sepals united, usually 5-lobed or toothed, occasionally 4 to 6, persistent; corolla hypogynous; petals **united, rotate or campanulate;** segments 5, rarely 4 to 6, folded or **twisted in the bud; stamens on** the corolla-tube equal and alternate **with its segments, sometimes united at** top; ovary 2-5-celled; **ovules many. Fruit varied in form, frequently a** many-seeded, pulpy berry, **sometimes a dry capsule. Herbaceous or** woody plants, with watery **juice.**

No. of genera, 66; species, 1,200; found in warmer parts of the old **world, and in temperate parts of America.**

LYCOPERSICUM. Tourn. Calyx 5- or 10-**parted,** persistent; corolla **wheel-shaped, tube short;** limb plicated, **with** 5 to 10 lobes; stamens, **5 or 6 in the throat of** the corolla, protruding; filaments short; anthers **oblong-conical, cohering** by an elongated membrane at top, opening **lengthwise on the inner** side. Fruit a berry, varying from a **prolate spheroid to subglobular,** usually an **oblate** spheroid or flattened globe, **pulpy, and many-seeded.** Annual.

L. esculentum, Mill. (Tomato. Wolf Peach.) Stem 2 to 5 feet high, branching, or straggling over shrubbery, prostrate when not supported. Leaves pinnately divided, alternate pairs of leaflets very small. Flowers in raceme-like clusters, common flower-stalk about 2 inches long, forked; sepals 5 to 10, linear-lanceolate, spreading; anthers united, pointed, with the points turned back. Fruit 1 to 5 inches in diameter, usually red, sometimes yellowish. Flowers June to August. Fruit ripens August to September.

The bruised leaves of the tomato emit a peculiar, disagreeable odor. Like other plants that grow from seed, it sports freely, producing varieties that differ mostly in size, shape, and quality of the fruit. Some of the larger fruit is very irregular, with deep grooves and alternate ridges; the favorites with market-gardeners and amateurs are the medium-sized, smooth-fruited varieties. Some are very small, from an inch to an inch and a half in diameter, and globular, called plum and cherry tomatoes; others small, pear-shaped. The last two varieties are esteemed for preserves and pickles.

LYCOPERSICUM ESCULENTUM (Tomato).

Geography. — It grows and fruits well in all southern Europe, especially in Italy and southern France and Spain, in Greece and northern Africa. It has been taken by Europeans to the islands of southern Asia and contiguous parts of the continent; is fruited in England, but under glass; and is an important crop in eastern United States, especially in the southeastern part of Virginia and the Carolinas, and is sent north as an early vegetable.

Etymology. — *Lycopersicum* is derived from the Greek words λύκος, a wolf, and περσικόν, a peach. The application is not apparent. *Esculentum*, the specific name, is of Latin origin, signifying "eatable." *Tomato*, the common name, is the native South or Central American name, carried to Europe by the Spaniards; its meaning is unknown.

History. — It is native to South and Central America, and is supposed to have been cultivated in Mexico at an ancient date. DeCandolle thinks it was first found by Europeans under cultivation in Peru. It was taken to Europe by the Spaniards, and introduced into the United States by Europeans; it came suddenly into pretty general use in the eastern United States after 1840, and is now one of the most popular and important table vegetables of the country. It is in very general use also in southern Europe.

Use. — The tomato is used raw as a salad, cooked as a sauce with meats, used to flavor gravies and soups, and is found very generally on the tables of both rich and poor throughout the United States and Europe.

SOLANACEÆ.

NICOTIANA, Tourn. (Tobacco.) Calyx tubular, hairy, divided into 5 parts, divisions narrow and pointed, half as long as the corolla-tube; corolla funnel-shaped, tube hairy; limb divided into 5 sharp segments, turned back, reddish; filaments 5, curved inwards, terminating in a lengthened, slender style, crowned with a round cleft stigma. Seed-vessel divided into 2 cells; seeds small, round, and numerous. Flowers in July and August in America.

1. **N. tabacum,** L. Stem round, terete, hairy, 4 to 6 feet high, stout and leafy (in cultivation). Root large and fibrous; bracts linear and pointed. Leaves 10 to 25 inches long, and 8 to 10 inches broad, entire, oblong, strongly veined, pointed and sessile. Flowers terminal, in loose panicles, rose-color. July.

2. **N. rustica,** L. Stem 3 to 5 feet high, round, terete, and stout. Leaves petioled, ovate, and shorter than in No. 1; segments of the corolla rounded. The whole plant is smaller and more hardy than No. 1, and is cultivated further north.

There are several other species, among which are:—

3. **N. macrophylla,** Lehm. (Great-leaved.) Leaves very large, clasping, ovate, and eared at the base; corolla inflated at the mouth.

4. **N. Persica,** Lindl. Lower leaves oblong, upper ones lanceolate and sessile; corolla salver-shaped; tubule elongated. Cultivated in Persia, and furnishes the celebrated Shiraz tobacco.

5. **N. repanda,** W. (Wavy Tobacco.) Leaves clasping, cordate, spatulate, repand; tube of corolla long and slender. Cultivated in Cuba.

NICOTIANA TABACUM (Tobacco).

6. **N. quadrivalvis,** Pursh. (Four-valved Tobacco.) Stem branched. Leaves petioled, oblong; corolla-tube twice as long as the calyx, segments obtuse; capsule 4-valved, round. Found under cultivation among the Indians in the Missouri valley.

7. **N. nana,** Lindl. (Dwarf Tobacco.) Leaves lanceolate, whole plant smaller than any other species. Found under cultivation among the savages of the valley of the Columbia river.

Geography —Tobacco arrives at the greatest perfection on virgin soil, or what the agriculturist calls "new land." It grows well in all parts south of the middle of the temperate zones and in the tropics. The best qualities are raised in Virginia, Kentucky, and the Carolinas, Venezuela, Cuba, and Brazil. It has been of late years an important crop in some parts of the Middle Atlantic States, especially Connecticut and Pennsylvania.

In Great Britain the growing of tobacco is prohibited by law, as a large portion of the national revenue consists of duty on its importation. It is an important crop in Holland, Flanders, France, Alsace, Hungary, and European Turkey.

Tobacco of a good marketable quality is raised in the Levant. Large quantities of an excellent quality are produced in the Indian Archipelago, in China, and Japan. The Dutch introduced its cultivation into south Africa, and the English have recently commenced its culture in Australia. The quality of that raised north of the middle of the temperate zones, as in Europe, is not so good.

Etymology. — *Nicotiana*, the generic name, was given to this plant in honor of John Nicot, a French statesman, who was instrumental in bringing it under cultivation in France. *Tabacum* is derived by some from *tabaco*, the name used by the American aborigines to indicate the instrument or pipe they used to smoke the dried leaves of the plant. Others derive it from Tobago, in the West Indies, others from Tabasco, in Mexico. The common name *tobacco* is derived from the same source. The other specific names explain themselves; as, *rustica*, of the field or the country; *macrophylla*, large-leaved, or long-leaved; *Persica*, grown in Persia; *repanda*, wavy, or sinuate-leaved; *quadrivalcis*, seed-vessel with four valves; *nana*, small, or dwarf.

History. — Soon after the permanent settlement of North America, learned societies and some of the sovereigns of Europe became interested in the natural history of the New World. They sent over men devoted to the study of nature to collect specimens of the animals and other objects of interest to be found in Virginia, the name applied at that time to the large tract of land claimed by the English. Among the naturalists sent out were enthusiastic botanists, who made large collections of plants and seeds, and conveyed them to the Old World.

In their explorations they found a plant, the dried leaves of which the aborigines smoked in an instrument called by them tabaco. The imperfect knowledge of the dialect of the savages possessed by the Europeans at that time led to the error that the substance they smoked was called tabaco, instead of the pipe through which they smoked it. Another history of the origin of the name is that a Spanish monk found the plant growing in Tobago, a province of St. Domingo.

It was introduced into Portugal in 1558, by Dr Fernandes, and thence into Spain in 1559, where it was grown as a medicinal plant. John Nicot, a French statesman, who was at that time minister to the court of Portugal, sent seed to Queen Catherine de Médicis, who caused it to be cultivated in France; and on account of the interest she took in its culture, it received the name of Queen's Herb. On account of the instrumentality of John Nicot in its introduction into France, Tournefort, a French botanist, named it Nicotiana.

Ralph Lane, the first governor of Virginia, and Sir Francis Drake brought to England in 1586 the implements and material for tobacco smoking, which they handed over to Sir Walter Raleigh. Lane is credited with having been the first English smoker, and through the influence and example of the illustrious Raleigh, the habit of smoking soon became rooted among the English. The custom was carried into Holland by young Englishmen who went there to prosecute their studies.

In less than fifty years after the tobacco-plant was first cultivated in Portugal the custom of smoking it spread over Turkey, Persia, India, Java, China, and Japan. This rapid spread is no doubt due in part to the ease with which the plant is cultivated throughout the temperate zones where rich soil is found.

It is claimed by the Chinese that they cultivated and used tobacco before the discovery of America; yet recent investigations have developed the fact that the plant cultivated there is a variety of the species most commonly cultivated in America, and has no characteristic difference, except such as different soils and modes of cultivation would induce. But the strongest argument is that the eastern languages have no name for it, but throughout the countries of Asia it is known by its American name, tobacco, which was no doubt obtained from the Portuguese who introduced it into China and Japan.

After careful examination we are forced to the conclusion that tobacco was brought to the notice of civilized man in the West Indies soon after their discovery, and that the practice of smoking it originated among the savages of the New World.

In the countries where the use of tobacco prevails, 27 of every 40 adult males are wedded to the custom. An account of the efforts to suppress the use of tobacco that have been made by medical men and by the highest authority constitutes an important chapter in its history.

The pope at one time lent his aid to its suppression. James I. of England, and the authorities of Russia and of Turkey, passed stringent laws forbidding its use, and executed them with savage barbarity. For using tobacco, men were whipped, their noses were slit, and sometimes cut off. In Turkey they were bastinadoed and beheaded. The pope thundered his bulls of excommunication at them.

James I. of England, in the beginning of the seventeenth century, published his counterblast against tobacco, in which he undertook to show how unworthy it is for a civilized nation to adopt customs from such barbarians as the American savages.

Notwithstanding this royal diatribe and many others that have followed all along down through the history of tobacco, its use has extended to every nook and corner of the world where civilized man has erected his habitation. Good men have written against it on the ground of its immoral effects. Political economists have attacked it on account of its entire uselessness. Physicians have fought it because of its mischievous effects on the health of the body; yet it has crept on and is still advancing.

Cultivation. — The plant is propagated from seed, and sports, forming varieties, many of which are under cultivation; but less attention has been paid to obtaining new varieties than the importance of the plant warrants.

Chemistry. — Nicotiana yields to the chemist eighteen different substances, the most important and characteristic of which are nicotine and nicotianin. Nicotine is composed of $C_{20} H_{14} N_2$. It is an oily liquid without color, and one of the most active poisons known; a single drop placed upon the tongue of a serpent causes death as instantaneously as an electric shock. It is an alkali which has the most intense affinity for acids; it is soluble in alcohol, and will mix with water. The disagreeable, sickening effect produced by the fumes of tobacco burned in an old pipe is due to the presence of nicotine.

Nicotianin contains the same substances, but in different proportions, and furnishes the odor of tobacco. The nicotic property appears in the leaf after fermentation.

These two substances are accompanied by sixteen others, the principal of which are *resin, potassa, chlorine, lime, silica, lignin,* and *glutin ;* and the whole plant is richer in nitrogen than any other vegetable substance that has been subjected to chemical tests. No one of the substances detected in tobacco possesses any nutritive qualities, and no animal is known to eat it. Man alone has found a use for it.

Use. — The dried leaves are used to chew or to smoke. They undergo various processes as preparation for chewing and smoking. The leaves are stripped from the stems, dried, and then rolled into cigars, or cut by machinery into shreds and put in the pipe; in that state tobacco is also chewed. It is also soaked in a liquor made of molasses and liquorice-water, after which it is pressed, and is then called " plug-tobacco."

The stems and other refuse parts are ground into snuff, flavored with substances to suit the fancy of buyers. The custom of snuffing is less common than formerly.

Effects upon the system. — When administered as a medicine it causes a sensation of heat in the throat, and a feeling of warmth in the stomach. The effect upon the system generally is to increase the activity and capacity of the absorbents, and dropsical affections sometimes disappear under its treatment. It produces nausea, causing vomiting and purging when administered in large doses, and gives rise to an indescribable sinking sensation at the pit of the stomach peculiar to itself. It sometimes acts as an anodyne, and occasionally promotes sleep.

Large doses bring on trembling of the limbs, faintness, dimness of sight, and cold sweats, succeeded by convulsions, paralysis, and torpor, terminating in death. The use of tobacco as a luxury is followed by effects which are very startling. It has been demonstrated that it has an injurious effect upon the mind, tending to enfeeble it and rob it of moral energy, impairing the memory, and producing a low grade of intellect.

Tobacco-oil is one of the most deadly poisons. It acts on the brain and nerve cord, first exciting, then lessening their activity; it induces convulsions and paralysis, and at length arrests respiration; it also injuriously affects the heart and contracts the intestines. Tobacco-poisoning in some degree is very common, and is probably experienced by every smoker when first acquiring the habit. In larger doses nicotine will kill like prussic acid; in small doses it causes tetanus, — one to two drops being enough to kill a dog or a rabbit in five minutes. Horses have been known to die from eating oats that had been kept in a granary along with tobacco. Tobacco-juice is distilled in smoking, and is very poisonous; some that was put in a student's beer caused his death; even contact of tobacco with the skin may poison the system. Thus a squadron of hussars who smuggled tobacco-leaf next the skin, were all made sick.

Statistics. — On the islands of Great Britain and Ireland there are about 300,000 shops in which tobacco is exposed for sale; the amount of sales exceeds 250,000,000 dollars per annum. In France a larger amount is sold in proportion to the inhabitants.

In Denmark the annual consumption is equal to 7 pounds to the person, including women and children. In the United States the proportion exceeds that of any other country of which we have statistics.

In New York City, Brooklyn, and Jersey City, it is estimated that about 800,000 people smoke; which is not far from one in every three, counting men, women, and children.

The entire consumption of the world is 5,580,000,000 pounds; and it is also estimated that about 300,000,000 of the whole human family smoke, and that the money spent for tobacco in the world would buy bread for the entire population of the United States.

SOLANACEÆ. 191

Most of the governments of Europe have either monopolized the sale or imposed a high excise or import duty upon it. In fact the influence it exerts upon the revenue of nations forms an important feature in its history. It is a special object of legislation in several governments of Europe; in some it is a royal monopoly, in others it is admitted from abroad under high import duties.

When we consider that this plant produces no nutritive substance, has no beauty of flowers nor delicacy of foliage, has no pleasant odors, has a disagreeable taste when a very little is eaten, producing vomiting and giddiness, and when taken in larger quantities causing death, we are surprised that it should play the part it does in the world, — that it should have become an important agricultural product; that it should be produced in such quantities as to require a large fraction of the tonnage of the world to freight it; that its commercial importance should compare favorably with any other single article; that it should be made the pet of empires and the object of special legislation; that thousands should be busied in its production and manufacture; that vast amounts of capital should be employed in its purchase and sale; and finally, that about one half of the male population of the globe should be addicted to its use, from which no good results can possibly flow.

ATROPA, L. (Deadly Nightshade.) Calyx leafy, 5-parted; corolla bell-shaped, limb 5-cleft; stamens distant, shorter than the corolla: anthers opening by slits; style longer. Berry globular, 2-celled upon the calyx, the size of a large cherry, black when ripe; the whole plant downy. Perennial herb. Leaves entire; pedicels solitary, recurved.

ATROPA BELLADONNA (Deadly Nightshade).

1. **A. Belladonna**, L. (Deadly Nightshade.) Stem trichotomous, branching, purple, 2 to 5 feet high. Leaves in pairs, unequal, oval, entire, pointed, from 8 to 12 inches long; petioles very short, radical Flowers large, axillary, and pendent, brownish purple. Flowers in June.

2. **A. Mandragora**, L. (**Mandragora officinarum**, L.) is a plant found in southern Europe, possessing similar properties to belladonna.

Geography. — The geographical distribution of the belladonna is not wide: it is found in the middle parts of the north temperate zone, in Europe, western Asia, and eastern North America.

Etymology. — *Atropa* was named for Atropos, one of the three Fates, daughter of Night, whose office it was to sever the thread of human life, — representing thereby the deadly character of this poisonous plant. *Belladonna,* the specific name, is from two Spanish words, *bella* and *donna,* and signifies "fine lady;" it is supposed to have been so styled because the fine ladies of Italy used it as a beautifier. The reason for the name *nightshade* is not apparent. *Mandragora* is derived from μάνδρα, pertaining to oxen or cattle, and ἄγριος, cruel, and relates to the poisonous effect this plant has upon cattle when it is accidentally eaten with their food.

History. — This plant was known to medical practice in the time of Theophrastus, and is supposed to be the plant whose fruit was eaten by Anthony's famished army (while retreating before the Parthians), by which so many were poisoned. It is also supposed to have played a part in the destruction of Sweyn's army when he invaded Scotland. A truce was agreed upon, and during the cessation of arms the Scots were to furnish food. They mixed the juice of the atropa berries with the bread and drink, which produced an intoxicating effect upon the Danes; and while in the deep sleep which succeeded, the Scots fell upon and slew them.

It is indigenous to Europe and western Asia, and was brought to North America by colonists, and in a few places, according to Dr. Gray, has escaped from cultivation.

Chemistry. — Atropa belladonna yields to the chemist two substances, atropine and belladonine.

These are the active principles of the plant, and are found in all its parts, — leaves, fruit, and bark.

Use. — Belladonna is administered in extract tincture, and in the form of plasters and ointment. All parts of the plant yield an active narcotic poison. It relieves inflammation, soothes irritation, and allays nervous excitement. It is used by oculists to cause insensibility or temporary paralysis of the retina. It is much used in homœopathic practice, especially in scarlet fever, and considered by this school a specific. When taken in large quantities it causes intoxication, accompanied by fits of laughter, ending in convulsions and death.

CAPSICUM, L. - Calyx, short, broad, bell-shaped, dentate, with 5 bristle-like teeth; corolla wheel-shaped, deeply 5-cleft, divisions or lobes valvate; stamens slender, rising from the base of the corolla; anthers short, opening lengthwise; disk inconspicuous; style thread-like; stigma club-shaped; ovary usually 3-celled; ovules numerous. Fruit a berry, with a thick or thin inflated pericarp, leathery or succulent, erect or nodding, globose, conical, linear, or oblong. Seeds flat, rough or smooth. Annual herb. Pedicels solitary or in pairs.

1. **C. annuum,** L. (Red Pepper.) Stem smooth, crooked branching, 2 to 3 feet high. Leaves egg-shaped, smooth, entire, on long foot-stalks. Flowers axillary, solitary, white; calyx tubular, with 5 small divisions; corolla wheel-shaped, in 5 divisions, plaited, and pointed; filaments short and tapering; anthers oblong; ovary egg-shaped; style slender, longer than the filaments;

SOLANACEÆ.

stigma blunt. Fruit drooping, conical, 1 to 2 inches long, pod with a thin shell, reddish-yellow when ripe, 2-celled, many-seeded; seeds yellowish-white, flat, kidney-shaped.

2. **C. fastigiatum**, L. (Cayenne Pepper. Bird Pepper.) Like the last, except that the flowers are in clusters of 2 to 3, and that the fruit is very small and in the shape of a cock's spur, and the pungency very much sharper.

3. **C. frutescens**. (Cayenne Pepper.) Much cultivated; is like the last, except that the pods are larger, and more pungent.

These three bear the most pungent fruit, and furnish the red pepper of commerce.

4. **C. grossum**, W. (Bell Pepper. Bull Pepper.) This is unlike any of the above, inasmuch as the fruit is not only very large, but the walls of the pod are thick and succulent, and very mild as to pungency. Pod from 2 to 6 inches long, and from 1 to 4 inches in diameter. Fruit used for pickling.

CAPSICUM ANNUUM (Red Pepper).

Geography. — The home of the capsicum is America. It is believed that it was first brought to the notice of Europeans as a condiment for food by the physician of the fleet on the second voyage of Columbus to the New World. It is now cultivated in almost every civilized country in the world where the climate admits of its cultivation. The southern Asiatics have names of their own for it; this goes to show that it is also indigenous there, but it has never been found in the Old World outside of cultivation; on the other hand, it has been found wild in South America. Capsicum is a tropical and subtropical plant, though it fruits in southern Europe, and in the United States as far north as the 43d parallel. It is grown in all tropical countries as a condiment or food. The West India islands, middle Africa, and southern Asia are all grateful regions of this fiery fruit. The outer skin of the fruit, as well as the seed, yields a red oil.

CAPSICUM GROSSUM (Bell Pepper).

Etymology. — *Capsicum* is supposed to be derived from the Greek word καπτος, strong, alluding to the pungent taste of the fruit pods. It has also been derived from κάψα, a coffer, box, or chest, referring to the pod which holds the seeds. *Annuum* is Latin, and means "yearly;" *frutescens*, having the appearance of a shrub; *fastigiatum*, tapering, or pyramidal; *grossum*, from the Latin *grossus*, thick, referring to the thick pod of this species. *Pepper* comes from the Greek πίπερι, and *Cayenne* refers to the country whence the best is brought.

Use. — As a medicine, capsicum is highly stimulant; a tea of red pepper is a specific in nausea, largely used in compounding the medicines of the Thompsonian practitioners. The ground fruit, or "pepper," is largely used as a condiment for food, especially in hot countries, and is much used in liniments. When mixed with the food of poultry it is said to preserve health and promote the laying of eggs.

SOLANUM, L. Calyx persistent, 5-parted; corolla rotate, subcampanulate; tube very short; limb plaited in the bud, 5-cleft, lobed or angular; anthers erect, connivent, but not united, opening at the top by 2 pores. Fruit a globular berry, 2-celled, many-seeded; seeds small. Flowering in August. Unarmed or prickly herb.

1. **S. tuberosum**, L. (Potato. Irish Potato. White Potato. Common Potato.) Stem herbaceous, erect, prostrate, or assurgent, 2 to 4 feet long, angular and branched towards the top, smooth or clothed with soft hairs at and near to the extremities. Leaves interruptedly pinnate; leaflets roundish, petioled, dark green, every alternate pair of leaflets very small; peduncles few-flowered; flowers blue or white. Rootlets give rise to tubers, globular or lengthened. Fruit a green berry, about the size of an ordinary marble; seeds numerous. Flowers in July and August. Roots giving rise to globular or reniform tubers. Tubers ripen July to October.

SOLANUM TUBEROSUM (Potato).

When propagated from the seed, the tubers of the first year are about the size of peas; the tubers produced by these are much larger, and the product of the third year is fit for the table. When propagated from the seed, the plant sports freely, and new varieties are thus produced. When the tubers are used to propagate from, the variety is constant.

There are other allied species, but they have not stood the test of cultivation, on account especially of the small size of their tubers, as follows:—

2. **S. Commersonii**, Dunal. Chile.
3. **S. immite**, Dunal. Peru.
4. **S. verrucosum**, Schlechtendal. Mexico.

These all differ in botanical features as well as in size and quality of their tubers, so much so as to entitle them to a specific character.

No plant in the whole range of table vegetables has had so large a share of attention from gardeners and amateurs as the potato, and every year new varieties are offered for sale by the seedsmen.

Geography. — The geographical zone of the potato is between the 35th and 50th parallels, and it even produces well up to 60°. It will grow and fruit well in any latitude where the cereals flourish.

It yields best in cool, damp climates; and the edible qualities are best where the plant is grown in rich sandy loam.

Etymology. — *Solanum* is of very doubtful derivation. The most plausible history of the word is that it is from the Latin word *solor*, solace, or comfort, due to the narcotic effect of some of the plants of this very large genus. *Tuberosum* is due to the tuberous character of its products. *Potato* is a corruption of *batata*, already explained. The French, from the same source, have made it *patate*.

History. — Humboldt believes the plant known to the Spaniards under the name *maglia* is the original of the potato. The potato is indigenous to Chile, and has been taken thence to neighboring parts of South America. Darwin, in his voyage to the south seas in 1840, touched at Chile, where he found the potato wild among the islands on the coast, growing in great luxuriance, with every appearance of being native to the soil, but saw nothing to lead him to believe that the natives cultivated it. In Peru, travellers report that the natives pay much attention to its growth. It is believed that the Spaniards carried it from Peru to Europe. This is the view taken by De Candolle. Another theory is that it was taken to Florida by the Spaniards, whence it found its way into the English settlements of Virginia, and thence to England. It has also been suggested that Raleigh, in warring against the Spaniards as a privateer, might have captured vessels with potatoes on board, and in that way have introduced the plant into Ireland. It is related that he had it planted on his estate in the south of Ireland, and that his gardener one day brought him a potato ball, or berry; tasting it, he was so disgusted with it that he ordered the gardener to root it out and destroy it. In attempting to do this, the gardener discovered the tubers attached to its roots; they were cooked, the value of this wonderful plant was revealed, and the cultivation of the potato dates from that incident. This occurrence is said to have taken place near the end of the sixteenth century. It was, however, known in other parts of Europe before taken to Ireland by Raleigh. It was for a long time confined to Ireland as a food-plant, but has now found its way throughout Europe.

Chemistry. — The potato yields to the chemist in 100 parts by weight: water, 68; starch, 17; wood, 09; mucilage, 06; = 100.

Compared with wheat, it is as 2–7; *i. e.*, a pound of wheat is worth 3¼ pounds of potatoes; or one bushel of wheat is worth 3¼ bushels of potatoes.

Use. — During the 300 years which have elapsed since its discovery, the potato has steadily advanced in value, until it has reached a point of importance in the economy of human life unequalled by any other food-plant outside of the cereals. As a root-vegetable for the table, nothing equals it. In all the countries of Europe, the United States of North America, and in Canada, it constitutes a large part of the food of the working classes, and is always upon the tables of the rich. The natives of Peru cook it by roasting it in the ashes. The most common mode of cooking is boiling; it is also fried, baked, and stewed with meat or milk. It is said that the Poles excel in cooking the potato, and that they practice thirty different methods of preparing it for the table.

Order XL. **PEDALINEÆ.**

Flowers perfect, irregular, axillary, solitary, racemed or in spikes, mostly 2-bracteolate; calyx 5-parted; corolla-tube cylindrical, throat swollen, 5-lobed; limb bilabiate, imbricate or subvalvate; stamens 5 on corolla-tube, 1 sterile, 4 fertile, 2 long and 2 short; ovary superior, 1-celled, rarely 2–4-celled. Fruit a capsule or drupe, 4-seeded. Leaves opposite or alternate, simple. Herbs.

No. of genera, about 10; species, 40; found in warm climates.

SESAMUM. L. (Oil Seed.) Calyx small, 5-parted; corolla bell-shaped, long and curved, oblique at base, 5-lobed, lateral lobes open,

DESCRIPTIVE BOTANY.

anterior shorter; stamens 4, attached to the base of the corolla, included; anthers arrow-shaped, attached at their backs, cells nearly parallel; ovary 2-celled, each cell divided into 2 dry false partitions; ovules many in each cell, arranged in a line; capsule oblong or ovate; seeds numerous, small, triangular in outline, differing in color; embryo straight, radical, short, testa crustaceous. Leaves opposite below, and alternate above, petioled, entire, incised or dentate, 3-parted or dissected. Flowers pale, violet, solitary, axillary, nearly sessile. Whole plant scabrous, erect, or prostrate.

1. **S. Indicum**, DC. (Sesame.) Stalk 2 to 3 feet in height, scabrous. Leaves ovate, oblong, or lanceolate, and alternate above, the lower ones 3-lobed or 3-parted and opposite, conspicuously feather-veined with yellow glands at the base of the peduncles. Flowers resembling the flowers of the foxglove in shape, color dingy white to rose; capsule velvety and pubescent, mucronate, 2-celled, developing in 4 cells; seeds numerous, ovoid, flat, varying in color, white, brown, or black, rather smaller than flaxseed. This species varies greatly in the form of its leaves and color of its seeds.

SESAMUM INDICUM (Sesame).

2. **S. orientale**, L. Larger leaves than in S. Indicum; flowers white, otherwise as in No. 1.

Leaves of both species abound in a gummy substance which they yield to water, forming a pleasant demulgent beverage.

Geography. — The home of the plant is believed to be the Sunda Isles, whence it was introduced into India, and thence to western Asia, southern Europe, and northern Africa, and from there' to the New World. To obtain the oil, the ripe seeds are first washed to remove all mucilage, and the oil is obtained by expression; the seeds yield 45 per cent of oil, as to weight. Its geographical zone is a tropical and subtropical belt extending both sides of the equator to the parallel of 35° all around the globe. It grows sparingly in higher latitudes.

Etymology. — *Sesamum* is derived from σησάμη, the Greek name of this plant; the specific name, *Indicum*, denotes the country where it is cultivated. The seeds are known in commerce as Til, Gingeli, and Tingili, Sesame, and Benne seeds. *Orientale* is from the Latin *orientalis*, and signifies "eastern."

History. — The S. orientale was known in the Levant and southern Europe and Egypt at least four centuries prior to the opening of the Christian era, being mentioned by Xenophon. It was taken to the West Indies by Europeans, and S. Indicum has found its way to Florida and other Southern States, and has crept along as far north as the vicinity of Philadelphia. It was detected by Judge Addison Brown in the ballast heaps at Communipaw, N. J. It is known as the thunderbolt flower. It is also the potent legendary opener of doors and caves. In the celebrated story of Ali Baba, the conjuror pronounces the words, "Open Sesame," and the door flies open.

Use. — The plant is used in Europe and India for a pot-herb, but it is most valuable for the oil it produces.

The oil resists putrefaction longer than any other fixed oil, and is considered the most delicate of all the sweet oils, except olive oil. It is used in India for food and for anointing the body, and for the fragrant oils used in religious ceremonies. It is used in all civilized countries for the same purposes for which olive oil is used. It is applied to illuminating purposes, to the manufacture of cosmetics and hair dressings, and especially to the manufacture of fine toilet soaps.

It congeals at a much lower temperature than other oils, and is hence well adapted to cold climates. It has of late years been largely employed in the manufacture of spurious butter, or oleomargarine. The natives of India make an article of diet of the seeds, and also of the refuse cakes after the oil is pressed out of the seeds, and they place the seeds in the graves with their dead relatives.

Order XLI. VERBENACEÆ.

Flowers perfect, irregular, or nearly regular, in a head, spike, raceme, or cyme, rarely solitary, usually bracteate: calyx with united sepals, persistent, tubular, limb toothed; corolla hypogynous; petals united and tubular, limb 4-5-lobed, frequently unequal, and lip-shaped, imbricate in the bud; stamens, 2 long and 2 shorter, rarely 5, attached to the corolla; ovary free; style terminal, forked at top. Fruit, a fleshy drupe, with 2-4 nutlets. Leaves opposite or whorled, rarely alternate, exstipulate, simple, or rarely compound. Herbaceous and woody. Plants and trees.

No. of genera, 59; species, 700; warm climates.

TECTONA GRANDIS (Teak).

TECTONA. L.f. Calyx bell-shaped, short, 5-6-cleft, sharp, tube swollen below, and contracted near the mouth; corolla-tube short, limb gaping, 5-6-cleft, lobes short, nearly equal, imbricated; stamens 5 to 6, attached to the corolla near its base, longer than corolla-tube; anthers ovate or oblong; ovarium fleshy, 4-celled, cells 1-ovuled; style elongated, 2-cleft, ovules attached to the side; drupe inclosed in the calyx; exocarp thin, endocarp fleshy. Seeds erect, oblong.

T. grandis. L.f. (Teak.) Trunk erect, 80 to 150 feet in height, and from 3 to 6 feet in diameter. Bark smooth, gray; branchlets 4-sided. Leaves opposite, or in verticils of 3, rough on the upper surface, downy beneath, entire, from 1 to 2 feet in length, and 6 to 18 inches wide. Flowers small, sessile, white, in terminal, compound, dichotomous panicles, bracts small. Fruit lens-shaped, in 4-celled drupes.

There are three species of this magnificent tree.

Geography. — The geographical home of the *Tectona* is tropical. One species is found in the East Indies, another in Burmah, and the third in the Philippines.

Etymology. — The generic name, *Tectona*, was altered from Tekka, the native name, whose signification is unknown. *Grandis*, the specific name, is the Latin for *large*, or *noble*, and is due to its magnificent size.

Use. — The teak is used in India in all structures where strength and durability are desired. It has been called the oak of the East. England has constructed some of her best ships of teak, some of which were built in India, and others in England, the timber having been taken home.

The leaf yields a red dye, and is also charged with an oil, the extraction of which forms an important industry. The oil is used for polishing and varnishing purposes.

Order XLII. LABIATÆ.

Flowers irregular, rarely regular, solitary or in pairs, or clustered in cymes in the axils of leaves or bracts, scattered along the extremities of the stems, or in densely crowded spikes; calyx persistent; sepals 5, connate, mostly lipped; corolla with united petals, hypogynous, tube occasionally twisted, limb 4-5-lobed, imbricated in the bud, usually 2-lipped, upper lip entire or notched, lower 3-lobed, sometimes 1-lipped and deeply cleft, sometimes bell-shaped or funnel-shaped, with four equal lobes and nearly equal stamens; stamens on corolla-tube usually 4, didynamous; ovary free; style simple, rising from the base of the ovarian lobes, 4-lobed; stigma usually forked. Fruit 4 akene-like lobes or nutlets, free or in pairs. Stems square; leaves opposite or whorled, exstipulate; subwoody and herbaceous plants, covered with glands containing an odoriferous volatile oil.

No. of genera, 136; species, 2,600; cosmopolitan.

LAVANDULA. L. Calyx spindle-shaped, 13-15 striate, 5-toothed, upper tooth longest; corolla-tube exserted, upper lip 2-lobed, lower one 3-lobed; stamens shorter than the corolla-tube; anthers 1-celled by confluence. Flowers small, spicate. Perennial herb.

1. **L. spica,** L. (Lavender.) Stem woody at the base, 15 to 20 inches high in a natural state, but reaching 5 feet under cultivation, branching near the ground. Leaves crowded about the base of the branches, whitish downy, oblanceolate, tapering to the base, sessile, edges revolute, leaves on the branches and upper part of the stem linear-lanceolate, uppermost ones shorter than the calyx bracts, and awl-shaped. Flowers in an interrupted spike, lilac-colored. July.

Var. **alba,** W. Has white flowers, otherwise as above.
Var. **latifolia,** W. Has broader leaves.

2. **L. vera,** DC., and **L. stœchas,** L. Were formerly in use, but are not used at the present day.

There are some twenty species in all; the above are the important ones.

LABIATÆ.

Geography. — The Lavandula is a native of Greece and the Grecian islands, whence it spread west, and was carried by Europeans to Hindustan; it was brought in the same way to the Atlantic States of North America, where it has become a common garden plant, cultivated for its delicate fragrance. It grows throughout the middle of the north temperate zone, has spread over the countries of the Levant, has been seen in gardens in Hindustan, and is common in the gardens of the eastern United States of North America. It is occasionally found growing wild near deserted dwellings.

Etymology. — *Lavandula* is from the Latin word *lavare*, wash, due to the circumstance that the spikes were used to perfume freshly-washed linen. The Romans perfumed their baths with it. The name *spica*, Latin, a spike, is due to the mode of inflorescence.

Preparation. — The flowers are distilled, the product being a delicate essential oil, and a coarser oil is obtained from the leaves and ends of the branches. The oil is largely manufactured in the department of the "Alpes Maritimes," in the southeast of France. The best, however, is produced in England, and brings in open market fifteen times more than the best French oil.

Use. — Lavender is stimulant and tonic. The pulverized dried leaves are used for a snuff to cause sneezing.

LAVANDULA SPICA (Lavender).

The leaves are laid among linen to perfume it.

The oil of lavender is a favorite perfume for manufacturing fragrant waters, and for compounding an aromatic spirit to remedy nausea, also to disguise the disagreeable taste of other drugs. The coarser oil produced from the leaves and stems is called *spike oil*, and is used by artists to mix their colors, and especially for painting on pottery.

MENTHA, L. Calyx 5-toothed; corolla 4-cleft, just a little longer than the calyx, nearly equal, one division broader and emarginate; stamens 4, straight, separate. Flowers in verticils, small, crowded, short-stalked, or subsessile.

1. **M viridis**, L. (Spearmint.) Stem 4-angled, from a creeping root, 12 to 20 inches high. Leaves wrinkled, subsessile or short-petioled, lanceolate acute, and cut serrate; bracts narrow, lanceolate or bristly; teeth of the

calyx hairy, spikes slender, interrupted. Whole plant possesses a strong agreeable odor. Damp places.

2. **M. piperita**, L. (Peppermint.) Stem quadrangular and grooved, sometimes hairy, from 1 to 2 feet high, slender, weak, purplish, and branching. Leaves on short petioles, ovate, pointed, and serrate, dark green above, smooth and shining, paler underneath, with purplish veins. Flowers in terminal spikes, lower parts interrupted, the lower whorl remote; bracts lanceolate and ciliate; calyx furrowed, with 5 dark purple ciliated striæ; corolla purple; filaments short, anthers included; stigma forked.

Var. **subhirsuta**, Benth. Has scattered hairs on the petioles and veins of the leaves.

There are many species to this genus, but those we have described are the most important, and the only ones that possess any economic or commercial value.

Geography. — The geographical distribution of the mint is a belt between 35° and 50° of north latitude, extending from the eastern side of the Mississippi valley to Japan. Loudon claims England as the home of the spearmint, as well as the peppermint. The *M. Piperita* is found wild in Hindustan, Japan, Persia, northwestern India, and in Egypt. In the days of Linnæus it was a well-known plant in gardens throughout central Europe.

Both these species were brought to the British colonies in North America by European emigrants, and have become naturalized throughout the northern and middle Atlantic States.

Etymology. — *Mentha*, the generic name, is said to have been given to this plant in honor of Mintha, the daughter of Cocytus, who according to the myth was turned into mint by Proserpine in a fit of jealousy. *Viridis* is the Latin word for *green*, due to the color of the plant. *Piperita*, from the Greek πίπερι, pepper, was given to this species on account of the biting pungency of the taste of the leaves. Spearmint is due to the spear-shaped spikes in which the flowers appear. Peppermint is merely a translation of *Piperita*. Mint is supposed to be a corruption of the word *mentha*, or *mintha*.

MENTHA PIPERITA (Peppermint).

History. — Mentha viridis, or spearmint, was under cultivation in the convent gardens in the ninth century. The exact locality which may be claimed as its home is not known, but it is a European plant.

It is claimed that the mentha piperita is a native of England, was discovered in Hertfordshire by a physician in 1696, and was described by Ray.

Cultivation. — It is cultivated for the production of oil of peppermint, which is obtained by distilling the green plant.

The cultivation for this purpose is carried on in England, on the European continent, and in the States of New York and Ohio in North America. The ground, which should be damp, is prepared in furrows, and roots are laid in them, after which they are lightly covered with earth and allowed to grow till the ground is covered. When in flower it is cut, and after lying for a part of a day, is placed into the stills. After the oil is extracted, the mint is cured and fed to cattle.

Use. — Spearmint is used to flavor sauces for meats, and to prepare a very popular beverage known as *mint julep*; also in the manufacture of essences and cordials; its medical qualities are carminative and anthelmintic.

Peppermint is largely used in the manufacture of the essential oil of peppermint, on which the medicinal character of the plant depends. Its medicinal qualities are anti-spasmodic, and it is used in domestic practice for pains in the stomach and lower intestines. In China and Hindustan it is largely used to relieve the flatulency incident to a vegetable diet. The taste is aromatic, warm, pungent, somewhat like camphor, and highly stimulant. The greater part of the oil, however, is used by the manufacturers of confectionery and cordials.

3. **M. pulegium**, L. (Pennyroyal.) Stem 6 to 10 inches long, bluntly 4-angled, much branched, usually trailing, slightly hairy. Leaves on short petioles, small, bluntly serrate, hairy underneath. Flower-stalks covered with short, thick hairs; flowers numerous, in sessile whorls; calyx greenish-purple, furrowed and hairy, with 5 unequal, ciliated, pointed teeth; corolla twice the length of the calyx, purple, 4-cleft, white at the base, clothed with long, soft hairs; stamens erect, longer than the corolla. Annual. Flowering from June to September.

Geography. — Found growing throughout Europe, Asia Minor, Persia, and northern Africa; was brought to northeastern America and introduced into cultivation; but the character and properties are so nearly allied to those of the American plant (*Hedeoma pulegioides*, Pers.) that it has disappeared, and the American plant is used in its place.

Etymology and *History.* — *Pulegium* is from the two Latin words, *pulex*, a flea, and *rego*, have power over, due to its reputation for destroying this lively insect. *Pennyroyal* is a curious corruption of the specific name. In England it is called *pudding grass*, because of its use for flavoring blood puddings; also *flea mint*, for the reason already stated.

The pennyroyal was known in the first century, and is mentioned by both Dioscorides and Pliny.

Use. — The medicinal properties of pennyroyal are carminative, antispasmodic, tonic, and stimulant. It is a well-known remedy in throat diseases. It is used as a gargle and as a liniment; also as an insecticide, or insectifuge, as the odor is said to banish various insects. The medical qualities reside in an essential oil obtained by distilling the entire plant.

Pennyroyal is administered by quacks for a blood-purifier, under the name of *Organs*.

ORIGANUM, L. Calyx ovate, bell-shaped, obscurely 13-nerved, 5-toothed, throat hairy; corolla 2-lipped, upper lip erect and notched, lower longer, and composed of 3 spreading lobes; stamens 4, ascending and spreading, mostly longer than the tube of corolla. Perennials; leaves nearly entire; flowers in crowded short or cylindrical heads; bracts usually colored; flowers purplish.

Species, 25 in number; mostly European.

1. **O. vulgare**, L. (Common Marjoram.) Stem purple, leafy, branched near the top, 1 to 2 feet high, and hairy. Leaves stalked, entire, or slightly serrate, sprinkled with resinous dots, paler underneath, ovate, lanceolate, an inch in length, petioles shorter than the blades, bracts purplish. Flowers in a terminal 3-forked panicle, in globular, compact heads; calyx hairy inside, with short, nearly equal, teeth; corolla twice as long as the calyx, with 4 broad, nearly equal, lobes, the upper one broadest and erect; stamens longer than corolla-tube. Flowers purplish-white. Whole plant highly aromatic to the taste. July to August.

2. **O. marjorana**, L. (Marjoram. Sweet Marjoram.) Stems numerous, woody, 12-18 inches high, much-branched. Leaves oval, obtuse, entire, on short petioles, blade hoary-pubescent, pale-green. Flowers small, white or pinkish, in crowded, roundish, compact, terminal spikes, bracts numerous and large; calyx tubular, 5-toothed, teeth sharp; corolla funnel-shaped, 2-lipped, upper one erect and rounded, lower one cut into 3-pointed segments. Leaves and flowers possess a pleasant aromatic odor. July to August.

These two species are the only ones whose medicinal or commercial value is of any importance.

ORIGANUM VULGARE (Common Marjoram).

Geography. — The geographical range of these two species is wide; they grow well all through the Levant and Mediterranean countries, and all over Europe as far north as the 50th parallel.

Etymology. — *Origanum*, the generic name, is from the Greek words ὄρος, "an elevation" or "hill," and γάνος, "beauty," hence the beauty of the hills. *Vulgare* is Latin for "common." *Marjorana* is supposed to be from the low Latin name, *majorica*. *Marjoram*, the common name, is a corruption of the Latin *majorica*.

History. — The Origanum was well known to the Greek and Romans, and was a favorite decorative plant at their marriage feasts, when it was woven into wreaths to crown the young married couple. It is mentioned by Pliny and Vergil. It was brought to North America by European colonists, and is a native of Europe and adjacent parts of Asia.

Use. — As a medicine it is stimulant, carminative, tonic, and sudorific, and it is prescribed for dyspepsia and other disorders of the stomach.

The Thompsonian practitioners use the dried leaves for snuff in cephalic difficulties. The leaves are used for tea as table beverage, and are preferred by some to teas from China.

The oil is very sharp, and used in liniments for sprains and rheumatism. On account of its beauty it is a favorite plant in the flower-garden.

THYMUS, L. Calyx bilabiate, with 10 to 13 striæ, 5-toothed, 3 upper teeth short, triangular, lower pair linear, subulate, ascending; corolla 2-lipped, upper lip notched, lower lip 3-lobed, lobes equal or middle lobe larger; stamens 4, mostly exserted; filaments straight, divergent. Leaves small, entire, strongly veined. Perennial herb, or woody.

1. **T. vulgaris,** L. (Sweet Thyme.) Stem a foot high, slender, woody, branched. Leaves sessile, linear-lanceolate or ovate, quarter of an inch long, edges revolute, hoary, especially beneath, and dotted with oil-glands. Flowers small, purple, in terminal globose heads, occasionally a few lower whorls. In a state of nature, clothed with a gray pubescence; under cultivation, greener, more luxuriant, and barely tomentose; fragrant, pungent, and aromatic. Perennial herb. July.

2. **T. serpyllum,** L. Stem slender, much-branched, procumbent, hard, approaching woodiness at the base, forming low, dense tufts, 6 to 12 inches in diameter, and covered with flowers. Leaves very small, ovate or oblong, fringed at the base by a few long hairs on each side; floral leaves smaller. Flowers usually 6 in a whorl, without any other bracts than the floral leaves, forming short, terminal, loose, leafy spikes; calyx hairy, and the whole plant sometimes clothed with hoary hairs.

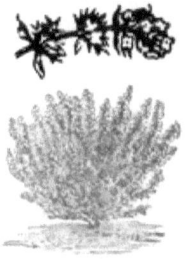

THYMUS VULGARIS
(Thyme).

Geography. — Thyme is indigenous to Spain and other parts of southern Europe, the coasts of the Mediterranean, the mountains of Greece and the islands of the Archipelago, the British Isles, and north to southern Siberia.

Etymology. — *Thymus* is from the Greek θυμός, signifying "courage" or "strength." *Vulgaris,* Latin *vulgus,* signifies "common." *Serpyllum,* Latin, means "creeping," and is due to the prostrate character of this species. *Thyme,* the common name, is a contraction of the generic name.

History. — Thyme was known to the ancient Romans, who used it in various ways to season food. It was in modern times described by Tournefort, and was brought to North America by European colonists. It was a favorite aromatic herb of the Greeks, and abounded on Mt. Hymettus, near Athens, famed for its thyme and honey. Ovid, Vergil, and Pliny, all speak of the thyme in connection with bees; and honey made from the thyme possessed the flavor of the plant, and was on that account highly esteemed.

Use. — Thyme is used in many culinary preparations for flavoring sauces, gravies, cheese, and meats, sausage, etc. The plant yields to distillation an essential oil, which is a powerful local stimulant. In combination with camphor and alcohol, it is used as a liniment for chronic rheumatism. It is an important ingredient in liniments, and is largely used for low grades of scented soaps. The plant is also employed for borders in planted grounds. The variegated varieties, T. platifolia, and T. aurifolia, are used for that purpose.

SALVIA, L. (Sage.) Calyx striate, 2-lipped, upper lip 3-toothed or entire, lower bifid, throat naked; corolla 2-lipped, gaping, the upper lip straight or curved, sometimes notched, the lower spreading and 3-lobed, middle larger; perfect stamens 2, the filaments have at their summits a cross thread, the transverse connective, one end of which bears a perfect half-anther. The other end has on it a defective half-anther. This transverse thread is an essential character of this large genus.

1. **S. officinalis**, L. (Common or Garden Sage.) Stem woolly, 1 to 2 feet high. Leaves elliptical-lanceolate, wrinkled, margins crenulated; calyx mucronate; whorls 6-10-flowered, in two opposite sets, of purplish-blue flowers. Whole plant strongly aromatic.

Var. **variegata** has leaves spotted with white. This genus has about 400 species, though S. officinalis is the only one of culinary importance.

2. The **S. pomifera**, L. (Apple-bearing Sage), is found upon some of the islands of the Grecian Archipelago. Upon its branches appear fleshy tumors from half to three quarters of an inch in diameter, supposed to be caused by the bite of an insect.

Geography. — The home of the sage is the countries of the Mediterranean. It was known to the most ancient writers on medicine and botany on account of its medical virtues and pleasant odor. From ancient time it has been cultivated throughout middle and southern Europe and the British Isles, and like similar plants was brought to northeast America by European colonists. It is also growing in British India, where it has been carried by the English.

The genus is very large, and some of the species are found in every quarter of the globe. The home of the S. officinalis and S. pomifera is the Levant and other Mediterranean countries. The S. officinalis grows well in all countries between 35° and 60° of the north temperate zone. It has been introduced into North America, but not naturalized; it is not growing except under cultivation.

Etymology. — *Salvia*, the generic name, is from the Latin *salvus*, "well," or "in good health," in allusion to its curative qualities. *Officinalis*, the specific name, means "of the shops." *Pomifera*, "apple-bearing," is due to the fleshy tumors on its branches. The common name, *sage*, is said to be due to properties of the plant, which tend to strengthen the mind, and thereby make men wise.

Use. — The sage is used to flavor sausage, for stuffings for roast poultry and other meats, and the pulverized dried leaves are used to flavor cheese. An infusion of the leaves is an astringent tonic gargle. The cold tea is administered for nervous headache; when sweetened with honey, the sage is a remedy for canker in the mouth and throat. Gerard says: "It is good for the head and brain. It quickens the memory and the senses. The juice of the leaves mixed with honey is good for those who spit blood. No man needs to doubt of the wholesomeness of sage." It was also used in his day as a hair-dye. It is used as a cooling drink in fevers, and is an important plant in the list of domestic remedies.

There are many species used for ornamental purposes. A variety of S. *officinalis*, var. *variegata*, is a favorite, and there are many with very showy flowers.

ROSMARINUS. L. (Rosemary, Dew of the Sea.) Calyx bell-shaped, slightly compressed, 2-lipped, upper lip concave, minutely 3-toothed, lower bifid, naked within; corolla gaping, downy, pale-blue, variegated with purple and white, tube longer than the calyx, upper lip erect and shortly bifid, the lower spreading and slit into three segments, the middle segment larger, concave, and declined; perfect stamens 2, longer than the upper lip, arched, bearing a flexed tooth above the base, supporting the blue oblong anther; style as long as the stamens, thread-like, 2-cleft, the posterior lobe small; nutlets, 4 at the bottom of the calyx. Shrub. Leaves narrow and entire. Floral leaves smaller. Only 1 species.

R. officinalis, L. (Rosemary.) Stem erect, 4 feet high, much-branched; branches hairy; branchlets 4-sided and downy. Leaves opposite, nearly sessile, an inch long, narrow, linear, obtuse, entire, revolute, dark-green, smooth and shining above, woolly, veined, and silvery beneath. Flowers axillary and terminal, blue. Both flowers and leaves have a strong odor, resembling camphor. A beautiful evergreen shrub.

Var. **variegata**, W., has variegated leaves. This, with other varieties, are under cultivation; but the R. officinalis is highly valuable for other purposes.

ROSMARINUS OFFICINALIS (Rosemary).

Geography. — The rosemary is no doubt native in all southern Europe. It is found without cultivation on the Greek islands in the Peloponnesus, and is under cultivation from western Europe to Japan, in the southern parts of the north temperate zone. It is a favorite in the gardens of Egypt, and is cultivated in Hindustan for medicinal purposes. As no native name is found for it in Asia, it is inferable that it has been introduced into those countries by Europeans.

Etymology. — *Rosmarinus* is compounded of the Latin words *ros*, "dew," and *marinus*, "of the sea," — dew of the sea, or sea-dew. *Officinalis*, Latin, means "of the shops." *Variegata*, Latin, means "variegated." *Rosemary* is without doubt a corruption of *rosmarinus*.

History. — When it was introduced into use is not known, but it was known to the ancients, Dioscorides, Pliny, Galen, and the Arabic physicians. It was known to the Saxons before the Norman conquest, and no doubt cultivated by them in England. Charlemagne ordered it planted in his garden; Vergil recommends it for the use of bees; and Lindley says the delicate flavor of Narbonne honey is due to access of the bees to the rosemary. It was brought to North America by colonists in the latter part of the seventeenth century.

Use. — It is difficult to find a plant to which greater and more numerous qualities have been attributed, or one that is more widely known or in greater repute than this humble shrub. The Greeks bound it on the head for garlands. The leaves and flowers decorated the bride to indicate fidelity. Anne of Cleves wore it at her wedding with Henry. It was laid upon the coffin, thrown into the open grave, or carried in the hand at funerals to indicate that the mourners would remember the dead. A sprig was worn to proclaim that the wearer had repented. Queen Bess adorned the walls of Hampton Court with rosemary. It was always found in the woman's department of the gardens of the high-born, and there was an English proverb, that "rosemary grows where the mistress is master." We have not space to relate half of what has been said of this little shrub. It is highly aromatic in all its parts. The flowers are distilled to procure the oil of rosemary, which is an ingredient in the perfumery of Hungary water and eau de cologne; it is used also to flavor spirits and liqueurs. It is used in making the well-known soap liniment, and the compound spirits of lavender. It is an ingredient in the manufacture of nearly all perfumery and of toilet soaps. The pulverized leaves are used for snuff.

NEPETA. 1. Calyx cylindrical, marked with 15 striæ or nerves, 5-toothed, generally oblique, barely 2-lipped; corolla slender below, swollen in the throat, upper lip emarginate, lower one spreading, 3-lobed, middle lobe largest, crenate, marked with crimson dots, margin turned over; stamens 4, ascending under the upper lip, near together. Perennial or annual herb. Leaves sometimes lobed or incised.

A large genus; some of the species are cultivated for ornamental purposes.

NEPETA CATARIA
(Catnip).

N. cataria. L. (Catnip or catmint.) Stem square, 3 feet high, branched; whole plant hoary. Leaves cordate, coarsely crenate-toothed and petioled. Flowers crowded, in large, hoary spikes, whorled, white or purplish. July to September.

Geography. — The home of the Nepeta Cataria is Europe and western Asia. It is found throughout the countries of the Levant. It was brought to North America by European settlers, from whose gardens it has escaped, and is found growing freely about dwellings and by the roadsides without cultivation.

Etymology. — *Nepeta,* the generic name of this plant, is said to have been given to it because it was first brought to the notice of naturalists at Nepet, a town in Tuscany. It is also claimed to be due to the fact that it cures the bite of the *nepa,* a scorpion. *Cataria,* the specific name, arises from the fact that cats delight in it, take it for medicine, and roll upon it when opportunity offers. *Catnep* or *catnip,* the popular name, was given for the reason that cats bite it or nip it.

Use. — Catnip is an important article in the *materia medica* of the matron. Tea made of the dried leaves, stems, and flower-buds is administered to infants to relieve pains in the bowels and to promote sleep. Herbal practitioners claim for it the qualities of a febrifuge, carminative, tonic, and soporific, and a *slight* narcotic.

MARRUBIUM, L. Calyx tubular, woolly, 5-10-toothed, with a corresponding number of striæ, teeth erect or spreading, setaceous, alternately shorter, erect or hooked; corolla 2-lipped, upper lip erect, flattish, sometimes divided, lower lip spreading, nearly flat or concave, 3-lobed, middle lobe largest and notched, tube the length of the calyx; stamens shorter than corolla, 4 in number; filaments parallel under the upper side of corolla; anthers 2-celled; verticils many-flowered, globose, with slender bracts. Leaves woolly and wrinkled, rounded at the base, serrate; floral leaves similar to the others in form. Flowers white or purple. Perennial herbs.

No. of species. 30.

M. vulgare, L. (Hoarhound.) Stem 12 to 18 inches in height, hoary, branching at the base, or numerous stems from the same root. Leaves ovate, rounded at the base, crenate-toothed, wrinkled, hoary, lighter underneath, on short stalks. Flowers white, sessile, in dense, globose verticils; nutlets dark-brown, obovate, and truncate. July.

MARRUBIUM VULGARE (Hoarhound).

Geography. — It is native in the countries of the Levant, the Peloponnesus, and other regions around the Mediterranean Sea. It is at the present day found growing in gardens all over Europe in the temperate zone. It was brought to the Atlantic States in America by European colonists, and escaping from gardens is found growing without cultivation about dwellings. It is found throughout the temperate zone in Europe, Asia, and America, keeping pace with civilization.

Etymology. — *Marrubium* is derived from *Marruvium*, an ancient city of Italy, on the shore of Lake Fucine. *Vulgare*, the specific name, signifies "common." *Hoarhound*, the common name, is due to the hoary appearance of the plant; *hound* is from *houn*, "bitter," or "disagreeable."

History. — This plant was known in the days of Theophrastus. Pliny, who lived 300 years later, also speaks of it. In those days the plant had a high reputation for curing pulmonary diseases.

Use. — As a remedy, the ancients attached great value to hoarhound, and more particularly for pulmonary diseases. At the present day it is a popular domestic remedy for coughs, colds, and affections of the lungs, administered in teas, syrups, and candies. The hoarhound has an aromatic smell, but a very bitter flavor.

ORDER XLIII. **CHENOPODIACEÆ.**

Flowers perfect, small, sometimes polygamous, sessile or pediceled, solitary or agglomerated, axillary or terminal, bracteate or naked; calyx 5–3-sepals, greenish and coherent at base, imbricate in the bud; corolla wanting; stamens hypogynous, or as many as sepals, or fewer at the bottom of calyx, opposite the sepals; filaments thread-like, mostly free, sometimes united just at the base in a cup; ovary egg-shaped, usually free, 1-celled; style 2–3-lobed or 2–3 styles. Fruit, a utricle; seed mostly free, lens-shaped or kidney-shaped. Leaves alternate, simple, sessile or petioled, entire, toothed or sinuate, frequently fleshy, without stipules. Herbaceous or suffrutescent, sometimes shrubby.

No. of genera, 80; species, 520; cosmopolitan; mostly in temperate climates.

BETA, Tournefort. Calyx hollow and contracted at the mouth, 5-cleft, persistent, becoming hardened at the base; stamens 5; ovary depressed, partly inferior; stigmas 2, the small bladdery fruit, with a thickish, hardened, depressed pericarp, enclosed in the calyx: seed horizontal. Leaves alternate; flowers in spikes. Herbs.

B. vulgaris, Moq. (Beet.) Stem 2 to 5 feet high, angled, branched in form of panicle, appearing the second year. Leaves of the first year 6 to 15 inches long, 4 to 8 inches wide, spatulate, edges wavy; radical leaves of the second year like those of the first; stem-leaves smaller, of dingy copper-color to dark-purple, ovate, lanceolate; root biennial, 3 to 10 inches in diameter, and 5 to 15 inches long, fusiform, tapering downwards to a slender fibrous point. Color, from dark-yellow to dark-red. Flowers greenish-white, in sessile, head-like cymes, forming slender spikes, arranged in leafy panicles; appearing in July. Seed rugose or wrinkled.

The beet is propagated from the seed, and sports freely, producing many varieties, the general forms of which are two,—the long beet, and the turnip beet.

BETA VULGARIS RAPA (Turnip Beet).

Var. **cicla**, the long, cylindrical-rooted beet.

Var. **rapa**, flat or turnip-rooted beet.

Var. **mangel-würzel**, large-rooted beet.

Under these forms there are many varieties, as may be seen by consulting the catalogues of the seedsmen, and the varieties under cultivation are very constant.

Geography.—The beet grows well in rich soil throughout the middle parts of the temperate zone, especially in Europe, north Africa, and the temperate parts of British India. It was brought to North America by British and Dutch colonists, and is largely grown here.

CHENOPODIACEÆ.

Etymology. — *Beta*, the generic name, is said to be from the Celtic word *bett*, "red," due to the color of the root. It is also claimed to be from *bete*, Anglo-Saxon, but derived from the Latin *beta*, used by Pliny. The specific name, *vulgaris*, is Latin, and signifies "common." *Beet*, the common name, is a mere corruption of the scientific name.

History. — The beet is a native of Europe and Western Asia, along the shores of the Mediterranean, and adjacent countries. The *Beta maritima* is believed to be the original plant, the seed of which is supposed to have given rise to the B. vulgaris of the gardens, out of which have grown by sports the numerous varieties that now exist. It was known to the Greeks and Romans at least 500 years before the Christian era. It is supposed to have been taken to Great Britain by the Romans in the time of Agricola, A. D. 79, and the spread into Gaul and Germany was no doubt due to the conquest of those countries by the Romans.

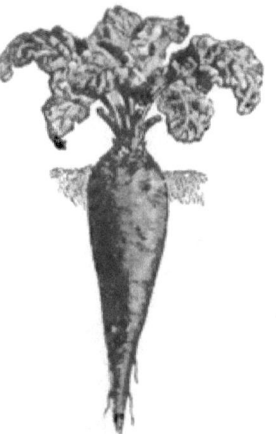

BETA VULGARIS CICLA (Long Beet).

Use. — The beet is an important table vegetable and pickle, and in Germany, France, and the British Isles is used as a salad. In central Europe, Germany and France especially, it is extensively used for sugar-making, and very largely for a feed for stock.

The sugar-bearing property of the beet although discovered in Germany, was first turned to practical use in France in the reign of Napoleon I., who, to render France independent of foreign nations for sugar, encouraged the manufacture of beet-sugar at home. From France it spread into adjacent European countries, and especially into Belgium, Germany, Austria and Russia; and at present beet growing and sugar making from the beet are ranked among the great industries of central and western Europe.

Statistics. — The annual product in Europe is about 600,000,000 bushels. An acre will produce 4,000 pounds of sugar. An acre of cane yields 7,200 pounds.

The crude beet-sugar contains a greater number of foreign substances than cane-sugar, and the refining process is more difficult and expensive. Its sweetening power is identical.

SPINACIA. Tourn. (Spinach. Spinage.) Flowers diœcious: bractlets; staminate flowers with a 4–5-parted calyx, segments equal, stamens on the receptacle, opposite the segments of the calyx; calyx of the fertile flowers tubular, swelled in the middle, 3-toothed; ovary egg-shaped, 1-celled, 1-ovuled, with 4 lengthened stigmas. Fruit 1-seeded, included within the hardened 2–4-horned calyx; seeds flattened. Leaves alternate and petioled. Flowers greenish and axillary. Annual herb.

S. oleracea, Mill. (Spinach.) Stem 18 to 24 inches high, slightly branched, more frequently simple. Leaves 3 to 5 inches long, tapering towards the base, frequently hastate and lanceolate; leaf-stalk varies from 1 to 4 inches in length. Flowers axillary, densely clustered or in raceme-like panicles, greenish. Fruit or ripened calyx generally armed with 2 or more strong, blunt spines; in some varieties the fruit is smooth or without spines. Flowers in June.

This plant, which is grown from the seed, sports, and many varieties have arisen. The most important ones are the round, smooth-leaved, and the prickly-leaved; the last is more hardy, and best for autumn planting.

A variety known as the *Flanders Spinach* has double-hastate and wrinkled leaves. The *Large Prickly* has very large leaves, rounded or spatulate. The *Lettuce-leaved* has rounded, deep green, smooth leaves on short stalks. These are the favorite varieties; there are others intermediate.

SPINACIA OLERACEA (Spinach).

Geography. — The spinach grows well throughout the middle of the north temperate zone, and is found in the kitchen and market gardens of the Old World, from Hindustan to the western shores and islands of Europe; and in the eastern United States of North America. It also has been carried to the islands of the South Pacific by European colonists.

Etymology. — *Spinacia* is derived from the Latin *spina*, "a thorn," on account of the armed condition of the seed-vessel. *Oleracea* is from the Latin *oleraceus*, "a pot herb." *Spinach*, or *spinage*, the common name, is a corruption of *spinacia*.

History. — The home of this plant is Persia, whence it has found its way by commerce, colonization, and travellers to all parts of the civilized world.

Just how or when it was taken to Europe is not known. It was brought to the United States from Europe, where it receives much attention by gardeners. It is fit for use early in the spring, to which circumstance its popularity and market value are largely due.

Use. — The leaves of spinach are boiled with salt meats, or alone, and served plain or with a vinegar-sauce, or eaten with salt only as a condiment. It is one of the most delicate of all the plants used for greens, and on this account justly ranks high as a table vegetable.

ORDER XLIV. **POLYGONACEÆ.**

Flowers perfect or polygamous, without involucre; perianth of 3–6 segments, inner ones or all often petaloid; stamens 1 to 9 or 6 to 8, sometimes 12 to 17; ovary free, occasionally adhering below, 1-celled,

1-ovuled, ovule basilar and erect. Fruit 3-angled, sometimes winged. Leaves alternate and entire, largely collected at the base of the stem, rarely opposite, mostly with clasping petioles and sheathing stipules. Herbaceous or frutescent, and sometimes climbing.

No. of genera, about 30; species, 600.

FAGOPYRUM, Tourn. Calyx composed of 5 colored equal sepals or parts; stamens 8, alternating with 8 honey-glands. Number of styles 3, capitate. Fruit 3-angled, in lax or dense cymes.

1. **F. esculentum**, Tourn. (Buckwheat.) Stem smooth, 1 to 3 feet high, branched, in an irregular panicle, furrowed, stout and hollow. Leaves varying, cordate, triangular, or hastate. Flowers in terminal and axillary cymose panicles, rose white. Fruit 3-angled, nut or grain inclosed in a dark-colored, coriaceous shell; kernel white.

Flowers in August; fruit ripens in September and October.

2. **F. Tartaricum** is a hardy species grown in Tartary and in northern Europe, endures light frost, differs from *F. esculentum* in the leaves, which are broader than long, with acute lobes. Fruit triangular, lance-shaped, with the angles sinuate-dentate; calyx very small.

3. **F. emarginatum** differs from the last in its fruit, whose angles are margined by a broad wing.

FAGOPYRUM ESCULENTUM
(Buckwheat).

Geography. — The zone of the buckwheat is not very wide; it is found in Russia, far north, and grows well in Canada and the northern United States. It will grow south of 35° of latitude, but does not thrive in hot climates. Though it does not endure the frost, it will not fill unless it has a temperature as low as 35° Fahrenheit.

Etymology. — *Fagopyrum* is derived from the Greek words φυγόs, "beech," and πυρός, "wheat," — beech wheat or grain. *Esculentum*, the specific name, is Latin, and signifies that the grain is eatable. The common name, *buckwheat*, is derived from the Anglo-Saxon word *boc*, "beech," and the word "wheat" signifying "beech-wheat." Linnæus called this plant *Polygonum fagopyrum*, "many-angled beech-wheat." Tournefort, the great French botanist, named it *Fagopyrum esculentum*, "eatable beech-wheat." *Tartaricum*, Latin, is derived from *Tartary*, the home of this species. *Emarginatum*, Latin, is due to the fact that the angles of the fruit are margined by broad wings.

History. — It is a native of central Asia and Tartary, and is found growing without cultivation in the valley of the Volga, and along the shores of the Caspian Sea. It was carried to Spain by the Moors, and has thence spread throughout northern and central Europe, where it has become naturalized. In France it is called *Blé Saracin*, or Sarasin Wheat.

Another account of its introduction into Europe is that it was brought by the Crusaders from Asia Minor, in whose northern fields it was found under cultivation, which would account for its **French name**.

Still another account is that it was brought into Russia from Tartary early in 1700, and thence spread through Europe.

It was introduced into northeast America by European colonists.

Cultivation. — In latitude 40° to 45° it is sown about the 1st of July, and ripens in September. It is planted when practicable upon new ground, and leaves the land in a good condition for next year's planting of other crops. The yield ranges in the northern and northwestern states from 20 to 50 bushels to the acre, but 30 bushels is a satisfactory yield. In the south Atlantic States the yield is less, ranging from 10 to 25 bushels.

Use. — Although this grain has not the importance of the cereals, yet it constitutes the bread of a large number of the people of central Asia and northern Europe. It is very popular as a material for griddle-cakes. In northern Europe it is used in making a dark bread, and there too the grain is hulled and used in the same manner as rice. It is an excellent feed for cattle and poultry.

RHEUM, L. Calyx colored, sepals 6, in double series, persistent; stamens 9, opposite the outer sepals; styles 3, very short and spreading; stigmas 3; fruit with 3 winged angles. Leaves large, mostly radical, on long stalks. Flowers in racemose or paniculate fascicles. Perennial herb.

1. **R. rhaponticum**, Linn. (Rhubarb.) Stem stout, hollow, 4 feet high, furrowed; stipules large and sheathing. Leaves entire, cordate, ovate, obtuse, and smooth, 12 to 24 inches long and 10 to 15 wide, cauline leaves smaller; sepals greenish, with white margins. Root large, fleshy, tuberous, and yellow within.

RHEUM RHAPONTICUM (Rhubarb).

2. **R. palmatum**, Linn. Leaves 5-lobed, palmate, rough; lobes sinuate-toothed, acute.

3. **R. undulatum**, Linn. Leaves oval-cordate, with undulated margins.

4. **R. compactum**, Linn. Leaves with a general heart-shaped contour, but interrupted with a number of deep sinuses.

The last three are the species which furnish the rhubarb-root of commerce, the R. palmatum supplying the larger part.

Several varieties have been obtained by sporting and hybridizing, which furnish very large leaf-stalks. There are some 10 varieties, that for one or another reason recommend themselves to the cultivator. The Victoria is admired for its gigantic size. The most delicate and desirable for the private garden is the R. compactum.

Geography. — Rhubarb will grow well as far north as the 50th parallel. The market is supplied by the products of regions lying between 35° and 45°. Unsuccessful attempts have been made in England to raise it for the druggist.

Darwin states that the root grown in England does not possess the medical properties of that raised in Tartary. It is found also growing in tropical India.

Etymology. — The name *Rheum* is from *Rha*, the ancient name of the Volga river, the plant having been first brought to the notice of man by specimens procured along the banks of the Volga. *Rhaponticum*, the specific name of the plant under garden cultivation, and from which the several varieties have been derived, is made up of the words *Rha*, the name of the Volga, and *Pontus*, a district of country where the plant was found. De Candolle gave it this name, which signifies "rhubarb from Pontus." It was formerly supposed to be the species which produces the root of commerce, and it has also been claimed that the R. palmatum is the true or parent species, from which all others have arisen, and that the R. compactum is the original species. But the properties of the different species and varieties are so similar that they may be substituted for each other. Rhubarb is a contraction of *rheum barbarum*, the rheum of the barbarians.

History. — The fine medicinal rhubarb known as Russian or Turkey Rhubarb entirely disappeared from commerce in 1863. It derived its name of Turkey from its being formerly brought into Europe through the Levantine ports, and in more recent times was named Russian from its being imported through the Russo-Chinese frontier town of Kiachta, at which place the Russian government maintained an establishment for its rigid inspection for nearly two hundred years, up to 1863, when all inspection was abolished. Neither the botanical source of this rhubarb nor the place of its origin were ever known. The present rhubarb of commerce, known as China or East Indian Rhubarb, is collected in the Chinese provinces of Shensi Kanshu and Szechuen, thence sent to Hankow, whence it reaches Europe and America by way of Shanghai. It is defined in the Pharmacopœia of the United States as "the root of Rheum officinale and of other undetermined species of rheum."

Use. — The root of the rheum furnishes one of the most reliable substances in the *materia medica*. It is cathartic, astringent, and tonic, and is largely used in alterative medicines and in preparations to correct and excite the digestive apparatus, as well as in medicines for reducing inflammatory conditions of the bowels. It contains much oxalic acid.

The leaf-stalk is used in culinary preparations, — tarts, pies, puddings, dumplings, and sauce. It is also preserved in sugar. It is now found in every well-kept kitchen-garden, and is an important article in the market-garden.

ORDER XLV. **PIPERACEÆ.**

Flowers perfect or diœcious, without envelopes, in long, peduncled spikes, each flower protected by a peltate or decurrent bract; stamens 2 or 3, sometimes 6 to many, filaments short; ovary sessile, globular, of 3–4-distinct carpels, with several ovules, or 1-celled, 1-ovuled. Berry dry or fleshy; seed globose, testa thin, cartilaginous. Leaves frequently succulent, opposite or whorled, occasionally alternate, entire; nerves obscure, reticulate; petiole short, sheathing at base. Herbs or shrubs.

No. of genera, 8; species, 1,000. Chiefly tropical; Asia and America.

PIPER, Linn. Flowers without perianth; stamens 2 to 4, occasionally 5, rarely more; filaments short; ovary 1-celled; stigmas 2 to 5. Fruit, a small berry, globular or egg-shaped; climbing shrubs, with alternate, entire leaves, which are strongly veined or nerved.

1. **P. betel**, Linn. (Betel Pepper.) Stem climbing, 5 to 8 feet high, or long, knotted at the nodes. Leaves opposite, 5 inches long, ovate, acuminate, uneven or obliquely cordate at base, 5-7-veined, leathery, glossy above.

There are over 600 species of the Piper, but this and the following are the most important.

Geography. — The Piper betel is a tropical and subtropical plant, and is distributed throughout the regions of southern Asia.

Etymology and History. — *Piper* and *pepper* are derived from the Greek πίπερι and πέπερι, pepper. *Betel*, the specific name, is the Malabar name of the plant. It is native to Java, and is cultivated wherever the betel-nut grows.

Use. — The leaves are sparingly sprinkled with shell-lime, and then wrapped around slices of betel-nut, and in that state they are used as a masticatory. They are also used for the same purpose without the betel-nut.

Medical practitioners among the Hindus recommend their use in the morning fasting, also after meals and on retiring. The properties of the betel-pepper are aromatic, carminative, stimulant, and astringent, and it is said to be a specific for headache.

In connection with the betel-nut, it is used by about one tenth of the whole human race, yet is not an important article of foreign commerce.

PIPER NIGRUM (Black Pepper).

2. **P. nigrum**, L. (Black Pepper.) Stem climbing, 20 to 30 feet long, with jointed or swelled processes at the nodes, branching in forks. Leaves broadly ovate, lanceolate, 5-7-veined, petioled, and 3 to 6 inches long, dark-green. Flowers in slender spikes, opposite the leaves, 3 to 6 inches long, greenish. Fruit, a globular, 1-celled berry, as large as a middling-sized pea, sessile, in loose clusters, to the number of 20 to 30 on a drooping stalk; the berries are first green, then red, and when ripe, yellow. A perennial shrub. This species produces both the black and the white pepper of commerce.

Var. *longum* produces a fruit similar to the *P. nigrum*, but inferior in strength and flavor, and is used to adulterate the *P. nigrum*.

Geography. — The home of the pepper-plant is southern Asia and the adjacent isles. Pepper has been an important article of commerce from the earliest times of communication between the East Indies and Europe, and was cultivated in southern India, Java, Sumatra, and Malabar, but especially at Bantam, in Java, whence the earliest shipments were made.

Etymology. — The specific name, *nigrum*, from the Latin, signifying "black," is due to the color of the berry when fit for market.

History. — It was known to the early botanists, Theophrastus, Dioscorides, and Pliny. It was in early times carried by caravans through lower Arabia and then across the Red Sea into Africa, and so to the shores of the Mediterranean, where a heavy duty was laid on it. The Arabian physicians of the middle ages used it as a medicine. The duties to which pepper was subject were so great in the middle ages that only the rich were able to procure it. In 1623 the impost-tax in Great Britain was five shillings per pound, and even as late as 1823 it was two-and-sixpence per pound, equal to 63 cents.

Cultivation.—The plant grows easily from slips, and is usually slipped beneath trees which it is to climb. It fruits the first year after slipping, but reaches its maximum crop in the fifth year. A plant in full bearing yields 10 pounds, and will last 20 years.

Chemistry.—The chemist obtains from the pepper-corn or berry an essential oil and a resin, on which its pungency depends. The oil known as the oil of black pepper gives the peculiar flavor. There is also present a tasteless alkaline substance known as piperine, $C_{17}H_{19}NO_{31}$, which forms four-sided prismatic crystals.

Preparation.—It is prepared for use by picking when in a semi-ripe state, and drying; in drying it turns black, hence its name. The white pepper is taken from the plant when nearly ripe, macerated in water, and the outer skin being removed, is bleached. The pungency of pepper is largely due to an oil which resides in the cuticle, hence white pepper is not so strong as the black.

Use.—Black pepper is used in civilized nations in almost all the preparations into which meat or fish enter, and in small quantities is supposed to promote digestion.

It was formerly used in medical practice as a stimulant and carminative in disorders of the stomach.

Order XLVI. MYRISTICACEÆ.

Flowers diœcious, inconspicuous, with simple perianth; male flowers with 3-15 monadelphous stamens; anthers often on the margin of a broad disk; female flowers with 1-celled ovary, having 1 erect ovule; capsule fleshy; seed erect, nut-like, enveloped in an aromatic, laciniated aril; testa hard. Leaves alternate, exstipulate, coriaceous, simple, entire, pubescent or scaly. Small aromatic trees or shrubs.

Only 1 genus, with 80 species; tropical.

MYRISTICA, L. Flowers diœcious, perianth leathery, tubular, somewhat bell-shaped, 2-4-parted. Staminate flowers, with 3 to 15 filaments, united into a column which terminates in a toothed disk; anthers attached to the filaments along the backs, and their dehiscence turned outwards, opening longitudinally. Pistillate flowers with a single style; stigma entire or lobed; ovule solitary, fruit appearing at the base or bottom of the pistillate flower; capsule fleshy, about the size of an apricot, and shaped like a short pear: the outer covering is half an inch thick. When ripe it opens by a smooth suture which extends from the stem all around to the opposite side, separating the shell into two equal parts. When partly open it exposes the nut enveloped in a network, which is the mace of commerce; inside the mace is the nut, inclosing a hard, black shell, inside of which is the kernel or nutmeg.

1. **M. fragrans**, Houtt. (Nutmeg and Mace.) Trunk 20 to 35 feet high, much-branched, branches erect, forming a tree which resembles the pear-tree; bark smooth, ash-colored, when wounded bleeding a glutinous red juice. Leaves alternate, on short petioles, elliptical, pointed, wavy, entire, obliquely nerved, bright-green above, grayish beneath, aromatic. Flowers and fruit present at the same time; flowers inodorous, small, axillary, 1 to 3 on a stem; calyx smooth, fleshy, 3-parted; segments spreading; corolla wanting; ovary oval; style short, with 2 stigmas. Fruit as described under the genus.

There are many species, about 80 in all; the *M. fragrans* is the most important, though several produce aromatic seeds.

2. **M. spuria**, Houtt., yields a mace, which is first yellow, turning red. From wounds in the bark it bleeds a red sap, sold as a substitute for dragon's blood.

3. **M. fatua**, Houtt., yields an inferior nutmeg, called long or wild nutmeg. It is used in India, but does not enter into commerce.

Geography. — The nutmeg requires a tropical climate, or at least a region of no frost. It has, by the appointed means for geographical distribution of plants, — the winds, waves, birds, and by human agency, — been carried from the Moluccas to the mainland of both hemispheres and their adjacent isles, and occupies at present a geographical belt all around the globe, extending just outside the tropics in both hemispheres.

Etymology. — *Myristica*, the generic name, is from the Greek μύρρα, myrrh, due to the aromatic fruit and leaves. *Fragrans*, the specific name, is Latin, also due to the odor-bearing fruit, leaves, and the oils obtained therefrom. *Nutmeg*, the popular name, is from the low Latin word *muscata*, a nutmeg, or a musk-like nut, referring to the perfume, which was supposed to resemble the odor of musk. *Mace* is from the Greek μάκερ, a spice.

History. — The nutmeg was introduced into European commerce by the Arabs through the Red Sea, early in the sixth century. An account of the tree and place of its growth was first given by an Arab traveller in the beginning of the tenth century.

The home of the nutmeg is the Molucca islands, but especially the Banda group of the Spice Islands. When the Dutch first came into possession of these islands, they attempted to confine the tree to a very few of them, in order the better to monopolize the trade as they did of the clove. When the English obtained possession of them, in the latter part of the eighteenth century, they made an effort to extend the area of growth, and introduced young trees into Sumatra, near Bencoolen, the Isle of Bourbon, Mauritius, and Madagascar, and also into the West Indies. The tree comes into bearing after ten years, but does not yield its largest crops till after 100 years, producing three full crops in a year.

Preparation. — When the nutmeg is harvested, the mace is removed and pickled in salt, afterwards dried in the sun and packed in boxes, and sent to market. The nut is dried either in the sun or by artificial heat, when it shrinks so that the shell may be broken without damage to the kernel. The kernel is then soaked in sea-water and lime, after which it is placed in heaps and left to heat sufficiently to destroy its vitality, when it is fit for market. The kernels that come to the United States take 100 to the pound; those sent to England, about 60, or about 4 to the ounce.

Use. — The medicinal properties of the nutmeg are tonic, digestive, stimulant, and intoxicating. In excessive doses it produces stupor, delirium, and death. It is used in the kitchen to flavor dessert sauces, and as a spice and an aromatic addition to cordials, and was at one time a favorite perfume.

The Hindu physicians pulverize it, and use it for a poultice for nervous headache, applying it above the eyes.

The odor or flavor, as well as the medicinal properties, depend upon the essential oil, of which the nuts yield about six per cent. The mace also yields an oil in somewhat larger quantities, which is very similar to that of the nut.

The mace also contains a fixed oil, which is obtained by crushing and pressure; used in pomades, soaps, and aromatic plasters.

Propagation. — It is raised from seed, and in its third year is ingrafted with branches from pistillate trees; and in the orchards a few staminate trees are planted to fertilize the others.

Order XLVII. LAURACEÆ.

Flowers perfect or declinous, regular, small, white or yellow, fragrant; perianth simple; calyx with united sepals, 6-lobed, herbaceous or petaloid, fleshy, usually inferior, rotate or urn-shaped; stamens, at the base or throat of calyx about 9 in number, 3 inner ones extrorse; filaments free or rarely monadelphous; anthers opening by uplifted valves; ovary free, 1-celled; style simple, stout, short; stigma obtuse, discoid, 2-3-lobed. Fruit, a berry, globose or ellipsoid. Leaves alternate, near each other, sometimes nearly opposite or whorled, exstipulate. Aromatic trees or shrubs.

No. of principal genera, about 34; species, 900; mostly tropical.

CINNAMOMUM. B. Flowers perfect or polygamous, receptacle funnel-form; perianth in 6 parts, with 12 stamens, 9 fertile and 3 sterile. Fruit, a berry, attached to the base of the receptacle; pericarp thin. Flowers in panicles, axillary or terminal, simple or in 3- to many-flowered cymes. Tree.

CINNAMOMUM CAMPHORA (Camphor-tree).

1. **C. camphora**, Nees. (Camphor Tree.) Stem 30 to 80 feet high, 1 to 2 feet in diameter; branches spreading horizontally, forming a symmetrical head; bark green, and on the young branches shining. Leaf-buds conical, glabrous, protected by stiff scales; leaves numerous, alternate, 3 to 6 inches long, on slender petioles, slender at each end, acuminate, entire, smooth, bright green, shining above and glaucous underneath, thick, stiff, and evergreen. Flowers small, on slender, spreading pedicels, forming small, spreading cymes, 2 to 3 in a long-stalked, axillary panicle, shorter than the leaves; perianth campanulate, smooth outside, very hairy within, greenish-white; ovary free, 1-celled, 1-ovuled; style slender, as long as the stamens, stigma small; seed egg-shaped, as large as a medium-sized pea, purplish, surrounded at the base by the enlarged tube of the persistent perianth; seed solitary. Flowers in July. This tree sometimes attains a very large diameter; it is reported that one in Japan has reached a circum-

ference of 50 feet. There are about 50 species of this genus, but they do not all yield camphor.

Geography. — The camphor-tree is found native in the edge and fringes of the tropic of Cancer. It is found in China, Japan, Borneo, and the island of Formosa. It also grows in regions of light frost, having been planted in southern Europe and in California.

Etymology. — *Cinnamomum* is derived from the Arabic word *kinamon*, cinnamon. *Camphora* is of Eastern origin; the signification is obscure, but it is supposed to mean "white," in allusion to the color of the gum.

History. — Camphor is a concrete volatile oil, obtained by distilling the wood with water. It yields to the chemist $C_{10}H_{16}O$, melts at 347° Fahrenheit, and boils at 400°; it is soluble in alcohol, and slightly so in water. It is purified by sublimation.

Camphor was not known to the ancients, and it is believed that it was carried into Europe by the Arabs in the sixth century, and by them introduced into the *materia medica*. The camphor of commerce is exported from the island of Formosa and Japan.

Preparation. — The mode of procuring camphor is to cut down the tree and reduce the trunk to fragments, and place the chips in a large still, the head of which is filled with rice-straw, the bottom of the still being filled with water. In the vaporization the camphor passes to the head of the still and crystallizes on the straw; it is picked from the straw, placed in packages, and sent to market. It is afterwards purified by sublimation in glass flasks, when the impurities are left at the bottom. In America the process is somewhat different. The gum is condensed in flat iron pans about sixteen inches square and one inch deep.

Use. — Camphor oil, obtained by draining the crystals in large vats before removing them from the straw and sticks that are taken from the heads of the stills, is used for medicine by the natives of Formosa.

The medicinal properties of the gum are stimulant, diaphoretic, anodyne, and narcotic. It is used to quiet restlessness, as a carminative, and a remedy in typhoid ailments.

It is dissolved in alcohol and in oil, and used as a liniment, and is an important medicine in domestic practice. It is poisonous and very inflammable. It is an insecticide, and used to preserve woollen and fur goods from the ravages of moths; also in herbariums, to prevent the destruction of dried plants by insects.

NOTE. — The Blumea balsamifera, an herbaceous plant of the order Compositæ, found in China, yields to distillation a camphor which is heavier than water, and more volatile than the ordinary camphor of commerce; it is used by the Chinese in medicine and to perfume the celebrated India ink.

The Borneo camphor is obtained from Dryobalanops Camphora, Colebr. This tree is a most magnificent object. The trunk rises to the height of 130 feet without a branch, the base is fortified with gigantic buttresses, and the top crowned with a cluster of branches clothed with large, shining leaves. Flowers showy, abundant, and fragrant. The camphor is obtained by felling the tree, cutting it into lengths, and then splitting it up, when the gum is exposed in layers in the wood, from which it is detached by means of a sharpened stick. The camphor is so pure as to need no process of refining, and is the precious camphor of the East, used in religious ceremonies and funeral rites.

2. C. zeylanicum, Breyn. (Cinnamon.) Second-class tree; trunk 20 to 30 feet high, forming a low, broad head; bark brown. Leaves opposite, oval, 4 to 5 inches long, rounded at the base, dark-green above, paler beneath, prominently 3-nerved. Flowers in large terminal panicles; flowers distant and bractless; petals inconspicuous; perianth divided into 6 oblong, rather blunt, equal lobes, imbricated in 2 rows, 9 stamens and 3 staminodia; anthers short; filaments hairy; ovary superior, 1-celled, with 1 ovule; style shorter than the stamens; stigma 2-lobed. Fruit ovoid, fleshy, half an inch long, smooth; seeds not filling the seed-vessel.

As this tree is propagated from seed, it sports freely, and many varieties are produced, but all constant as to quality of bark.

Geography. — The Cinnamomum zeylanicum is a tropical and subtropical tree, requiring a mean temperature, not below 70°. It grows throughout the East Indian Archipelago. It was formerly largely cultivated in Ceylon, but the cultivation of tea and coffee are rapidly supplanting it. It has been taken to the West Indies, South America, and the isles of the Pacific.

Etymology and History. — *Zeylanicum* is the Latinized form of *Ceylon*. When cinnamon was first introduced into use as a spice or a medicine is not known; it is spoken of in the Bible as one of the substances of the anointing oil used in the installation of the priesthood; it was known in England in the eleventh century, and has always been a favorite spice. When the Dutch came into possession of Ceylon, they limited the supply to sustain the price, and the English did the same. In the beginning of the last century the oil of cinnamon sold in London for its weight in gold.

Cultivation. — The Dutch began the cultivation in Ceylon. The plants are cut back to about six buds; from these the shoots are allowed to grow to the height of ten feet, at which time they will have reached a diameter of about an inch or more; these are cut at the time the sap begins to flow, and divided into lengths a foot long, which are split in halves, and the bark removed. The pieces are laid one inside the other and tied together in small bundles, which are left a few days to dry; the strips are then laid upon a rounded stick, and the cuticle scraped off. The pieces are then allowed to dry and curl up; when dry, the smaller are inserted into the larger, and in that way made into a sort of solid to prevent breaking. These are tied together into bundles weighing about thirty pounds each, covered with gunny cloth, and sent to market.

CINNAMOMUM ZEYLANICUM
(Cinnamon).

Use. — The chips and pealed sticks, with the bark upon the twigs, are distilled, and yield the oil of cinnamon; the leaf also yields an oil. The root yields a peculiar camphor.

Cinnamon is used as a spice in confectionery, cakes, and pastry, cordials and prepared liqueurs, and perfumery. In medicine it is stimulant, aromatic, carminative, and a remedy for nausea.

3. C. cassia, Bl. Stem 20 to 30 feet in height, symmetrically branched. Leaves like those of C. zeylanicum, but more obtuse; when young, flame colored. This tree, when clothed with its flame-colored leaves, interspersed with pure white flowers, is an object of enchanting beauty.

Besides the *C. cassia*, there are three other well-marked species, as follows:—

4. **C. obtusifolium**, Nees, with very blunt leaves.
5. **C. pauciflorum**, Nees, few-flowered.
6. **C. iners**, Reinw., feeble-wooded.

These all produce bark resembling the *Cinnamomum zeylanicum*, but less pungent and delicate. The bark of the cassia is sold in the American market for cinnamon; very little true cinnamon reaches the United States. It is easily distinguished by the druggist and dealer. The taste and flavor are higher, and the bark is not thicker than good writing-paper. The barks of the shops of America are cassia, and not the true cinnamon, and the oils are likewise mostly cassia oils.

Geography. — The cinnamon-producing cassia trees are tropical and subtropical. Found in southern India, Java, Sumatra and Ceylon, eastern Africa, and Australia.

Etymology. — *Cassia* is the Latinized Hebrew word, *ketzioth*: and *gatsa*, to cut, is also given as the root of this word, alluding to the mode of obtaining or harvesting the bark, *i. e.*, a bark which is cut.

Cultivation. — The modes of growing the tree and harvesting the bark are precisely like those for the true cinnamon. Cassia buds of commerce are the unripe fruit of the *C. cassia* and other species.

Use. — The properties of cassia products are about the same as those of true cinnamon, only less intense. The cassia buds are used in confections, and to flavor bitters and cordials. The bark is used in all cases the same as cinnamon; most consumers do not know what the true bark is.

Order XLVIII. **SANTALACEÆ**.

Flowers perfect or polygamous, white, green, yellow, or red, lateral or terminal; perianth single, tubular, variously 5–3-lobed, valvate in the bud; stamens equal and opposite to the perianth lobes, inserted on their middle; filaments short; ovary inferior, 1-celled, mostly 3-ovuled. Leaves opposite or alternate, entire, narrow, usually sessile. Trees, shrubs, or herbs.

No. of genera, 28; species, 220; tropical and temperate regions.

SANTALUM. L. (Sandal-wood Tree.) Perianth bell-shaped, 4–5-parted; lobes spreading; petals 4, spreading; stamens 4; filaments thread-like; anthers ovoid; style conical or cylindrical; stigmas 2 to 4; drupe globose, truncate, or crowned. Leaves alternate. Trees.

1. **S. album**, L. Trunk 20 to 30 feet high, branches numerous, opposite, drooping; bark smooth, grayish-brown; twigs glabrous. Leaves without stipules; petioles slender, half an inch long; blade 1 to 3 inches long, oval or lanceolate, tapering at the base, sharp or blunt at the extremity, entire, smooth both sides, glaucous underneath. Flowers small, numerous, short-stalked, in small pyramidal, erect, terminal, and axillary 3-forked, panicle-shaped cymes, without odor; bracts small; perianth bell-shaped, smooth,

short; segments 4, triangular, sharp, spreading, fleshy, straw colored, changing to purple; stamens 4, opposite the segments of the perianth; filaments short, inserted on the mouth of the perianth, alternating with the erect lobes; anthers short, 2-celled; style thread-like; stigma small, 3-4-lobed, on a level with the anthers. Fruit the size of a pea, bearing a sort of crown, which is the remains of the perianth, nearly black, not hard and bony, with 3 ridges extending half way down from the top.

There are eight species, three of which, besides the S. album, yield an oil and perfume, but of an inferior quality to that of the S. album.

2. **S. myrtifolium**, East Indies.
3. **S. yasi**, Fiji Islands.
4. **S. Freycinetianum**, Sandwich Islands.

Geography. — The geographical range is tropical and subtropical, in Asia, Malaysia, and the islands of the Pacific Ocean. Farther India and China produce most of the wood that reaches England and America.

Etymology. — *Santalum*, the generic name, is the Latinized form of the Persian name, said to come from the Sanskrit *chandana*, sandal, the tree. The word *chand* signifies "shine," hence the shining tree. *Album*, the specific name, is the Latin for white. *Sandal-wood*, the popular name, signifies "shining wood," due to the light color of the sap-wood.

SANTALUM ALBUM
(Sandal-wood Tree).

History. — This tree is a native of the East Indies, and is highly prized on account of its fragrance. The use of sandal-wood dates as far back at least as the fifth century B.C., for the wood is mentioned under its Sanskrit name "chandana" in the *Nirukta*, the earliest extant Vedic commentary. It is still extensively used in India and China, wherever Buddhism prevails, being employed in funeral rites and religious ceremonies; comparatively poor people often spend as much as fifty rupees on sandal-wood for a single cremation. Until the middle of the eighteenth century India was the only source of sandal-wood. When it became known to the western or European nations we have no means of knowing; most likely the wood was introduced into commerce very early, increasing as the means of conveyance improved. The discovery of sandal-wood in the islands of the Pacific led to a considerable trade of a somewhat piratical nature, resulting in difficulties with the natives, often ending in bloodshed, the celebrated missionary John Williams, amongst others, having fallen a victim to an indiscriminate retaliation by the natives on white men visiting the islands. The loss of life in this trade was at one time even greater than in that of whaling, with which it ranked as one of the most adventurous of callings. About the year 1810 as much as four hundred thousand dollars is said to have been received annually for sandal-wood by Kamehameha, King of Hawaii. The trees consequently have become almost extinct in all the well-known islands, except New Caledonia, where the wood is now cultivated.

Preparation. — The tree is propagated by seeds, which must be placed where they are intended to grow, since the seedlings will not bear transplantation, probably on account of deriving their nourishment parasitically by means of tuberous swellings attached to the roots of other plants. The trees are cut down when between eighteen and twenty-five years old, when they have attained their maturity, the trunks being about one foot in diameter. The felling takes place at the end of the year, and the trunk is allowed to remain on the ground for several months, during which time the white ants eat away the valueless sap wood but leave the fragrant heartwood untouched. The heart-wood is then sawn into lengths two to four feet long. These are afterwards more carefully trimmed at the forest depots, and left to dry slowly in a close warehouse for some weeks, by which the odor is improved, and the tendency of the wood to split obviated. An annual auction of the wood takes place, at which merchants from all parts of India congregate. The largest pieces are chiefly exported to China, the small pieces to Arabia; and those of medium size are retained for use in India.

Use. — The fragrance resides in the heart-wood and the root. The heartwood also is valuable in the arts, the sap-wood being too soft for the construction of furniture. Sandal-wood is hard, close-grained, takes a fine polish, and is employed for making musical instruments, toilet-boxes, fans, and fancy articles; drawers and boxes for preserving furs, silks, and woollens from the depredations of insects are also constructed from it, and it is largely used by engravers, for whose purposes it excels the famous box-wood. It is also used by the Chinese and other Asiatic pagans as a perfume to burn before their idols.

Oil of sandal-wood is obtained from the chips, sawdust, and raspings of the wood, by slow distillation. The oil is employed for adulterating the attar of roses, and for compounding medicines. It is an ingredient in the favorite handkerchief extracts, colognes, and fancy soaps of the shops. The seed of the sandal-wood yields a fatty oil by decoction, used for illuminating purposes.

ORDER XLIX. **EUPHORBIACEÆ.** (SPURGE FAMILY.)

Plants of various habits, generally with milky juice. Leaves mostly alternate, stipulate, and often undivided; inflorescence usually compound, sometimes with a calyx-like involucre inclosing several reduced declinous flowers; perianth single, of united sepals, or none, or double (when double it consists mostly of small, distinct petals); stamens 1–1,000; ovary superior, usually 3-celled (rarely 2–many-celled); cells 1–2-ovuled; ovules pendulous from the inner angle, anatropous. Fruit capsular, separating from the axis into cocci (sometimes a drupe); embryo in axis; fleshy or oily endosperm; radicle superior.

No. of genera, 195; species, 3,000. Habitat. tropical and temperate zones.

EUPHORBIA, L. (Spurge.) Flowers monœcious, without floral envelopes, several in a cluster, inclosed in a calyx-like involucre, with 4 to 5 lobes, frequently with 4 to 5 glands; staminate flowers, 9 or more in a cluster, each with 1 stamen and bract; pistillate flower central, with a 3-celled, 3-ovuled ovary on a long pedicel; styles 3 in number, bifid; capsules 3-lobed, with 3 seeds or nuts. Juice milky.

E. Ipecacuanhæ, L. Stems usually short and in clusters, slender, and diffusely, bifurcately branched. Leaves opposite, oblong, linear-lobed, or slit, variable, sessile, heads on thread-like pedicels; seed white, compressed, pitted; root very large, forked, and perennial. Sands of New Jersey and south.

Etymology. — *Euphorbia* is named in honor of Euphorbus, physician to King Juba of Mauritania. *Ipecacuanha* is from the Brazilian *ipecaaquen*, road-side sick-making plant. Spurge is from the Latin *expurgare*, cleanse.

Use.— The E. Ipecacuanhæ is seldom used commercially. True ipecacuanha is obtained from Cephaelis (see Cephaelis of Rubiaceæ, p. 162).

HEVEA, Aub. Calyx 5-toothed, valvate; petals wanting; stamens 5 to 10, united, forming a tube; anthers extrorse; styles 3 in number, short and emarginate; inflorescence a raceme, made up of a number of few-flowered cymes; pistillate flowers above, and the staminate ones below; ovary ovoid, 3-celled; capsule large; exocarp somewhat fleshy, endocarp slightly woody; seeds large, oblong, smooth; testa dry and brittle. Large trees, with 3-foliate leaves.

H. Braziliensis, Mill. (Caoutchouc.) Trunk 50 to 60 feet high; bark rough, grayish-brown; branching near the top; branches and branchlets covered with a rough bark, the branchlets disfigured with tumors or swellings. Leaves on long petioles, branching at the end into three parts, each division terminated with a fleshy evergreen obovate-acuminate leaflet, dark-green above and light beneath; seeds oval and spotted.

There are 18 genera and 44 species of plants from which the gum elastic of commerce is obtained, mostly large trees. The most important of those that yield the largest quantity of the best quality are, first, the Hevea Braziliensis, order Euphorbiaceæ; Ficus elastica, order Urticaceæ; and Castilloa elastica, order Urticaceæ.

At the exposition at Philadelphia in 1876, a gum was exhibited procured from an undescribed plant of the Compositæ, found in Durango, Mexico.

Geography.— The hevea is a native of the region of South America drained by the Amazon and its tributaries. The *ficus* is distributed over southern Asia, middle Africa, and northern Australia. These, as well as the other trees and vines that yield the gum-elastic, are tropical or strictly subtropical plants. The castilloas are found on the Pacific slope in northern South America, and extend into central America. Castilloa elastica was sent from America in 1875 to Kew Gardens, England, and thence to India, and in 1876 Hevea Braziliensis reached India by the same route, and is thriving there and upon the island of Ceylon.

EUPHORBIA IPECACUANHÆ (Spurge).

Etymology.— The meaning of *hevea* is obscure. *Braziliensis*, the specific name, denotes the country where it is indigenous; *elastica* is from the Latin *elasticus*, elastic or pliable. The name *India rubber* has been applied to this substance because early in its history it was used to erase or rub out pencil marks, and as it was brought from the West Indies it was called India rubber. *Gum elastic*, another popular name, means "pliable gum." *Caoutchouc* is the native South American name.

History.— In 1493 the Spaniards, on the second voyage of Columbus, saw the natives of Haiti amusing themselves with elastic balls which they threw and caught. More than a century after that (in 1615) Torquemada published an account of the New World and its products, in which he says the natives smear the juice of a tree over their bodies, and when it coagulates, they **scrape** it off and knead it into masses, and work it into vessels to hold water and other substances. More than a century after the appearance of Torquemada's work, a scientific description was read before the French Academy by La Condamine, and was published in the transactions of the Academy. Oublet, in 1755, described the hevea botanically.

At the beginning of the present century the usefulness of the gum had not been developed; in its native state it was soft and sticky in a high temperature, hard and inelastic when exposed to the cold, and hence of very little practical value. In 1842, three hundred and fifty years after it was first seen by Europeans, experiments showed it to possess the power of uniting with sulphur. When thus combined, it is said to be vulcanized; it is not affected by temperature, and resists solvents. At the will of the manufacturer it may be made very pliable and more elastic than in its natural state, or very hard.

Preparation.—The mode of harvesting this remarkable substance is to wound the tree by piercing the bark to the down-flowing sap, which is a yellowish-white, milky fluid. This, as it flows out, is caught in vessels of clay or bamboo, and carried to the camp, where it is smeared over an instrument resembling a wooden shovel, and held over a smoky fire. When the first smearing is coagulated or hardened, another coat is put on, and the process is repeated until the successive layers have produced a coating of eight inches or more in thickness, when it is slit down the side, slipped off, and hung up to dry and harden. It at first presents a silver-gray color, soon changing to yellow, and finally to a dingy black, as seen in commerce. It has been suggested that the vinegar of wood, or pyroligneous acid of the smoky fire, plays an important part in its coagulation, as the gum prepared in this manner is superior to that of any other preparation.

HEVEA BRAZILIENSIS (CROUTCHOUC).

The South American gum was formerly hardened upon clay moulds in form of bottles and lasts, and when in a proper state the clay was picked out.

In India the mode of collecting the sap destroyed the trees. The government has taken it in hand, and formed plantations, and less destructive modes of harvesting are now practiced.

Use.—The uses to which India rubber has been applied since the vulcanizing process has been put into practice are so numerous that only a few can be mentioned here; in fact it would be difficult to enumerate all the uses to which gum elastic has been put. It enters into the manufacture of every sort of waterproof clothing used for man or beast. The messenger boy encases himself in India rubber, and defies the pelting storm. The soldier in the camp spreads down his rubber blanket, and the cavalryman on the march, in a storm, protects himself and his horse by a covering of gum elastic cloth. It is manufactured into toys for the infant, bands for holding papers and packages together, boots, shoes, hats, beds and cushions, life-belts, garden-hose, door-springs, roller-skates, bumpers for railroad cars, etc.

EUPHORBIACEÆ.

BUXUS, L. Flowers monoecious; calyx of the staminate flowers 3-sepaled; corolla 2-petaled; stamens 4; pistillate flowers with 4 sepals, 3 petals, and 3 pistils; seed-vessels surmounted by 3 beaks, 3-celled, 2 seeds in a cell. Shrubs and small trees. Evergreen.

B. sempervirens, L. (Box. Box-tree.) Stem from 10 to 30 feet in height, and from 3 to 8 inches in diameter, with densely crowded branches. Leaves oval, about half an inch in length, dark-green, glossy; the flowers in axillary clusters, male and female flowers on the same plant, greenish and inconspicuous. April.

Var. **angustifolia**, W., is the same as the above, except that it has narrow leaves.

Var. **suffruticosa**, W., is a dwarf variety, and is used for borders in the garden and parterre.

Geography.—The geographical range is not great. It is a native of southern Europe, western Asia, and especially the countries of the Mediterranean Sea, Syria, Persia, and the regions south of the Black Sea.

Etymology.—*Buxus* and *box* are from the Greek πύξος, box-tree, and our word *box* comes from the same source, because boxes were made of the wood of this tree. *Sempervirens*, the specific name, is due to its evergreen foliage. *Angustifolia* is from the Latin *angustus*, narrow, and *folium*, leaf. *Suffruticosa* is from the Latin, *fruticosa*, bushy.

BUXUS SEMPERVIRENS (BOX).

History.—The box was known to and used by the ancients. The wood was used by them to make combs and musical wind-instruments, and the Romans used it to ornament their planted grounds; and as it bears the knife well, they clipped it into various shapes, making it represent lions and other wild beasts among their shrubbery. It was brought by colonists to North America, and is found in planted grounds throughout the middle States and Virginia; it is not hardy north of 41° north latitude.

Use.—The dwarf form is used for ornamental purposes in planted grounds, and as borderings for gardens. The wood of the box-tree is used by turners and musical-instrument makers, also for mathematical instruments, scales, rulers, sheaves, pestles, screws, inlaying for tables and other cabinet ware, and especially for wood-engraving, for which purpose, on account of its close grain, it is invaluable.

It also enters into the *materia medica.* An extract is prescribed for intermittent fevers and for kidney complaints. Its seeds are purgative, and all parts of the plant are poisonous, especially the leaves. It is said that in those parts of Persia where it grows, camels cannot be kept, as they are sure to feed upon it and poison themselves.

Marts.—The supply of the wood of the box comes from Smyrna, Constantinople, and the islands of the Grecian Archipelago. It is sold by weight. England imports about 80,000 pounds annually.

CROTON, L. Flowers monœcious; calyx of the staminate flowers 1-5-parted, cylindric, and valvate in the bud; corolla, with petals corresponding to the division of the calyx, small; stamens 5 to 20; calyx of the pistillate flower 5-, sometimes 8-parted; corolla minute, sometimes wanting; styles 3-forked, sometimes compound-forked; seed-vessel 3-lobed, with 3 cells; cells 1-seeded. Plant downy and aromatic.

C. tiglium, L. (Croton-oil Plant.) Stem 15 to 20 feet in height, erect. Leaves alternate, acuminate, serrate, smooth, with 2 glands at the base; petioles shorter than the blade. Flowers in terminal racemes. Fruit 3-celled; cells 1-seeded.

There are about 500 species.

CROTON TIGLIUM (Croton-oil).

Geography. — The C. tiglium has a narrow geographical range; it is tropical and subtropical. It is under cultivation in southern Hindustan and the East India islands.

Etymology. — Croton is from the Greek word κροτών, a tick, — because the seed in shape resembles the sheep-tick. *Tiglium*, the specific name, is of obscure signification.

History. — The Croton tiglium was described by Acosta in 1578, and the seeds came in use as a purgative medicine under the name of *grana tiglii* in the seventeenth century, but fell into disuse. In 1812 the English physicians in India revived its use, and a few years afterwards the expressed oil was brought to the notice of medical men.

Chemistry. — The oil is about the consistency of castor oil, and is obtained from the seeds by pressure. Chemists have thus far failed to obtain the substance which gives efficacy to the oil. Its chemical formula is $C_9H_{14}O_2$.

Use. — Croton oil is used sparingly, and administered with great caution in very small doses as a purgative after other remedies have proved ineffective. One drop mixed with olive or some other sweet oil is a full dose; a drop placed upon the tongue produces irritation throughout the entire intestinal canal; a few drops mixed with olive oil applied to the surface of the body cause an almost immediate eruption of the cuticle. It is used for a counter-irritant.

EUPHORBIACEÆ.

MANIHOT, Plum. Flowers monœcious and apetalous; calyx of the staminate flowers 5-parted and imbricated; stamens 10, in two series; filaments free, slender; anthers attached by their backs to the filaments; cells along the sides covered with cracks, running lengthwise; calyx of the pistillate flowers 5-parted, deciduous; receptacle below the base of the ovary sometimes furnished with stamen-like processes; style 3-lobed; ovary 3-valved, 3-celled; ovules 1 in each cell. Fruit in a capsule, 5-berried; berries 2-valved; seeds smooth. Leaves alternate, digitate; stipules small, deciduous; inflorescence a branched raceme, terminal or axillary. Fertile flowers occupy the lower part of the raceme. Root fusiform and fleshy.

M. utilissima, Pohl. (Tapioca. Bitter Cassava.) Stem slender, 5 to 9 feet high, woody below, branched above, smooth, bark whitish. Leaves large, on long purplish foot-stalks, warty near the base, falling off early; stipules narrow, triangular, acute, smooth, falling, blade divided into 3 to 7 oblong, acute, narrow lobes, 2 to 4 inches long, smooth above, glaucous beneath. Flowers monœcious, in axillary or terminal racemes; perianths bell-shaped, deeply cut into 5 acute segments; pistillate flowers, larger and occupying the lower part of the raceme; ovary surrounded at the base by a ring-like receptacle, smooth and purple, 3-celled, an ovule in each cell; style short, surmounted by 3 stigmas. Fruit on short stalks; fruit-vessel half an inch long, globular, glabrous, embossed with protuberances, 6-winged, 3-celled, separating into 3 berry-like divisions, with a single seed in each, one eighth of an inch long, oblong, smooth, and gray. Root fleshy, 3 feet long and 6 to 9 inches in diameter, weighing 10 to 30 pounds, fusiform in shape, and charged with a milky juice.

MANIHOT UTILISSIMA (Tapioca).

There are 30 well-marked varieties of this species grown in Brazil.

Geography. — The home of the tapioca plant is tropical and subtropical South America, whence it has been carried to other warm countries, as southern Asia and western Africa.

Etymology. — The history of the names of the tapioca plants is somewhat obscure; the names have been frequently changed, and there are a number of synonyms. Loudon places the species under the genus Jatropha, and makes the cassava plant *J. manihot;* but Pohl places the two species under manihot, as follows: *Manihot utilissima*, *M. api*, regarding all others as forms or varieties of these two.

Manihot is a name applied by the natives. *Utilissima*, the superlative of the Latin adjective *utilis*, signifies "very useful." *Tapioca* is a name in the Brazilian native dialect for the starch or substance prepared from the root of the plant. *Jatropha* is derived from two Greek words: ἰατρός, physician, and τροφή, food, due to the fact that physicians order it as food for invalids. *Api, Aypi,* or *Aipi,* is an ancient native name, whose meaning is unknown. *Cassava,* like *tapioca,* is a native name, whose signification is obscure.

Cultivation. — The root grows best in loose, dry, well-fertilized, sandy loam. During the first month or six weeks it needs rain or irrigation; after that, it grows well without either, and is the most productive and valuable crop to be made in its region, far exceeding either coffee, sugar, or cotton. One acre will produce about 4,000,000 of pounds when the root is full grown, that is, from a year to eighteen months after planting, which is done by plunging 'slips of the plant into small prepared hillocks of sand, or by burying a section of a stem containing a bud. The M. api varieties mature in about eight months.

Preparation. — The roots of the varieties of the M. api species are not poisonous; they are prepared by roasting, and have the taste of roasted chestnuts. The varieties of the M. utilissima produce the larger roots, which are prepared as follows: The natives peel the roots, then reduce them to pulp by rasping them upon a rude grater made by inserting rough fragments of stones in a piece of bark; the juice is then forced out by allowing it to drain in loosely made baskets, and then baking the pulp in ovens; or it is made into flat cakes baked or dried upon hot stones.

The tapioca brought to market is largely made by reducing the roots by circular rasps or graters turned by water, placed into coarse bags, the juice being removed by a press. The pulp is then subjected to heat in open ovens or on iron plates, and constantly stirred till dry.

Farina is the coarse meal made from the root, universally used by the Brazilians. The fine siftings make the tapioca.

Use. — The fresh-grated pulp, and the juice which is expressed from it, is charged with a substance analogous to prussic acid; yet when the pulp and the juice are subjected to heat, the poisonous character disappears, and the pulp is turned into a wholesome starchy food that sustains life in large and densely populated districts, and is used for puddings for dessert, and for invalid food, all over Europe and the United States.

The root, cut in slices and exposed to heat, is an excellent food for cattle. The natives of Brazil make a fermented liquor of the juice, which is highly intoxicating.

The fresh juice has been administered to cats and dogs, which die with convulsions in about twenty-five minutes. Thirty-six drops administered to a criminal caused death in six minutes.

RICINUS. Tourn. Flowers monœcious, valvate in the bud; sepals narrow and reddish, 3 to 5 in number; no corolla; stamens numerous; filaments repeatedly branching; anthers from the tops of the branches of the filaments; ovary globose, 3-celled, 1 ovule in a cell; style short; stigmas 3-bifid, plumose, and colored; capsule large, 3-celled, covered with blunt, rough spines; seeds large, oily, somewhat in shape of a tick that infests sheep. Annual.

R. communis, L. (Castor-oil Plant.) Stem round, stout, frosted or glaucous, white, shining, purplish, red towards the top, 6 to 12 feet in height in the middle United States, reaching 15 to 20 feet in hot climates, where it is perennial. Leaves alternate, on long petioles, 6 to 10 inches in diameter; subpeltate divided into 7 lanceolate, pointed, serrate segments; staminate flowers on the summit, fertile ones on the lower part of the spike.

Geography. — The geographical range is broad. Though a subtropical plant, it grows well and matures its seeds in a latitude as high as 42°.

Etymology. — *Ricinus*, the generic name, is from the Latin *ricinus*, a sheep tick, which the seed resembles. *Communis* is Latin, signifying "common." The common name, *castor-oil plant*, is believed to have been due to the resemblance of the oil to a liquid contained in little sacks in the groin of the *castor* (*beaver*). Another supposition is that it was called in the West Indies *agno casto*, and that *casto* has been corrupted into *castor*.

History. — The home of this plant is southern Asia. De Candolle believes its home to be eastern Africa. It was known to the ancients, and is spoken of by Dioscorides, who states that the seeds are violently cathartic. When it was first cultivated is not known. It was grown in southern Europe and the countries of the Levant at least 300 years before the Christian era. It is cultivated in Japan, Bengal, eastern and northern Africa, southern Europe, and the United States.

It was brought to America by European colonists. It is largely cultivated in the southwestern United States to produce the oil of commerce. Since the invasion of Kansas by the grasshoppers, the cultivation of the castor-oil plant has been largely practiced.

Preparation. — The mode of procuring the oil is by expression. The coats of the seeds are first removed; the

RICINUS COMMUNIS (Castor-oil).

oil is squeezed out and then boiled with water; the foreign substances are thereby coagulated, and the clear oil floats upon the top, whence it is drawn off and put into suitable casks and vessels for market. The best is made in Italy, where it is manufactured in large quantities and with great care, in the vicinity of Verona, from seed grown in Italy and imported from Africa and India.

Use. — The plant is grown for ornament, and among tropical plants makes a desirable object.

The oil is a mild, speedy, and safe purgative, and for children is the best cathartic medicine in use.

It is used in the preparation of salves, liniments, and ointments, and dissolved in alcohol is largely used in preparations for hair-dressings.

It mixes with alcohol in all proportions, becomes solid at 0° Fahrenheit, and boils at 265°. It is the heaviest of all oils. The purgative principle is not known to chemists.

Order L. URTICACEÆ. (Nettle Family.)

Trees, shrubs, or herbs. Leaves generally alternate and stipulate. Flowers unisexual or polygamous, and axillary; perianth single, sometimes adnate to the ovary in the female flower, or wanting; stamens, when present, usually equal and opposite the lobes of the perianth; ovary 1-celled, 1-carpelled; style simple or 2-cleft; ovule solitary. Fruit an akene or drupe, or many fruits coalescing into a syncarp, with fleshy accrescent torus.

Genera, 108; species, 1,500; habitat, warm and temperate countries.

MORUS, Tourn. (Mulberry.) Flowers unisexual, usually monœcious, but in some plants diœcious or polygamous; male flowers spiked and axillary. Sepals 4, equal, imbricate in æstivation, but expanded in flowering; stamens 4. Pistillate flowers, with a 4-sepaled calyx, in opposite pairs, 1 pair larger, all upright and persistent, and becoming pulpy and juicy. Leaves alternate, simple, exstipulate, deciduous, lobed, rough, 3-4 inches broad. Flowers greenish-white. Fruit the aggregate of the ovary, called "mulberry."

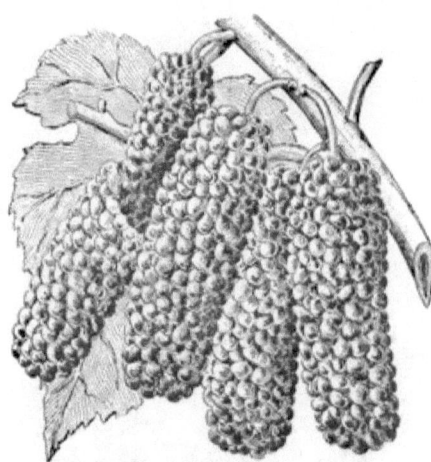

Morus rubra (Red Mulberry).

1. **M. rubra**, L. (Red Mulberry.) Leaves ovate, cordate, serrate, rough above, soft and downy beneath, frequently lobed on young shoots. Flowers often diœcious. Fruit dark-purple (when ripe turning black), cylindrical, half to three quarters of an inch in diameter. Fruit ripe in July. Tree 15 to 30 feet high, but sometimes grows to a much greater height; head 12 to 15 feet in diameter.

2. **M. alba**, L. (White Mulberry.) Leaves obliquely ovate, heart-shaped, acute, serrate, frequently lobed, smooth and shining above. Fruit whitish, soft, sweet, insipid.

Geography. — M. rubra is a common tree in western New England, the southern parts of Upper Canada, the Dakotas, Kansas, and the South. M. alba flourishes in southern Europe and Asia. It is also found in the United States south of the forty-third parallel of latitude in the older parts of the country.

Etymology. — *Morus* is from the Greek μόρον, a mulberry. *Rubra* is Latin for "red," and *alba* is Latin for "white." *Mulberry* is from the Old English *moolberg*, Anglo-Saxon, *morberie*, from the Greek μόρον.

History.— The white mulberry is a native of China and Japan; it was introduced into Europe about the middle of the fifteenth century, and soon became naturalized in Italy and Greece, as well as in Asia Minor and Armenia. Cortes procured its introduction into Mexico in 1522; it was sent to Virginia in 1619 by order of James I., and its cultivation was commanded by law.

Up to the time of the Revolution the British government paid a bounty on raw silk produced in the North American colonies; some of this silk was, in 1772, presented to the queen and to the wives of the proprietors of Pennsylvania by Dr. Franklin, then the agent of Pennsylvania in London. In the same year a colony of Saltzburgers, who were familiar with the mulberry and silk culture, sent to England from their settlement of New Ebenezer, in Georgia, nearly 500 pounds of silk. Shortly afterward the settlement was broken up, and the industry was obliterated during the Revolution. Upon the restoration of peace, efforts were made in many of the states, dating from 1783 in Connecticut to 1866 in California, to revive the industry by legislation. In 1838 a mania for the importation of and speculation in the plants of M. alba, var. multicaulis, of the South Sea Islands, broke out in the United States, hundreds of thousands of dollars having been invested during a single week in Pennsylvania alone. The passion for the cultivation of the tree subsided within a year, but the species remains, and has become naturalized.

On account of the absence of cheap and abundant skilled labor in North America, all stimulations by government aid have failed to establish the mulberry and silk culture there, while China, Japan, western India, Italy, and the Levant, where cheap skilled labor can be found, have become the sources of the world's supply.

Use. — M. rubra, or the red mulberry, has a hard, light, durable wood, much used in ship-building in the southern United States for the light timbers of vessels and boats. The fruit is blackish-red, with an agreeable subacid taste. It is wholesome and refreshing when eaten directly from the tree before the acetous fermentation sets in. The leaves are rough, and are used for feeding silk-worms, but for this purpose are inferior to the leaves of the M. alba.

M. alba, or the white mulberry, is the tree so extensively grown for the culture of silk-worms, and the finest and strongest silks are produced from worms fed upon its leaves. It is a much more rapid grower than the other species of this genus, and is distinguished by its many slender white-barked shoots. Its leaves are slender, and highly charged with a glutinous, milky substance resembling the sap of the hevea. This substance gives strength to the silk produced by the silk-worms fed on its leaves. Trees grown upon high, gravelly, poor soil, in hilly localities produce better silk than trees grown upon generous soil.

ULMUS. (Elm.) **Flowers in lateral** groups, appearing in advance of the leaves, bisexual **and monœcious;** calyx campanulate, 5- or 4-8-lobed, imbricate **in the bud,** persistent; stamens equal in number to segments; style **short, or** wanting; stigmas 2. Fruit a samara, with a membranous **wing.** Leaves alternate, **stipulate,** serrate, feather-**veined,** usually **unequal at** the base, **rough to** the touch. Flowers **reddish-white.**

1. **U. Americana.** L. (American Elm. White Elm. Weeping Elm.) Trunk 2 to 5 feet in diameter near the base, usually dividing into 2 or more branches

within 15 to 50 feet from the ground, which when growing near each other interlace and form graceful curves and arches; the leading branches sometimes reach the height of 120 feet.

2. **U. fulva**, Mx. (Slippery Elm. Red Elm.) Trunk 40 to 60 feet high, 12 to 25 inches in diameter; wood reddish-yellow, tough, inner bark highly charged with mucilage. Branches rough, not forking as in No. 1. Leaves oblong-ovate, acuminate, more nearly equal at base than in No. 1, unevenly serrate, pubescent, rough; buds, before expansion, clothed with soft hairs, large. Flowers at the end of young twigs; calyx downy and sessile. Stamens short, 7. Flowers in April.

3. **U. racemosa**, Thomas. (Corky Elm.) Trunk 50 to 80 feet high, 1¼ to 3 feet in diameter, in habit and appearance like No. 1; branches, when young, slightly pubescent; scales of the buds ciliate; bark of the branches roughened by corky lines. Leaves as in No. 1, with more regular veins. Flowers in racemes.

4. **U. alata**, Mx. (Winged Elm.) Trunk 29 to 30 feet high, 10 to 15 inches in diameter. Branches here and there roughened with corky ridges, or wings, otherwise smooth; scales of the buds and young branchlets glabrous. Leaves unequal at base, downy beneath, oblong-ovate or lanceolate, sharp, thick, small, and doubly serrate; petioles short.

ULMUS CAMPESTRIS (English Elm).

5. **U. campestris**, L. (English Elm.) Trunk 60 to 80 feet high, branching irregularly, branches extending sub-horizontally; bark of a dull lead-color, smooth when young, cracking into irregular strips with age. Leaves rough, doubly serrate, rather small. Flowers rusty-brown; samara oblong, deeply cleft, smooth, and yellow. Flowers in March; seed ripens in May.

The English elm is an important timber-tree, and much attention has been given, especially in England, to its cultivation. No tree sports more freely than the elm, and English writers describe about 20 well-marked varieties to be found in Europe.

There are about a dozen species of the elm, with a large number of varieties. Linnæus went so far as to advance the belief that all the elms are varieties of *one species*.

Geography.—The geographical range of the elm extends, in Europe, from the Mediterranean countries to the middle of European Russia; in America, from the southern banks of the St. Lawrence River to the Gulf of Mexico; and west to the foot-hills of the Rocky Mountains.

Etymology.— *Ulmus* is from the German *ulm*, Latinized into *ulmus*, supposed to come from the base *ol*, grow, and believed to have been applied on account of the rapid growth of this tree. *Elm*, the common name, is from the old European name, *ulm* or *ilm*. The specific names of the American species are

Americana, derived from America, its native country; *fulva*, Latin, yellow, due to the color of wood; *racemosa*, the flowers being borne in racemes; *alata*, from the Latin *ala*, a wing, due to the ridges in the bark of the branches, which are wing-like; *campestris*, Latin *campester*, pertaining to a plain or field, due to the fact that the plant grows in open places.

History. — The elm is mentioned by Pliny. Little is known of its early history, and at what date it began to assume importance as an ornamental tree is not recorded. There are some remarkable specimens mentioned for size and age. One planted by Henry IV. of France was standing in 1790. One in England, planted by Queen Elizabeth, was cut down in 1745; this was more than four feet in diameter.

The elm is the most majestic tree we have in planted grounds; the most desirable are the U. Americana of America, and the U. campestris of Europe.

Use. — The Ulmus Americana is a favorite ornamental tree, on account of its majestic form; it forks into large branches, and when planted in parallel rows along walks and drives, the branches interlace, forming graceful curves and pointed arches. The timber of this tree has not been used much in America, because an abundance of better timber is to be found. The European U. campestris is not only prized as an ornamental tree, but it yields an excellent lumber, which is remarkable for its durability, especially in water.

The American species, U. fulva, has a thick inner bark, which is highly charged with mucilage, and is used in medical practice for throat and bronchial troubles, and for poultices.

HUMULUS, L. Flowers diœcious; calyx of the staminate flowers 5-petaled, with 5 stamens; anthers with 2 pores; pistillate flowers axillary in short strobiliform spikes; bracts leaf-like, laxly imbricated, 2-flowered, each floret sessile at the base of a scale-like involucre, embraced by its involute margin; calyx urceolate, truncate, with small teeth; ovary ovoid, compressed ovule, single and pendulous; strobile membranaceous, made up of the enlarged imbricated bracts and scales. Fruit roundish, egg-shaped, inclosed in the truncated calyx; cotyledons linear, spirally involute. Perennial twining herb, with opposite leaves.

H. lupulus, L. (Common Hop.) Root branching; stem 20 feet long, a number from the same root, twining with the sun, striate or angular, twisted; slender branches near the top, upon which the flowers and fruit appear. Leaves opposite, and lobed near the root, alternate and entire above, scabrous on the upper surface; petioles long; stipules elliptical-lanceolate and wedge-shaped below, the scales sprinkled with resinous dots, which resin produces the peculiar odor and taste of the hop. July.

There are several varieties, which differ very little from each other.

Geography. — The hop grows wild throughout middle Europe and Siberia, as far north as the 62d parallel, the Levant, and Asia Minor, and has been introduced into Egypt. It is also indigenous to southern Japan and to North America, along the foot-hills of the Rocky Mountains, along the upper Arkansas, Missouri, and Mississippi rivers, and near the shores of Lake Winnipeg, also throughout the Atlantic States north of Virginia. The variety in cultivation was brought to northeast America by European colonists.

Etymology. — The name *humulus*, from the Latin, *humus*, earth, was given to this plant because it delights in new rich earth. Its native haunts are in the deep soils of swamps and low grounds. The specific name is a diminutive of *lupus*, the Latin for wolf, "a little wolf." As it grows among the willows it twines about them and chokes them, as the wolf does a flock of sheep. The common name, *hop*, is from the Old English *hoppen*, which signifies "climb," hence the climbing plant.

History. — When this plant was introduced into cultivation we have no means of knowing. It is said to have been brought into England from Flanders, and to have attracted the attention of gardeners and agriculturists first in the reign of Henry VIII. more than 400 years ago. Malted liquors had been formerly called *ale;* but the use of hops made them *beer*. This plant was known to the Romans before the Christian era. Pliny speaks of it as a garden-vegetable.

HUMULUS LUPULUS (Common Hop).

Use. — The hop plays an important part in the manufacture of beer and ale. The plant furnishes the substance known as *lupulin*, in which the virtues of the hop reside in part. It is said to clarify these liquors, and to prevent acidification.

It is aromatic, astringent, tonic, sudorific, and anodyne, and promotes sleep in some cases of insomnia when other remedies fail. A pillow filled with hops is said to have been used by George III. of England in his severe illness in 1787, by direction of his physician, Dr. Willis.

It is an important ingredient in the domestic *materia medica*. It is used also in making yeast. Young shoots were formerly prepared for the table and eaten as a substitute for asparagus. In Sweden a strong cloth is made of the fiber of the stem.

FICUS. Tourn. (Fig Tree.) Flowers monœcious, lining the interior surface of a hollow globular or pear-shaped fleshy receptacle, at the top of which is an opening, which is shut by small scales, staminate flowers above, and the fertile ones beneath: calyx of the staminate flowers 3-parted; stamens 3; pistillate flowers with 2 stigmas and a 5-cleft calyx. Leaves simple, alternate, stipulate, deciduous, and lobed. Fruit in shape of a little bottle, edible.

1. **F. Carica,** L. (Common Fig.) A small, irregularly branched tree, 5 to 20 feet high; or an irregular straggling bush branching near the root, forming an irregular head. Branches cylindrical; bark pale-reddish, young branches

showing scars, from which the leaves and stipules have fallen; twigs downy. Leaves alternate, on long, thick, curved, and downy petioles; blades 4 inches long, nearly as wide, stiff and rough on the upper side, soft, woolly underneath, cordate at base, 3 to 5 palmate, broad, blunt lobes, irregularly and coarsely toothed; stipules large, clasping the whole stem or branch, falling off early. Fruit axillary and solitary, on short stalks, varying from 1 to 3 inches in length, smooth, purplish, turning to a dingy yellow when ripe, soft and fleshy, with numerous seed-like nuts, 1-celled.

2. **F. elastica.** (Indian Fig. India Rubber.) Trunk from 80 to 120 feet in height, and 5 to 10 feet in diameter. Like the F. Bengalensis it produces aerial roots, which it throws to the ground, where they frequently take root in the soil or in the crevices of the rocks among which it delights to grow. A singular feature of this tree is its enormous roots, which lie upon the surface of the ground, coiling and curling over and about the rocks like great serpents. Branches large and irregular; bark gray. Leaves ovoid or elliptical, dark-green, thick, leathery, and regularly veined, and acuminate and glossy. Flowers in axillary panicles, crimson. Fruit small, and not edible.

This tree yields the India rubber of the East Indies.

3. **F. Bengalensis,** W. (Banyan Tree.) This tree is the most remarkable of the genus Ficus, and seems to deserve a place here.

The trunk is from 5 to 9 feet in diameter, and rises to the height of 100 feet. The branches extend horizontally, and send down vertical branches to the ground, which take root, become stems, and branch throwing down other branches, which take root in the same way, until the whole presents the appearance of a vast leafy canopy, supported in some cases by more than fifty pillars, covering a space from 300 to as much as 400 feet in diameter. It is stated by travellers that these strange unions of trees sometimes rise in pyramidal form to the vast height of 150 feet.

Ficus Carica (Common Fig).

The *ficus* genus is large, including some curious and interesting trees. The edible figs are confined entirely to the species *carica* and its numerous varieties.

The varieties of F. Carica are as follows: 1st, growing without cultivation, with small and nearly entire leaves; 2d, under cultivation, with large leaves, deeply cut and lobed, fruit white or dark; 3d, under cultivation, with large leaves, nearly entire.

These three divisions separate into several varieties, each depending upon the size, shape, and color of the fruit.

Geography. — The fig grows well in all subtropical countries; and while it will endure the temperature of 40° north latitude, and with slight protection fruits sparingly, it flourishes best just in the edge of the region of no frost.

Etymology. — The word *ficus* has been derived from the Latin word, *fecundus*, fruitful, on account of its heavy bearing; also from *fag*, a Hebrew name; as well as from the Sanscrit *feg*. *Fig* is a mere translation of the word *ficus*. *Carica* is from Caria, from which town fine figs were exported in ancient times. *Banyan*, Hindu for "merchant," is applied to these trees on account of their frequent use as market places. The other names are self-explanatory.

History. — The fig is spoken of frequently in Scripture. Greek tradition carries back the use of the fig to remote antiquity, leading to the inference that it was used prior to the cereals, and figured as largely in the support of human life as the plantain family. Even so late as after the Exodus we find the Israelites deploring the failure of the fig crop as a great calamity. In the days of Vergil the cultivation of the fig near Rome was carried on to greater perfection than that of the vine.

The home of the fig is believed to be western Asia, perhaps Persia, whence it has worked its way both eastward and westward. In very early times it had spread throughout the Mediterranean basin, along both shores, and as far west as the Canary Islands. In the days of Theophrastus, who lived about 300 years before the beginning of the Christian era, it was a well known fruit. It has spread through all the subtropical countries where European colonization has been established. The tree endures the climate of southern England, but does not fruit well there. Some trees, carried to England in 1525, and planted in the garden of Cardinal Pole, are now in good health; some of them being 50 feet in height and 10 inches in diameter. The fig is supposed to have been taken into England in the first century by Agricola.

Use. — The fig is a favorite dessert fruit, both in a natural and a preserved state. In southwestern Asia, southern Europe, and northern Africa it constitutes the principal food of a large number of people, and is eaten just as it is taken from the tree. When figs are dried in the sun or in a kiln and packed tightly, they keep well, and endure long voyages without damage.

Medicinally, they are laxative, and roasted they are used as a poultice for boils, and applied to the gums to allay inflammation.

Some of the Chinese and Indian fig-trees are the abodes of the lac insect, *Coccus lacca*. The females make their homes upon the ends of the twigs of the fig-tree, and deposit thereon a resinous substance, which enters into the manufacture of sealing-wax, varnish, waterproof hats, etc.

Marts. — The markets of the world are supplied by Turkey and Greece, Spain and Egypt. About 7,000,000 pounds are taken from these countries to the United States annually, and about 20,000,000 pounds are taken to England.

CANNABIS, Tourn. Flowers diœcious; staminate flowers in a raceme; calyx with 5 nearly equal sepals; stamens 5, nodding; pistillate flowers spicate, clustered, single-bracted; calyx urceolate, 1-sepaled, and membranous; ovary globose, 1-celled, inclosing a single ovule; style terminal; seed hanging. An erect annual. Leaves alternate above, opposite below, digitate.

C. sativa, L. (Hemp.) Stem 5 to 18 feet high, roundish, angular, sulcate, and rough-branched. Leaves opposite below, alternate above, digitately divided; leaflets 5 to 7, linear-lanceolate, and toothed, the two at the base smaller and frequently entire, the stipulate foot-stalks 1 to 3 inches long.

URTICACEÆ.

Staminate flowers green, pedunculate, axillary, crowded at the summit of the stem and branches; pistillate flowers sessile, usually in pairs.

Geography. — The geographical range of hemp is very wide. It flourishes throughout the edges of the tropics, and all through the temperate zones, to about the 50th parallel.

Etymology. — *Cannabis* is said to be derived from the Arabic word *cannab*, made up of *can*, a reed, and *ab*, small; hence, a little reed. *Sativa*, the specific name, is from the Latin *sativus*, sown or planted. The meaning of the word *hemp* is obscure.

History. — The home of the hemp is supposed to be Chinese Tartary, northern India, and southwestern Siberia, whence it is supposed to have been carried into Europe by the Scythians about 1,500 years before the Christian era. Herodotus states that the ancient Scythians burnt the seed, and were intoxicated by breathing the fumes.

Hemp is an important crop in China, Chinese Tartary, Japan, Persia, Hindustan, Egypt, southern Africa, most of the states of Europe, and especially Russia. It has been introduced into America, and is cultivated in Canada and the United States. Hemp of a superior quality is raised in southern Russia and Poland. The hemp of the North produces the best fabrics and cordage. The plant when grown in hot countries possesses qualities wholly unknown to it when raised in colder regions. That grown in the tropics and subtropical regions yields substances that are narcotic and intoxicating.

Use. — The fiber of the hemp plant is among the most important of all the textile products. It is made into cloth, and furnishes material for the coarse clothing of a large part of the people of northern Europe and Asia. The bagging-cloths,

CANNABIS SATIVA (Hemp).

and the sails and cordage for vessels all over the world, are made of hemp. The Russians obtain an oil from hemp seeds, which they use in their culinary preparations and to mix their paints, and in the manufacture of soft soap. The seed is also fed to caged birds, and is said to change the color of their plumage from red to black.

The leaf, when grown in warm climates, is smoked, and produces a narcotic and intoxicating effect upon the smoker, which is said to alleviate pain, increase the appetite, and give rise to mental cheerfulness. It also produces violent coughing and spitting of blood. From the whole plant also exudes a resinous substance, which, when smoked, produces intoxication; when taken internally in small doses, it produces furor and imparts wonderful strength; when taken in larger quantities it produces hilarity and stimulates the appetite, but the patient finally becomes insensible, and his limbs will remain in any position they may be placed. After a time the person recovers without any apparent ill effects to either mind or body. The resinous substance is the celebrated hashish of the Arabs. *Bhang* is a narcotic intoxicating drug, prepared from the leaves and seed-vessels of hemp. It is a favorite drink in the East Indies.

Those who frequent places in Egypt where hashish is sold are of the lowest class, and the term has come to convey the idea of disorderly or riotous people. The plural, hashshasheen, is believed to be the origin of our word "assassin," because the Arabs in the time of the Crusades used the drug to produce insensibility in their victims.

Order LI. **JUGLANDACEÆ.**

Flowers monœcious, staminate ones small, and often in hanging catkins; perianth single, attached to the inner face of a bract which is 6-lobed, sometimes 2–3-lobed; stamens 3 to 40, inserted at the base of the bract; filaments very short, free or coherent at the base; pistillate flowers terminal, solitary or few, and clustered; calyx tube ovoid, limb 4-toothed; styles 2, very short; stigmas 2, elongated, recurved. Fruit drupaceous, rarely nut-like, containing a single nut; epicarp fleshy, fibrous within, indehiscent; nut woody, rugose, and irregularly grooved lengthwise. Trees or shrubs, with watery, resinous, aromatic juice. Leaves odd-pinnate, exstipulate; staminate catkins from last year's branches, or at the base of the younger branches.

No. of genera, 5; species, 3; temperate regions, and mountains in the tropics.

JUGLANS. L. Bract of the pistillate flower with its bractlets closely adhering to the ovary, irregularly toothed at the perianth limb so as to resemble an outer perianth; exocarp of the drupe closely adhering to the wrinkled endocarp, or at last coming away irregularly. Staminate and pistillate flowers separate, but upon the same plant; staminate flowers in solitary, drooping catkins; calyx composed of 5 to 6 scale-like sepals; stamens 18 to 36, usually about 20. Pistillate flowers, 1 to 5 in a group, terminal on the new wood; calyx ovate, and 4-toothed; petals 4 in number; styles short, 2 in number; stigmas 2. Fruit a drupe, nut rugose, hard, globose, a little compressed laterally, 2-valved. Covering of the nut a fleshy husk, indehiscent. Kernel large, oily, sweet.

JUGLANS CINEREA (Butternut).

1. **J. cinerea,** L. (Butternut.) Stem from 10 to 30 feet in height, irregularly branched, and from 6 inches to 1 foot in diameter. Leaf made up of 7 or 8 pairs of leaflets, and a terminal one; leaflets rounded at the base, elliptical-lanceolate, serrate, and pubescent underneath; aments cylindrical. Flowers greenish, appearing in April and May. Fruit cylindrical, 2 inches long and

1 inch in diameter, ending with an acuminate tip. Exocarp or outer shell like that of J. nigra, but thinner; nut corrugated in the direction of the longer axis; kernel sweet and buttery.

Geography. — Geographical range between the parallels of 48° and 36° north latitude, from the Atlantic to the Mississippi River.

Etymology. — *Juglans* is derived from the words *Jovis*, Jupiter, and *glans*, Latin for "nut," — that is, nut or fruit for Jupiter or the gods, on account of its delicate quality as a food. The common name, *butternut*, is due to the delicate buttery taste of the kernels. The specific name, *cinerea*, comes from the Latin *cinereus*, ash-colored, due to the ashy-gray color of the bark.

History. — The home of the butternut is northeastern North America. It is sometimes planted for ornament, but delights in rocky places, and loves the hills; has been planted in middle Europe and in England, and grows and fruits well there.

Use. — The butternut is a favorite dessert nut. A delicate salad oil is obtained from the kernels by expression, and the fruit in an unripe state is used for pickling.

The wood has a coarse grain, but takes a good polish, and is used for cabinet ware and for wainscoting. The chips are also used to manufacture beer.

The outer shell of the fruit furnishes a dye, which is largely used to color home-made fabrics in the western part of Kentucky and Tennessee.

A decoction of the bark is boiled down to a mass and made into pills, which are administered as a gentle cathartic.

JUGLANS NIGRA (Black Walnut).

2 **J. nigra**, L. (Black Walnut. Black-wooded Walnut Tree.) Stem straight, 40 to 80 feet high, and 1 to 4 feet in diameter; branches crooked and straggling, forming an open and picturesque head, from 15 to 30 feet in diameter. Leaves compound, with 7 to 8 pairs of leaflets and a terminal one, odd one frequently wanting; leaflets slender, cordate, acuminate, unequal at the base, on short petioles. Flowers greenish, appearing in May in the northern limits, and in April further south. Fruit ripens in October.

Geography. — The black walnut is indigenous in southwestern New York, and further south to the Gulf of Mexico, and west some distance beyond the Mississippi River. It has been introduced into the eastern middle States and southern New England, where it grows and fruits well.

Etymology. — *Nigra* is from the Latin *niger*, black, from the color of the wood. The word *walnut* was given in England, and means "a strange or foreign nut," signifying that it came from abroad, from Anglo-Saxon *wealh*, strange, and *nut*.

History — The black walnut has been introduced into England and southern Europe by seeds from America. It grows well in England, but is becoming very scarce in the United States.

Use. — The wood of the black walnut is hard, very dark, and takes a good polish; it is strong and tough, and is largely used in the manufacture of cabi net ware, especially tables, bedsteads, bureaus, and chairs; also in joining, wainscoting, and for floors, panels, and doors. In the southwest it is used largely for lumber and for fencing. It takes the place of mahogany with us in the manufacture of furniture and cabinet ware.

The fruit is highly esteemed as a dessert, and is used when in an unripe state for pickles and catsups. A good salad oil is expressed from the kernel, and the shells are used for dyeing purposes.

3. **J regia**, L. (English Walnut.) Stem 20 to 40 feet in height, and from 10 to 20 inches in diameter; branches rather straight, head symmetrical; leaves consisting of 3 to 5 pairs of leaflets, increasing in size towards the top, terminating with a single one; leaflets ovate, acute, margins wavy, on short petioles; catkins oblong, 2 to 2½ inches long, peduncle short. Fruit subglobose, mucronate, about 2 inches in diameter; exocarp leathery, smooth, ovoid; shell or endocarp wrinkled; kernel large and sweet. Flowers in early summer; nut matures in October.

Like other trees grown from the seed it has many varieties, the most important of which are: —

Var. **maxima** (large fruited). Nut twice the size of the J. regia, but perishable.

Var. **tenera** (tender-shelled). The shell is thin, so that small birds pierce it before it is ripe; very delicate to the taste, but not a prolific bearer.

Var. **serotina**. Endures the frost, and can be cultivated in higher latitudes than the J. regia.

There are many other forms, differing from the species only in the size or quality of the fruit.

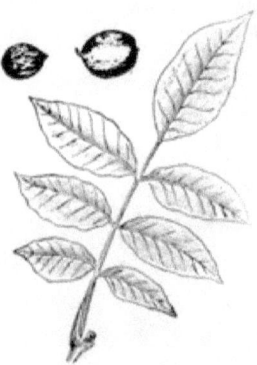

JUGLANS REGIA
(English Walnut).

Geography. — Though a native of the southern parts of the temperate zone, the English walnut fruits in a latitude of 45° in Europe and Asia, and grows well in the Atlantic States of North America, but does not fruit freely north of Virginia. It is extensively cultivated in southern California. Its home is a region below 40°, extending from the country southeast of the Black Sea eastward to Japan.

Etymology. — The specific name, *regia*, is the Latin for "royal" or "kingly," due to the high esteem of its quality.

History. — Food plants necessarily attracted the attention of man in the earliest period of his existence; and nuts, on account of their edible character in an uncooked state, have always been favorites. At the present day nuts form an important part of the food of the laboring classes, and with them the walnut holds high rank. London, in his work on trees, states that between Heidelberg and Darmstadt the walnut is the principal tree, not only for the fruit, but for shade. In that region when a young farmer desires to marry, he is obliged to furnish proof to the intended bride's father that he has planted with his own hands a stated number of walnut trees, which are already in an advanced stage of growth.

Use. — In the Levant, where the English walnut reaches perfection, it constitutes a large portion of the food of the masses. It is highly prized in

Europe and America for a dessert nut. A table oil is expressed from it; and in a green state it makes an excellent pickle. It is also used to flavor sauces, and is an important article in the celebrated walnut sauce. The wood takes a good polish, has a brown color, is richly veined, and is highly prized by cabinet makers, rivalling mahogany. The plain kind is used for gunstocks. The root is gnarled, and when sawed into thin slices makes valuable veneering.

HICORIA, Raf. (Hickory Nut.) Flowers unisexual; both staminate and pistillate flowers and leaf developed from the same bud; the pistillate flowers terminal, few in number; bract small or none; bractlets none; staminate flowers in pendulous catkins in the axils of the lower leaves; 3 on a peduncle, their perianth irregularly 2–3-lobed; stamens 3 to 10; pistillate flowers shortly spicate, few, with a minute bract or none; perianth enfolding and adhering to the ovary, with a free 4-parted tip; stigma sessile upon the ovary, 2–4-lobed. Husk or outer shell of fruit fleshy, 4-valved nut, somewhat 4-sided, smooth or slightly wrinkled. Flowers greenish.

1. **H. ovata**, Mill., Britton. (Carya Alba, Nutt.) (Shell-bark. Hickory nut. White Walnut.) Stem from 40 to 60 feet high, and from 1 to 2 feet in diameter, rather regularly branched, forming a symmetrical head. Bark gray, and falling in strips. Leaves composed of two pairs of leaflets and a terminal one, lateral ones sessile, terminal one petioled, all oblanceolate, the lower pair smaller, subacuminate, sharply serrulate, downy beneath. Fruit flattish, globose, with four grooves extending along the length of the husk, which, when ripe, separate into four sections, freeing itself from the nut, which is marked by four seams or ridges extending lengthwise; shell thin; kernel delicate. Ripens in November; flowers in April and May.

HICORIA OVATA (Hickory Nut).

2. **H. sulcata**, Britton. (Carya sulcata, Nutt.) (Thick Shell-bark.) This species differs from the last in the size of the fruit, which is much larger than that of the C. alba, and the leaf has from 3 to 4 pairs, and the nut has an acuminate tip; in other respects it is well described in C. ovata. A larger tree than C. alba.

Geography.—The geographical range of these last species is the northern and middle States, from the Atlantic to the Mississippi River, and it bears well in corresponding latitudes in Europe.

Etymology.—The generic name, *hicoria*, is of unknown origin, supposed to be an aboriginal name of the tree or its fruit, probably the latter. The old generic name, *carya*, is from the Greek word κάρυον, the walnut tree, said to have been given in honor of Carya, daughter of Dion, king of Laconia, who, according to the Greek myth, was changed by Bacchus into that tree. The specific name, *ovata*, is from the Latin *ovum*, an egg, referring to the shape of a plane of the fruit parallel to the axis of growth. *Sulcata*, from the Latin *sulcus*, a furrow, derives its name from markings on the fruit.

History. — The home of these species of hicoria is North America; they have been introduced by seed into Europe, where they grow and fruit well. Seeds were first planted in Europe in 1629. These are the most important of the hickories. There are several other species whose wood is similar, but which bear inferior fruit.

Use. — The fruit of the H. ovata and H. sulcata is very delicate, and is valued for a table dessert. The shell is full, and the kernel sweet. Both yield an excellent salad oil, which is obtained by expression.

The wood splits easily, but is hard and tough, and is used largely in the manufacture of agricultural instruments, axe and hammer handles, and hubs and spokes of carriage-wheels. It is a very valuable material for fuel.

3. **H. olivæformis**, Nutt., Britt. (Pecan Nut.) Stem 80 to 90 feet high, and from 1 to 2 feet in diameter; bark rough and shaggy. Leaves with slender petioles; leaflets in 6 to 7 pairs, and a terminal one, lanceolate-falcate, acuminate, and sharply serrate, on short petioles. Flowers greenish. Fruit oblong, 4-angled, with distinct valves, the green husk inclosing an olive-shaped nut with a thin shell; kernel fills the entire shell, and possesses a delicate, pleasant flavor.

Flowers appear in May. Fruit ripe, October and November.

Geography. — The Hicoria olivæformis is indigenous to southern North America, and delights in a damp rich soil; it will grow and fruit in the latitudes south of 40°, and north of 20°.

Etymology. — The specific name was suggested by its shape: *olivæformis*, olive-shaped.

HICORIA OLIVÆFORMIS (Pecan Nut).

History. — Nuttall, the English botanist, first described this tree, whose fruit Pursh sent to Europe. It was planted in Prussia, and found its way into England in 1766.

Use. — The pecan nut is a favorite dessert nut, and has become an important article of commerce. It is shipped to the West Indies, also to Europe. In its native forests it is highly valued as a mast upon which droves of swine fatten, which are allowed to run at large while the nuts are falling.

The wood of this tree is white, tough, and durable, and used largely in the manufacture of agricultural implements, and, like other species of the genus, makes excellent fuel. It is characteristic of North America, no wood equally tough, elastic, and suitable for these purposes being known in Europe.

Order LII. CUPULIFERÆ.

Flowers monœcious; staminate flowers in pendulous, bractless catkins, on last year's branches, or at the base of this year's branches; calyx usually 5, occasionally 5-12-parted; stamens 2-20; anthers 2-celled. Pistillate flowers solitary or clustered, terminating few-leaved branches; calyx attached to the ovary, 6-toothed or wanting; ovary 2-3-celled; ovules 1-2 in a cell, pendulous. Fruit, a nut, 1-seeded by abortion, 1-3 in a cup or shell. Leaves alternate, pinnately veined, simple, falling or persistent, stipulate, with an involucre of accurrent woody bracts.

Number of genera, 10, species, 400; chiefly in north temperate regions and in tropical mountains.

CASTANEA. Tourn. (Chestnut.) Male flowers in clusters of long, slender, cylindrical, erect aments; calyx 6-parted; stamens 5 to 15 in number. Fertile flowers in 3's, surrounded by a 4-lobed involucre, which when ripe is leathery and beset with weak prickles about half an inch long; calyx 5-6-lobed, the tube adhering to the 3-6-celled ovary; Number of stigmas equal to the number of cells. Involucre 4-valved; nuts usually 3 in number, sometimes 1; when the involucre contains 1 nut, it is top-shaped; when there are 2, the nuts are plane on one side and convex on the other; when there are 3, the outside ones are plano-convex, and the middle one flattened into a wedge shape. The nuts are from three quarters to an inch in length, and sometimes as wide as long. Covering shell thin and horny. Leaves simple.

1. **C. vesca.** Gaert. (Chestnut.) Trunk from 50 to 70 feet in height, ranging from 1 to 5 feet in diameter, throwing out branches nearly horizontal, which extend 20 to 30 feet, sometimes forming a head 50 feet in diameter. Leaf oblong-lanceolate or oval, mucronately serrate, glabrous on both sides. Flowers yellowish, appearing in May. Fruit in October.

Var. **Americana** is the American chestnut, and differs from the European chestnut only in bearing a smaller and more delicate fruit. The tree grows to the height of 80 feet, and when in the forest reaches the height of 40 or 50 feet without a branch, but when standing alone branches low.

2. **C. pumila,** Mx. Stem 6 to 15 feet in height, branching low and profuse, shrub-like in appearance. Leaf oblong, ovate, or obovate, mucronately serrate, hoary, tomentose on the under side, 3 to 5 inches long and about 2 inches broad, smooth above, acute at the apex, and obtuse at the base; petioles long. Flowers axillary; nut solitary, small, and very sweet.

CASTANEA VESCA (Chestnut).

In Europe great efforts have been made to improve the chestnut; and as the trees are produced from seed, the varieties are numerous. The American tree is believed to be identical with the European Castanea vulgaris, Lam. In America no efforts have been put forth to improve the fruit, hence no varieties have arisen.

Out of about twenty varieties grown in England, four are considered as greatly improved. In France also much attention has been given to the improvement of the chestnut, and many varieties have been produced, differing only in the size and quality of the fruit.

Several species of chestnuts have been found in eastern Asia recently, which were formerly classed under the oaks, and there has been a species discovered in Nepaul, northern India, recently.

Geography. — The geographical range of the chestnut is very broad. It grows well and is indigenous all along the eastern coast of North America, from 40° to 43° north latitude, extending west to eastern Kentucky and Tennessee. The C. pumila is found between 30° and 40° north latitude, — from southern Pennsylvania to northern Georgia. The C. vesca grows well throughout the middle and southern counties of England, in all the countries of middle and southern Europe and northern Africa, and in the countries of the Levant and southern and eastern Asia, wherever it has been planted.

Etymology. — *Castanea* was named for *Castane*, a city of Thessaly, famous for chestnuts. *Vesca* is from the Latin *vescor*, eat, referring to the edible character of the fruit. The common name, *chestnut*, is due to the fruit or nut being inclosed in a box or chest.

History. — The home of the chestnut is not exactly defined. De Candolle says it forms natural forests from the Caspian Sea westward to Portugal. It has also been stated that its home is the country between the Black and Mediterranean seas, and that it was carried west after the Roman conquests. Pickering says it is native to China, and Thunberg saw it near Jeddo, in Japan. It is believed that the Emperor Tiberius took it to Italy from Asia Minor, and that it thence spread all over southern Europe. It is also indigenous in North America.

There are some very remarkable chestnut-trees in the world, some of which have reached a greater diameter of trunk than any other tree. The most noted is the celebrated Mount Ætna chestnut, under which a hundred mounted horsemen took shelter. The enormous size of this tree has led to the belief that it is the union of a group of trees that stood near each other in their youth. M. Jules Houel, a French scientist, nearly a hundred years since, made a journey to measure and make a drawing of it. He found it to be 160 feet in circumference, and on the closest and most careful examination could find no evidence that it is not a single trunk. There are other large trees in the neighborhood, measuring from 36 to 40 feet in diameter.

There are three large chestnut-trees in the southern suburb of the city of Yonkers (just outside the northern limits of New York City.) Two of these measure respectively 24 feet 10 inches, and 19 feet 6 inches in circumference. These are of the American variety, and are in an advanced state of decay. By a calculation from the best known data, the largest of these trees is about 210 years old.

Use. — The chestnut is a favorite nut in many parts of Europe; and in the countries of the Levant it constitutes a very important article of food. It is roasted or boiled, and eaten with salt. It is also eaten raw. A flour is made of the nuts in a dried state, which is used for various culinary purposes, principally for griddle cakes.

The wood is not hard, but is very durable, and takes a high polish. It is used for cabinet work, bedsteads, tables, etc., and by reason of its durable character is very highly valued for fencing material. It is also a strong and valuable timber for building purposes.

The nut forms an important article of commerce. The best European nuts come from Spain.

QUERCUS, L. (Oak.) Staminate flowers in groups of slender hanging catkins; stamens 5 or more, surrounded by sepal-like **bracts**, 6 to 8 in number. Pistillate flowers axillary and erect; ovary surrounded by an adnate calyx, the limb of which is toothed; style short; stigma 3-lobed; ovary 3-celled, rarely 4–5, containing 6 ovules, 5 of which are abortive. Fruit oblong, somewhat in the form of a modern musket-cartridge, with the base inserted in a cup, which is clothed with imbricated scales. Flowers greenish, appearing in regions of frost during the month of May. Leaves simple, alternate, stipulate, deciduous, but persistent. A few evergreen in the southern fringe of the north temperate zone.

1. **Q. alba**, L. (White Oak.) Trunk 60 to 80 feet high, 4 to 5 feet in diameter; bark grayish-white; much-branched. Leaves 3 to 5 inches long, sinuate-lobed, in opposite pairs, 1 to 4 pairs of lobes, with a terminal one; lobes coarsely and irregularly toothed, pubescent underneath; acorn on short peduncles, large and sweet, edible.

Var. **pinnatifida**, Mx. Like Q. alba, except the leaves, which have 3 to 4 pairs of well marked lobes.

Var. **repanda**, Mx Leaves with a wavy margin.

Geography.—The oak thrives best in the temperate zones, above 35°, and is found in a zone between 30° and 60° quite around the globe. It is found in the mountain-top as well as in the valley and the plain below, and is indigenous in the above-named parallels in both hemispheres. The white oak is common in the United States and Canada.

Etymology.— *Quercus*, the generic name, is derived from Latin *quercus*, an oak. The popular name of the oak among the Celts was *drew*, from which the word Druid was derived, signifying "priest of the oak." *Alba*, the specific name of this species, is from the Latin *alba*, white, and is due to the grayish-white bark of the trunk. Oak comes from the Anglo-Saxon name of the tree, &c.

History.—The oak is famous in all ancient writings in which trees are mentioned. Among the Gauls it was held sacred. Oak groves were the abodes of priests, and no religious ceremony was complete without oak-boughs or oak-leaves. The Greeks and Romans also dedicated the oak to their gods; and the Roman peasants initiated the harvest by a festival, in which their heads were adorned with wreaths woven with the leaves of the oak. It was upon the oak that the Druid priests found the mistletoe, which figured so largely in their religious ceremonies.

Many oaks are noted for historical events. Less than a hundred years ago the oak was still standing in the New Forest against which the arrow glanced that killed William Rufus. The Royal Oak at Boscobel concealed the person of Charles II. after the disastrous battle at Worcester. The oak at Torwood, at the place where Wallace convened his followers, still stands. Alfred's Oak, at Oxford, which was in existence when the university was founded, may still be seen. And in our own country an oak in the city of Hartford, Conn., concealed the charter of that colony, and was known afterwards as the Charter Oak.

Abraham's Oak (Q pseudococcifera), near Hebron, in Palestine, is many hundred years old; it measures twenty-three feet in circumference, and its branches extend forty-five feet from the stem, forming a head ninety feet in diameter

Use. — The white oak furnishes a hard, durable timber for frames of buildings, axles of carriages, floors, tables, chairs, handles for axes and hammers, wainscoting, panelling, church furniture, shipbuilding, and mill-gearing. The bark is highly charged with tannin, and is a valuable material in the manufacture of leather. The fruit of the Q. alba, which is the sweetest of all the species, is excellent for fattening swine; the pork thus fatted is said to produce the most delicious bacon. The delicate flavor of the Virginia hams is said to be due to the feeding of swine upon acorns.

2. **Q. robur,** L. **Q. pedunculata,** Willd. (British Oak.) Trunk 50 feet high and upward to 100 feet; when standing in open grounds it branches low, spreading out so as to form a head whose diameter is greater than the height; branches crooked, gnarled, and very large; bark gray and rough; leaves on short petioles, blade oblong, made up of 3 to 5 unequal pairs of lobes, and a terminal one; sinuses narrow, lobes rounded. Fruit sessile or on long peduncles, oblong, elongated, brown, buried to one fourth of its length in the hemispherical cup, which is clothed with rough imbricated scales. Flowers greenish-white, appearing in April. Fruit ripens in September.

QUERCUS ROBUR (British Oak).

Like all trees that propagate themselves by means of their seeds, the Quercus robur has run into a great variety of forms, of which the following are the most prominent, and may be found growing in the public grounds in Washington:—

Var. **sessiliflora.**
Var. **pubescens,** Lodd. Leaves downy beneath.
Var. **fastigiata,** Lodd. Branches compact and upright.
Var. **pendula,** Lodd. Branches decidedly pendulous, or weeping.
Var. **heterophylla,** Loudon. Leaves varying greatly in size and form; some lobed, others lanceolate and entire.
Var. **foliis variegatis,** Lodd. Leaves variegated with white and red streaks. A beautiful specimen is growing in the public grounds at Washington.
Var. **purpurea,** Lodd. Foot-stalks of the leaves tinged with purple, and the leaves when young entirely purple.

There are many other varieties, but less striking.

Geography. — The British oak is indigenous to the continent of Europe, and most likely to England. It grows in the south of Europe, the Levant, and northern Africa.

Etymology. — The many names by which the British oak has been known to botanists constitute not the least of its features. *Robur* is from the Latin word *robur*, and indicates strength; it is also an old name for the oak tree. *Pedunculata* refers to the long foot-stalks of the fruit. The variety names, with the exception of *heterophylla*, are all derived from the Latin, as follows: *sessiliflora*, sessile-flowered; *pubescens*, covered with down; *fastigiata*, sloping to a point; *pendula*, hanging down; *foliis variegatis*, variegated leaves; *purpurea*, purple-colored. *Heterophylla* is from the Greek ἕτερος, different, and φύλλον, a leaf, hence varying leaves.

History. — The celebrated character of the British oak seems to call for a special notice. Some of the most remarkable specimens of this tree are interesting for their age and size. The Framlingham oak, used in the construction of the "Royal Sovereign," squared four feet nine inches, and yielded four square beams, each forty-four feet in length. An oak felled at Whitney Park, Shropshire, in England, in 1697, was nine feet in diameter without the bark, and yielded from the trunk alone twenty-eight tons of timber. The head of this great tree was one hundred and forty-four feet in diameter. Another English oak, in Holt Forest, Hampshire, measured, seven feet from the ground, thirty-four feet in circumference. Another, at Newbury, measured forty-five feet around. Still another, in the vale of Gloucester, was fifty-four feet in circumference; and one in Dorsetshire gave a girth of sixty-eight feet.

Use. — The wood of the British oak is hard and tough, and resists great force without fracture; these qualities make it rank very high as a material for shipbuilding. Its acorns formerly took a high place in European history as food.

The oak forests of central Europe furnished food for swine and other domestic animals, and the people themselves subsisted largely upon acorns. It was regarded as one of William the Conqueror's most oppressive acts that he deprived the people of England of the use of the oak forests, where they had been accustomed to collect the acorns for their swine.

3. **Q. bicolor**, Willd. (Silver-leaved Oak. Swamp White Oak.) Trunk 60 to 70 feet high, 4 feet in diameter; bark scaly, and greenish-white. Leaves nearly sessile, downy, white underneath, bright-green above; obovate, coarsely and bluntly toothed, entire near the base. Acorns in pairs, peduncles longer than the petioles; nut long, dark-brown; cup shallow, and fringed with short, slender, thread-like processes.

Geography. — It is well distributed throughout the eastern and northeastern United States.

Etymology. — *Bicolor*, the specific name of this oak, is from the Latin word *bicolor*, two colors, and refers to the contrast in the colors of the two sides of the leaf, one of which is a bright-green, and the other a silvery white. *Silver-leaved* arises from the color of the under side of the leaf. The name *swamp white oak* is due to its fondness for wet ground.

Use. — The lumber is valuable for building purposes; it is hard, durable, and takes a good polish; it also makes excellent fuel. The bark is highly charged with tannin, but is thin, and is not profitably obtained for markets where thicker bark is available.

4. **Q. coccinea**, Wang. (Scarlet Oak.) Trunk 60 to 80 feet in height, sometimes 4 feet in diameter; bark thick, gray outside and red within.

Leaves divided into 3 to 4 pairs of lobes, much like the leaves of the Q. palustris; petioles longer than in Q. rubra, deep-green, shining on both sides; lobes cut, toothed and acute, turning scarlet with the early frosts. Acorn ovate, half buried in the scaly top-shaped cup.

Geography. — Indigenous to southeastern North America; the northern limit is southern New England; common in the middle and southern Atlantic States.

Etymology. — *Coccinea* is from the Latin *coccineus*, scarlet, and has reference to the color of the leaves after frost.

Use. — The wood is largely used for making barrels, and the bark is a favorite with tanners. The tree is also used in planted grounds.

Var. **tinctoria**, Gray. (Black Oak. Yellow-barked Oak. Dyers' Oak.) Trunk 70 to 100 feet in height, and 3 to 4 feet in diameter; bark furrowed, dark without, and yellow within. Leaves downy beneath, obovate, oblong, broad-lobed, broadest near the end, sinuses not deep, lobes coarsely toothed, teeth pointed. Acorn flat, globose, half buried in the flat, thick cup.

Geography. — The Q. tinctoria is a native of eastern North America, and is widely distributed throughout the eastern and middle States.

Etymology. — *Tinctoria* is from the Latin *tinctor*, a dyer, because the bark furnishes a dye. The popular name, *black oak*, is due to the color of the bark.

Use. — The wood of this tree is sometimes used for cooperage and construction, and is excellent fuel. The bark is largely used for dyeing; it yields the querciton, which is much used in calico printing, to give the yellow color to cotton fabrics. It is also used for tanning.

5. **Q. falcata**, Mx. (Spanish Oak. Sickle-leaved Oak. Downy-leaved Oak.) Trunk 60 to 70 feet high, 4 to 5 feet in diameter. Bark thick, black, and furrowed. Leaves on long petioles, blade 6 inches long, downy beneath, obtuse at the base; in the northern limits of the tree the leaves take on a slender entire form, widening towards the upper end, where they terminate in three lobes; further south the usual form of the leaf is in 1 to 2 pairs of pointed, mucronate, scythe-like lobes, entire or irregularly and coarsely toothed sinuses, deep and wide. Acorn globular, small; cup shallow.

Geography. — It is native from southern New Jersey (where it seldom attains a height greater than 40 feet) to the shores of the Gulf of Mexico, where it grows to its full size, 60 to 80 feet. It is a subtropical tree, and flourishes best below the parallel of 35°.

Etymology. — *Falcata*, the specific name of this tree, is from the Latin word *falcatus*, scythe-like, from the supposed resemblance of the lobes of the leaves to the shape of a scythe. *Spanish oak*, the common name, is obscure in its origin and meaning.

Use. — The Spanish oak is a beautiful, well-formed tree, used for ornamental purposes. The wood is an excellent fuel. The bark is highly charged with tannin, and extensively used in the manufacture of leather.

6. **Q. macrocarpa**, Mx. (Moss-cup Oak. Burr Oak. Mossy-cup White Oak.) Trunk 50 to 70 feet high, and 2 to 3 feet in diameter, branching into a symmetrical head. Bark grayish, rough, the bark on the branches roughened by longitudinal corky ridges. Leaves downy beneath, lyrate, larger than those of any other species, frequently a foot long and 8 inches broad, made up of 3 to 5 pairs of lobes and a terminal one, the terminal lobe greatly expanded and notched. Fruit larger than the fruit of any other species in America.

Acorn subglobular, two thirds inclosed in the cup, the orifice of which is fringed with long, flexible, thread-like processes.

Geography. — It is found sparingly in western New England and in New York, but abounds in western Virginia, Kentucky, and Tennessee, and west and south.

Etymology. — *Macrocarpa*, the specific name, comes from the two Greek words μακρός, long, and καρπός, fruit; hence, long-fruited.

Use. — This oak is a beautiful, symmetrical tree, and for that reason is to be found in all large collections of trees in planted grounds.

The wood is strong, tough, and durable. The bark is used for tanning hides. The fruit is highly prized for food for swine.

7. **Q. obtusiloba**, Mx. (Post Oak. American Turkey Oak. Iron Oak. Upland White Oak.) Trunk 40 feet high, 12 to 18 inches in diameter. Bark thin, grayish-white, branching irregularly. Leaves leathery, dark-green above, and grayish beneath, blade cut by deep sinuses into two pairs of lobes and a terminal one; lobes rounded. Acorns small, and deeply buried in the roughish gray cup.

Geography. — It is seldom found in the northeastern United States, but abounds in the middle Atlantic and southern and southwestern states, east of the Mississippi.

Etymology. — *Obtusiloba* is due to the blunt, rounded lobes of the leaves. As to the common names, — *iron oak* alludes to the hardness and durability of the wood; *turkey oak* is so called because turkeys feed on the acorns; *upland oak* takes its name from the localities where the tree grows; *post oak* is so named because the wood is used for posts.

Use. — The wood is hard and durable, takes a good polish, and is much used for timber where exposure to the weather is required, — especially for bridges, fence-posts, and railroad ties. The fruit is sweet, and is used for feeding swine; also turkeys and other poultry.

8. **Q. nigra**, L. (Black Jack Oak. Oak of the Barrens.) Stem from 20 to 30 feet in height, and 6 to 15 inches in diameter. Bark very dark; branching irregularly. Leaves on short petioles, blade firm in texture, wedge-shaped, sometimes 3-5-lobed; lobes abruptly pointed and terminating in spines. Acorn globular, half covered by the cup. Not abundant.

Geography. — Geographical range is from Massachusetts to the southern States, and west to Illinois. This tree is an important feature in many barren regions.

Etymology. — *Nigra*, the specific name, is from the Latin *niger*, black, and refers to the color of the bark, as does the common name.

Use. — The wood of this species is too small to be valuable for lumber, but makes excellent fuel. The bark is rich in tannin, but on account of the small size of the tree, it cannot well be obtained in large quantities, and is but little used.

9. **Q. palustris**, Du Roi. (Pin Oak. Swamp Spanish Oak. Water Oak.) Trunk 50 to 70 feet high, and 2 to 4 feet in diameter, branching low and forming a graceful head. Bark smooth and dark. Leaves divided into 3 to 5 pairs of lobes, separated by deep, broad sinuses, distinguished from Q. rubra by more narrow lobes, and a color and consistency more delicate. Nut subglobose; cup flat.

Geography. — It is found growing in southern New York, New Jersey, and west in the same latitude to the Mississippi; it is not common in northern New York and New England.

Etymology. — The specific name, *palustris*, is from the Latin word *palustris*, boggy, wet, marshy, alluding to the favorite locality of this species, which is generally found in wet places. This is also indicated in the common names: *swamp Spanish oak*, *water oak*, *meadow oak*. The name *pin oak* arises from the circumstance that the knots are slender, and sometimes on splitting the wood they draw out, appearing like pins. It on this account splits with difficulty.

Use. — Q. palustris is used sparingly in planted grounds for ornament. It forms a beautiful head, and its abundant delicate foliage makes it a rival of the Q. rubra as an ornamental tree. The lumber is coarse and poor, and not as good for fuel as the Q. rubra. The bark of this tree is sometimes used in tanning.

10. **Q. Phellos**, L. (Willow-leaved Oak) Trunk 30 to 60 feet in height, straight, 10 to 20 inches in diameter. Bark smooth and thick. Leaves light-green, about 4 inches long, and 1 to 2 inches wide; linear-lanceolate, pointed; when young toothed; light-green. Acorn subglobose; cup, saucer-shaped.

Var. **sylvatica**, Mx. Leaves on the young tree lobed.

Var. **latifolius**, Lodd. Leaves like those of var. sylvatica, but broader.

There are several other forms, all shrubs.

Geography. — It abounds in southern Virginia and farther south, and is found in New Jersey as far north as Monmouth County. It has been reported as growing in Suffolk County, New York, in planted grounds.

Etymology. — *Phellos*, the specific name of this species, is from the Greek word φελλός, a cork; but why applied to this species is not apparent.

Use. — The Quercus phellos is a beautiful object in the lawn, and is always found in the southern states of the United States in planted grounds. The wood is soft, and not used in building where better lumber is obtainable. The bark is charged with tannin, but thus far has not been largely used by tanners.

11. **Q. Prinus**, L. (Swamp Chestnut Oak.) Trunk 75 to 90 feet high, and 2 to 3 feet in diameter. Bark dark-gray, branching regularly. Leaves on long petioles, blade 7 to 8 inches long, 3 to 4 wide, conspicuously veined beneath, oblong-ovate or elliptical, coarsely and deeply crenate toothed, resembling the leaf of the chestnut. Acorn large, sweet, oval, and brown; cup shallow and scaly.

Geography. — The chestnut oak or chestnut-leaved oak is found throughout the northern United States, and as far south as Virginia, and west to the Mississippi. It attains its full size in southern Pennsylvania and northern Maryland and Virginia.

Etymology. — *Prinus* is from the Greek πρῖνος, ever-green oak. The name *chestnut oak* is due to the shape of the leaf, which resembles that of the chestnut.

Use. — The wood of this tree splits easily, is hard, durable, and takes a good polish. It is used for frames of buildings, planks, etc., and is highly esteemed for fuel. The fruit is sweet, and greatly valued as food for swine. In Virginia, southern Pennsylvania, and the eastern parts of the Carolinas large droves of swine are fattened on acorns, principally of this species.

Var. **acuminata**, Mx. (Chestnut Oak.) Trunk 40 to 70 feet in height, 1 to 2 feet in diameter. Bark whitish and furrowed, irregularly branched.

Leaves on longish petioles, blade oblong, lanceolate, obtuse at the base, sharply toothed, green above and pubescent underneath, resembling the leaf of the chestnut tree. Acorn egg-shaped, deeply set in the hemispherical cup, subsessile.

Geography. — It grows in southern Vermont, where it is a second-class tree. It increases in size southward to the southern part of Virginia, where it reaches its full height. It extends westward to the Mississippi, along the ridges that trend through Virginia, Kentucky, and Tennessee.

Etymology. — *Acuminata* is Latin for pointed, and alludes to the sharp-pointed leaves.

Use. — The wood of this variety is hard and durable, making timber that endures the weather, and it is much used for rails and shingles. It splits freely, and is highly prized for fuel. The bark is well charged with tannin, and is used in the manufacture of leather. The fruit is sweet, large, and abundant, and is used for fattening swine in Virginia, Kentucky, and Tennessee.

Var. **monticola**, Mx. (Rock Oak. Rock Chestnut Oak.) Trunk 30 to 40 feet in height, the top made up of straggling, irregular branches, especially in the rocky, hilly localities, where it is found in the northern and middle states of the United States. Leaves smaller than in Q. prinus, much the same in form, but the teeth are more regular and blunter; when very young, covered with a white down. Acorns in pairs, on short peduncles, and deeply inserted in the cup, which is clothed with loose scales.

Etymology. — *Monticola* is from the Latin *mons*, a mountain, and *colo*, inhabit, and is due to the localities in which this tree delights. It is found on rocky hills and mountain sides, — sparingly in southern New England, more frequently in southern New York, commonly in southern Pennsylvania and Virginia. The names *rock oak* and *rock chestnut oak* are also due to the localities of the tree, and the latter to the shape of the leaf.

Use. — The bark of this species is rich in tannin, and is largely used in the manufacture of leather. The wood is excellent fuel.

12. **Q. rubra**, L. (Red Oak.) Trunk 50 to 80 feet in height, and 3 to 5 feet in diameter. Branches long and spreading. Bark smooth, and dark gray. Leaves smooth, oblong, divided into 3 to 4 pairs of sharply toothed, acute, mucronated lobes, separated by deep and rounded sinuses. Flowers greenish-white, appearing in May. Nut ovate; cup flat, and saucer-shaped. Ripe in October. The foliage varies considerably with the age of the plant and conditions of locality and soil.

Var. **runcinata**, Engl. Sinuses shorter; lobes more upright; fruit very much smaller; cup top-shaped at base.

Geography. — The Q. rubra is emphatically an American tree, and in planted grounds where it has room it forms one of the most graceful objects of the lawn; it is not exceeded in beauty by any of the oaks. It grows throughout northeastern America, and west to the Mississippi, and south down the Mississippi valley.

Etymology. — The specific name of this oak, *rubra*, from the Latin *ruber*, red, was applied on account of the color of the leaf, which after the appearance of frost turns a deep red; hence also the common name *red oak*.

Use. — The tree is highly prized as an ornament in planted grounds. The wood is strong, but has a coarse grain, and does not take a fine polish. It splits easily, and is valuable for barrel staves; it makes an excellent fuel. The bark is prized by tanners.

13. **Q. virens**, Ait. (Live Oak.) Trunk 40 to 60 feet in height, much branched above, forming a broad, picturesque head. Bark thick, very dark. Leaves subsessile, blade thick, elliptical, oblong, varying in form, — entire, lobed, or irregularly toothed, — downy in star-like spots underneath. Fruit peduncled; acorn long, ovate, about one third inclosed in the cup.

Geography. — Its home is North America; its geographical range is narrow. It abounds in the regions of the southern Atlantic and the Gulf States.

Etymology. — *Virens*, the specific name, is from the Latin adjective *virens*, green, and is due to the evergreen leaves.

Use. — On account of its great strength, it is highly prized for use in naval architecture; it is also excellent fuel. The geographical range is so small, and the mode of lumbering is so wasteful, that a speedy exhaustion of the supply is to be apprehended, and legislation is suggested to protect the live oak forests of Florida and Georgia.

14. **Q. suber**, L. (Cork Tree.) Trunk 20 to 35 feet in height. Bark spongy and cracked. Leaves on short petioles, ovate-oblong, leathery, remotely dentate, occasionally entire, downy underneath, and evergreen. Flowers greenish-white, appearing in May. Acorn long and subcylindrical; cup hemispherical, clothed with overlapping scales.

The products of the cork tree are so valuable that the tree has not only been protected, but large plantations have been made; and as the trees are raised from seed, a number of varieties have arisen, the most important of which are the following: —

Var. **latifolia**, Bauh. Leaves broader than those of the species.
Var. **angustifolium**, Bauh. Leaves narrow.
Var. **dentatum**. Leaves large and toothed.

Geography. — The Quercus suber is indigenous in southern Europe and northern Africa; it does not flourish north of the middle of France.

Etymology. — *Suber* is the old Latin name for the cork tree. Linnæus placed it under Quercus, and made *suber* the specific name. *Cork*, the common name, is derived from the Latin *cortex*, cork, and signifies the outer thick bark between the epidermis and the cuticle.

History. — When cork was first applied to its present uses is not known. The Romans were acquainted with its use during the first century.

Preparation. — Harvesting the bark is begun when the tree is from twenty-five to thirty years old. Removing the cork does not injure the tree; on the contrary this is said to be conducive to its growth. The first crop is of poor quality. The second stripping occurs ten years after the first, but the third stripping yields the best bark. It is taken from the tree by making an incision with a sharp instrument around the tree near the base, just deep enough to avoid wounding the liber. Three feet above, a parallel incision is made, and so on up to the branches, making in all three or at most four incisions. It is then slit vertically in widths convenient to handle. The pieces are then forced off with a flat piece of wood, which is introduced between the liber and the cortex. It is held over live coals till the surface is slightly charred, to close the pores. It is then subjected to pressure to take the curve out of it, after which it is piled under cover to dry, and when dry it is fit for market.

Use. — Cork is applied to many uses, the most important of which is in the manufacture of corks for bottles, for which purpose it is especially adapted.

The Romans used it for buoys for fishing-nets and anchors; also for life-preservers. Camillus wore one when he swam the Tiber during the siege of Rome by the Gauls. It is used now for the same purpose; also for cushions and mattresses, and soles of shoes; and it is worked into a sort of felt floor-cloth. In Spain the wealthy line their houses with it. The ancient Egyptians used it for making coffins. The wood is durable, but is not largely used in the arts. The fruit is sweet.

15. **Q. infectoria**, Oliv. (Gall Oak.) Stem 4 to 6 feet high, much branched, forming a straggling shrub. Leaves ovate, oblong, and smooth on both sides, pale beneath, deeply toothed. Fruit sessile; nut elongated, cylindrical; cup tessellated; fruit appearing next year after the appearance of the flowers.

The insect *cynips quercus galli* punctures the leaves, and deposits its eggs in the wounds; these wounds become tumors, from an eighth of an inch to an inch in diameter, subglobular in form, and armed with blunt, spine-like processes. When these tumors are dry and hardened, they constitute the nut-galls of commerce.

Several varieties appear in market, the principal of which are blue and white. The blue gall is gathered before the young insect has gnawed through, and the white afterwards. The blue gall is by far the most valuable.

Geography. — The gall oak is found in all the eastern Mediterranean countries, especially in Asia Minor and northwestern Syria. The best galls come from Aleppo.

Etymology. — *Infectoria*, the specific name of this plant, is from the Latin word *infector*, a dyer, alluding to the circumstance that its products are used in dyeing.

Use. — The nut-gall figures largely in the manufacture of black ink. A solution of copperas, mixed with a decoction of nut-galls, produces a jet-black dye. The nut-gall is also an important article employed in the tanning of hides.

16. **Q. ægilops**, L. (Ægilops. Valonia Oak.) Trunk 20 to 50 feet in height, and 1 to 2 feet in diameter. Bark grayish, sprinkled with brown spots. Branches spreading, forming a hemispherical head. Leaves on short petioles, blade 3 inches in length, coriaceous, ovate, oblong, pale-green above and downy underneath; coarsely toothed, teeth pointed. Flowers greenish-white, appearing in May. Fruit large, nearly inclosed in hemispherical cups, which are covered with long, spreading, lanceolate scales.

Var. **pendula**. Branches long, slender, and drooping.
Var. **latifolia**. Leaves broader.

Geography. — It is native in the countries of the Levant, and abounds throughout Greece and the Grecian Archipelago; it is found sparingly in Italy, but does not grow in middle or western Europe.

Etymology and History. — *Ægilops*, the specific name, is from the Greek αἴξ, αἰγός, a goat, and ὄψ, the eye, goat's-eye, due to the circumstance that an infusion of the shells of the half-grown fruit is used as a remedy for the disease of the eye known as *goat's-eye*, thus named because goats are afflicted with it. This tree was known to Dioscorides and to the ancient Greeks.

Use. — The fruit of this oak formerly constituted the food for a large number of people. The shells or cups are highly charged with tannin, especially when half grown; but on account of the expense of procuring them in an unripe state, they are allowed to ripen. The shells of the ripe fruit are called *ralonia*, the half-ripe ones *camata*, and those gathered in a still earlier state are called *camatina*. The camatina are most highly charged with tannin, the camata next, and the valonia least. Ordinarily the tree is not large.

The wood is excellent for the manufacture of furniture, takes a fine polish, and is durable.

As a dye, the shells of the acorns are in great demand.

CORYLUS, Tourn. (Hazelnut.) Catkins of the male flowers cylindrical. Pistillate flowers, with an involucre of imbricated scales, 2 in number, attached at the base, and fastened to the under surface of the bract; stamens inserted upon the scales near their base, 8 in number; anthers tipped with beard. Female flowers in a flattened, bud-like catkin; bracteal scale entire and ovate; calyx membranous, inclosing the whole ovary, terminating in a short fringed tube. The two stigmas long and filiform. Fruit, a nut, egg-shaped and bony.

1. **C. avellana**, L. (Hazelnut. Filbert.) Stem shrubby, 3 to 5 inches in diameter near the base, made up principally of ramifications, rising from 3 to 8 feet high. Leaves somewhat roundish, cordate, acuminate, and irregular serrate; stipules lengthened. Fruit-covering bell-shaped, ragged at the margin. Nut brown. Ripe in October.

CORYLUS AVELLANA (Hazelnut).

2. **C. colurna**, L. (Constantinople Hazelnut.) Trunk 40 to 50 feet high, 12 to 18 inches in diameter; stipules lanceolate, acuminate. Leaves as in No. 1. Fruit larger and longer.

A dozen varieties are arranged under these two, differing in size and form of the fruit and leaf.

3. **C. rostrata**, Ait. (Beaked Hazelnut.) Stem much branched, forming a straggling shrub from 4 to 8 feet high. Leaves ovate, irregular, serrate, and slightly lobed; stipules narrow-lanceolate. Fruit-envelope tubular, bell-shaped, 2-parted, divisions cut, toothed. Fruit excellent, but smaller than the European species.

4. **C. Americana**, Walt. (American Hazelnut.) Stem branching, forming a shrub like the last, but somewhat larger, 3 to 8 feet high. Leaf rounded at the base, sometimes slightly cordate. Envelope of the fruit globular, bell-shaped; edges coarsely toothed. Nuts as in the last.

The last two species are natives of North America, and abound in thickets and along fence-rows, in the northern and middle States, as far south as Virginia. Of these there are no varieties. The European varieties have doubtless arisen from attempts to improve the fruit by cultivation.

Geography. — The hazelnut grows well between 35° and 55° latitude in the northern hemisphere, but is confined to the eastern parts of the western hemisphere, and to the western parts of the Old World.

Etymology. — The name *Corylus* is said to be derived from the Greek κόρυς, a helmet, referring to the manner in which the calyx enwraps the fruit. The specific name, *avellana*, is derived from Avellino, the name of a city in southern Italy. *Colurna* is from the Greek words κόλος, mutilated, and οὐρά, a tail, referring to the lacerated fringe of the fruit-envelope. *Rostrata*, Latin, signifies "beaked," and refers to the beak-like extremity in which the fruit-covering of this species terminates. The word *Americana* explains itself. The common name, *hazelnut*, is from the Anglo-Saxon *hœsil*, a head-dress, that is, a nut with a head-dress. *Filbert* has been regarded as a corruption of the word "full-beard," referring to the fringed envelope.

History. — The history of this plant is very obscure. It is indigenous to the countries of the Levant. It was originally brought into Italy from Pontus, and was called by the Romans *nux Pontica*, which name was changed in process of time into *nux avellana*, because the plant was first cultivated near the city Avellino, in the kingdom of Naples. It now grows all over middle, southern, and western Europe.

Use. — The hazelnut is a common dessert nut, and in parts of western Asia and Europe it constitutes an important article of food.

The wood of the C. colurna is white, and of a fine grain; it is used for hoops and fishing-rods. The wood of the other species is worthless for timber. The fruit is an article of considerable economic and commercial importance.

FAGUS. L. (Beech.) Staminate flowers in drooping, globose, head-like catkins, 3 to 4 in a group or head, accompanied by minute deciduous bracts; calyx bell-shaped, 5-7-parted; stamens 8-12, and sometimes 16 in number, attached to the bottom of the calyx, and extending above its mouth; filaments slender, with 2-celled anthers. Pistillate flowers, in groups of 2 to 6, usually in 2's, inclosed within a pitcher-shaped 4-lobed involucre, made up of a number of scale-like processes, interior united; calyx with 6 awl-shaped lobes; styles 3 in number, slender. Fruit, an edged, three-faced nut, dark; shell tough; kernel white and sweet, edible. Leaves simple and alternate. Flowers apetalous, presenting a green hue.

1. **F. ferruginea**, Ait. (American Beech.) Stem 40 to 60 feet high, and 1 to 3 feet in diameter, regularly and densely branched, forming a symmetrical head; bark ashy-gray, smooth. Leaves oblong-ovate, pointed, toothed, veins extending into the teeth. Fruit-covering armed with spreading and crooked prickles.

2. **F. sylvatica**, L. (Beech of Europe.) Trunk from 60 to 80 feet high, 2 to 4 feet in diameter. Leaves ovate, dentate, glabrous, with margins ciliate. Fruit inclosed in a rough envelope, armed with blunt prickles, otherwise as F. ferruginea. Under this species there are several well marked varieties or subspecies.

3. **F. obliqua**, Mx. (Oblique-leaved Beech.) Leaves oblique, otherwise like F. sylvatica.

4. **F. colorata**, DC. f. (Cuprea Copper Beech.) Leaves copper-colored.

5. **F. betuloides**, Mx. (Birch-leaved Beech.) Leaf like the birch, and evergreen; forms forests in Tierra del Fuego; also native in **Van Dieman's Land.**

Remarkable for the production of an edible fungus, which appears on its branches. There are three other well marked species in South America.

Geography. — The beech grows well in the temperate zones up to 60° north latitude, and as far south as 50°, but does not flourish in the tropics.

Etymology. — *Fagus,* the generic name, comes from the Greek word φαγεῖν, eat, because the Greeks used the nuts for food. The specific names are derived from the Latin: *ferruginea,* iron-wooded; *sylvatica,* from *sylva,* growing in the woods; *betuloides,* from *betula,* birch-like; *obliqua,* oblique-leaved. The common name, *beech,* signifies "eat."

History. — The beech was a well-known tree in ancient times, esteemed for its fruit and for its shade by both the Greeks and the Romans. Vergil immortalized it. He describes Tityrus in his First Eclogue as reclining beneath the shade of a broad-spreading beech tree. All the species have been introduced into the gardens and planted grounds of Europe.

Use. — The nut of the beech in the north of Europe is a highly prized dessert nut, and constitutes an important part of the food of the inhabitants of northeastern Poland and western Russia. A delicate oil, rivalling that of the olive, is obtained from it, which is used for the table, and also for illuminating purposes, and large droves of swine are fattened upon it.

The wood is hard, and is prized for fuel, and used in cabinet ware for chairs, bedsteads, screws, and wooden shovels; also for shoemakers' lasts. The F. ferruginea of North America and F. sylvatica of Europe are the most important for ornamental purposes.

Order LIII. SALICACEÆ.

SALIX, L. (Willow.) Catkins with entire imbricated scales, subcylindrical; stamens 1 to 5 or more, with 1 to 2 little glands. Fertile flower, with a little gland at base of ovary; pistil stalked or sessile; stigmas 2, short, each occasionally 2-lobed. Leaves simple, alternate, mostly stipulate, usually lanceolate and serrate. Trees and shrubs. A large genus; 170 species.

SALIX BABYLONICA (Weeping Willow).

1. **S. Babylonica,** L. (Weeping Willow.) Stem 50 to 70 feet high, branching low and irregularly; young twigs slender and weeping. Leaves exstipulate, lanceolate, acuminate, finely serrate, glabrous and glaucous beneath; catkins appearing with the leaves.

Geography. — Western and southern Asia.

Of this species there are three well marked varieties, as follows: —

Var. **vulgaris.** Young shoots pale-green, slender, with an annular or wing-like twist just above the axil of the leaf. Leaves furnished with large stipules. Flowers appear in June.

SALICACEÆ.

Geography. — This **variety abounds in the** southern parts of England, especially about London.

Var. Napoleona. Shoots reddish-green, leaves as in var. vulgaris, but exstipulate.

Geography. — This variety is supposed **to have arisen from cuttings** carried **from England and** planted in St. Helena. **In** 1823 cuttings were **taken from a tree which stood** near Napoleon's grave, **and planted** in England; **from these plantings cuttings were** brought to America, **and the weeping willows of the eastern parts of the** United States agree in botanical characteristics **with this variety.**

Var. crispa, syn. S. annularis. (Curled Willow, Ring-leaved Willow.) **Leaves lanceolate, acuminate, serrate,** glabrous and glaucous **underneath, curled into a ring.**

Geography. — **The variety native to England has been** brought by nurserymen **to America, and is found frequently in planted** grounds throughout **the eastern United States.**

Etymology. — *Salix*, **the generic name of the willow, is derived from the two Celtic words,** *sal*, **near, and** *lis*, **water, due to the circumstance that it delights in wet places.** *Willow*, **the common name, comes from the** Low German **word** *wichel*, **give way, or bend, and is due to the pliancy of the branches.** *Babylonica*, **the specific name, is said to have been suggested to Linnæus by the following passage from the 137th Psalm;** —

> " By the rivers of Babylon there we sat down;
> Yea we wept when we remembered Zion;
> We hanged our harps upon the willows in the midst thereof."

History. — **It is generally believed that the first tree of** this species grown **in England was the celebrated Twickenham tree, which has** the following **history: Alexander Pope, the poet, received from the Levant** a box of fresh **figs, packed in willow leaves. The poet noticed that the** small twigs to which **the leaves were attached were alive, and hoping to get a new** plant, rooted **one, which grew into the renowned tree. Another account says** that the poet **was present when a** package **of merchandise from** Spain **was** opened; **the hoops that bound it were** willow, one of **which Pope planted, and thus obtained the tree. This is said to have taken place in** 1730. Another date is also **claimed for the advent of the weeping willow** into England, 1692, thirty-eight **years previous. If it was planted at the** earlier date, Pope could have had no **agency in its planting, for he was then in** his fourth year.

In the Magazine of American History, it is stated that **Sir Henry Clinton brought cuttings of the weeping** willow to America, **which were planted on the estate of John P. Custis, in Virginia.**

If Sir Henry Clinton brought the *first*, he could **not have** brought the var. **Napoleona, for that was first** planted in England in **1823, and** Clinton came to America **in 1775, and died in** 1795. If he brought **the willow, it** was a cutting, **most** likely from **the** variety **vulgaris,** which is very **common in** England, and **not the Napoleona, which is the variety that** prevails among our nurserymen, **and has** been **brought to** America since 1823. It cannot be true that no **weeping willows** grew here previous to **1823, for** it seems improbable that **some of the large** willows should have grown **to** the **size** they have attained **in so few years. In 1848 a tree** in a garden in Flushing measured **14 feet in circumference, making it** about 4½ **feet in diameter.**

Use. — The weeping willow is used as an ornamental tree; it is a very rapid grower, and forms a beautiful and picturesque object in planted grounds. The wood is soft and light, and does not make valuable lumber nor good fuel. Its charcoal is used by chemists in some blow-pipe experiments.

2. **S. viminalis**, L. (Osier Willow.) Stem 20 to 30 feet high, slender, shooting up 10 to 12 feet in a single growing season. Leaves linear-lanceolate, acuminate, sparingly toothed, long, silky beneath; stipules small, aments appearing before the leaves; scales roundish and very hairy; silky down on the under side of leaf, white. The shoots are best for wicker-work and baskets when two years old.

Geography. — Native all over Europe and northwestern Asia; grows easily from slips, and is grown largely in middle Europe.

Etymology. — *Viminalis*, the specific name, is from the Latin *viminalia*, an osier.

Use. — The osier willow is used for wicker-work, baskets, chairs, settees, hoops, etc. There are several other species used for the same purposes.

The charcoal of willow is used in the manufacture of gunpowder.

SALIX VIMINALIS (Osier Willow).

ORDER LIV. **ORCHIDACEÆ**.

Flowers perfect, or imperfect by arrest, terminal, solitary, or in a spike, raceme, or panicle, bracteate, occasionally springing from the middle of a leaf. Perianth superior, petaloid irregular, made up of 6 parts in two series, free or coherent, persistent or falling; the outer whorl or sepals 3 in number, 2 lateral and 1 inferior, mostly superior by torsion peduncle. Petals 3, alternating with the sepals, the 2 lateral ones similar, the other lip-shaped, mostly inferior by torsion, varying greatly in form and color, limb of the lip 3-lobed or entire. Stamens usually 1, opposite to the odd sepal, accompanied by 2 rudimentary stamens, adnate to the style in gynandrous columns, pollen-grains often united into masses. Ovary inferior, 1-celled, or occasionally 3-celled, with 3 parietal placentæ. Capsule membranous or leathery, cylindric, ovoid, winged, or a dehiscent pod. Seeds small and very numerous, testa crustaceous and black. Perennial herb,

ORCHIDACEÆ.

terrestrial or epiphytic. Stem or scape usually simple, cylindric or angular. Leaves mostly radical, those on the stems close together and equitant, alternate or opposite sheathing, and glabrous, cylindrical or linear-lanceolate, entire or emarginate, sometimes cordate; veins usually parallel.

No. of genera, 334; species 5,000; in warm and temperate regions.

VANILLA, Swz. (Vanilla.) Calyx composed of 3 sepals, outer side greenish and petaloid. Petals of the same size and similar to the sepals; lip entire, its claw adnate to the gynandrous column, its limb broad and concave around the column. Anthers terminal, forming a sort of lid; pollen granular. Fruit a pod, cylindrical, 6 to 10 inches long, and half an inch in diameter, fleshy; seeds numerous, imbedded in a soft black pulp. Natives of the tropical countries of both hemispheres. They are lianes, climbing over lofty trees.

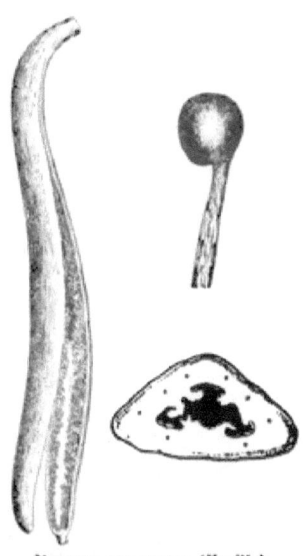

VANILLA PLANIFOLIA (Vanilla).

1. **V. aromatica**, Swz. (Spicy Vanilla.) Stem 4-sided, climbing, 15 to 30 feet long, epiphytic, but not parasitic, fastening itself to the bark by rootlets given off at every node. Leaves from 10 to 15 inches long, narrow and fleshy. Flowers greenish.

2. **V. planifolia**, Andrews. (Flat-leaved Vanilla.) Stem cylindrical and slender, very long. Leaves alternate, sessile, 4 to 6 inches long, oval and pointed, sheathing at the base, persistent, dark-green above, paler underneath. Flowers 2 inches in diameter, pale yellowish-green, sessile, in loose axillary spikes, with 8 to 10 short, triangular, green bracts. Perianth made up of 6 fleshy deciduous leaves, in two rows. Ovary inferior, 2 inches long, cylindrical and stalk-like, fleshy and 1-celled. Fruit a fleshy pod, 5 to 9 inches long slender, filled with small black lens-like seeds.

These two species produce the vanilla of commerce. It is propagated by cuttings; hence there is little opportunity for sporting, and the number of varieties is accordingly limited; there are a few varieties however.

The cuttings are fastened to trees upon which they are to grow, and the fertilization is artificial, by placing the pollen upon the stigma with a splinter of bamboo, or some other delicate instrument.

Geography. — Vanilla is a tropical and subtropical plant, found native in the damp, wooded districts of southern Mexico, and in the coast regions of Vera Cruz and adjacent provinces. It is also cultivated in Guatemala, Guadaloupe, Brazil, Mauritius, Bourbon, Madagascar, and Java.

Etymology. — *Vanilla* is the diminutive of the Spanish word *vaina*, sheath, due to the resemblance of the pod to the sheath of a knife. *Aromatica*, the specific name, signifies "spicy." *Planifolia* means "flat-leaved."

History. — Vanilla is the only orchid whose product is of commercial importance. It is a native chiefly of tropical America, and was brought to the notice of Europeans by the Spaniards, who found it in use among the Aztecs when they invaded Mexico.

Chemistry. — The flavoring substance of vanilla is called by chemists *vanillin*. It has been produced artificially in the laboratory.

Preparation. — The best vanilla is the product of the V. planifolia. The beans are removed from the vine as soon as they have attained their full size, but before they are ripe, and are placed in a situation that exposes them to the heat of the sun or to artificial heat, — a part of the time wrapped in cloths, and a part of the time exposed. Under this treatment they attain their aroma and dark color, and are made fit for the market. The pods are fleshy, flexible, compressed or cylindrical, five to ten inches long, dark-brown, slender, hooked at the stalk end, the surface furrowed lengthwise, with a greasy, shining appearance when properly cured. After a few months the surface becomes covered with a bloom of fine crystals of vanillin. The pods are prepared for market by sorting into lengths and qualities, tying up in parcels of fifty each, and packing the bundles in tin boxes that hold twenty to fifty pounds each, which are then incased in cedar boxes, holding four to six tins.

Use. — Vanilla was formerly used as a medicine for stomach disorders and indigestion. Its use at present is for flavoring. It is added to chocolate, ice-cream, sauces, syrups for soda-water, tobacco, snuffs, liquors, and perfumery.

Order LV. ZINGIBERACEÆ.

Flowers irregular, in a raceme, or a terminal or lateral bracteate panicle. Perianth superior, formed of 3 to 4 whorls. Calyx green, 3-leaved, imbricate. Corolla of 3 imbricated divisions, alternating with the divisions of the calyx, colored and tubular at the base; stamens 1–6. Ovary inferior, 1–3-celled; style sometimes dilated, petal-like, straight or curved, sometimes slender. Stigma terminal; capsule 1–3-celled. Seeds globular or angular; testa leathery. Leaves alternate, simple; petioles sheathing; blade plane large, entire, veins extending from the midrib. Herbs, frequently gigantic.

No. of genera, 36; species, 450; in warm climates.

CURCUMA, L. Calyx tubular, 3-toothed. Corolla tube dilated upwards; divisions of the limb equal; upper lip open. Filament petal-like, dilated, fleshy at the top; lateral staminodia united with it below, 3-lobed. Ovary inferior, 3-celled. Style thread-like; stigma capitate. Capsule 3-celled; cells 3-valved. Seeds numerous, arillate. Root tuberous and palmate, perennial. Leaves herbaceous; petioles sheathing; scope simple. Flowers in a simple, erect spike, yellow. Stemless herb.

C. longa, L. (Turmeric.) Stem wanting. Leaves sheathing each other, lanceolate, a foot long, springing from the crown of the root; the scape passes through the center of the leafy tube formed by the sheathing petioles; scape leafy and crowned with a simple spike of small flowers, creamy yellow.

Of this genus there are about twenty-five species, several of them yielding arrow-root; but the C. longa is the only one that yields the coloring-matter known as curcuma, or turmeric.

Geography. — The curcuma is indigenous to the south of Asia and the Malay peninsula; it flourishes only in the regions of no frost. It is cultivated and made an article of commerce in Hindustan, Cochin China, southern India, Bengal, Java, and the isles of the Pacific.

Etymology. — *Curcuma*, the generic name, is the Latinized form of the Arabic name *kurkum*, signifying "yellow." *Longa*, the specific name, is Latin, signifying "long," and refers to the length of the scape. The common name, *turmeric*, is of unknown origin.

History. — The home of the curcuma is Farther India and the Asiatic islands. Where its products were first introduced into domestic economy or the arts is not recorded. The drug was known to the people of the countries of the Levant and southern Europe prior to the commencement of the Christian era. Dioscorides speaks of it under the name of *Cyperus Indicus;* and it is believed to be the "saffron" spoken of in the Scriptures.

Chemistry. — It yields to the chemist a substance to which the color is due, called *curcumin;* the aromatic taste and smell reside in another substance known under the name of *turmerol*.

Use. — The young tubers do not contain a coloring-matter, and are largely used for food in the islands of the coast of Asia and in the Pacific. The young root, dried and reduced to a powder, is the arrow-root of the East Indies. The pulverized *ripened* root is used principally for coloring.

CURCUMA LONGA
(Turmeric).

In India it enters into curry powder and other culinary preparations. Its dye is not permanent, yet it is largely used.

The medicinal properties are stomachic, slightly tonic, and cordial. It is used in coloring foods, medicines, and cosmetics. In the East it is made into a paste with oil, and used to anoint the body at marriage festivities.

Paper colored with turmeric is used for testing the presence of alkalies in liquids; if alkali is present, the paper becomes brown.

MARANTA, Plum. (Arrowroot.) Calyx of 3 distinct lanceolate sepals; staminodia petaloid, united at base with stamens. Flower perfect, bracted; bracts under the branches narrow, appressed; corolla tubular at the base, curved, 3 imbricated divisions at the limb, alternating with the sepals.

M. arundinacea, W. (Arrowroot.) Stem 2 feet high, branched; rhizome fleshy. Leaves lanceolate, hairy. Flowers in clusters; 2 flowers on a stalk, peduncles short. Fruit globular, one eighth of an inch in diameter; rhizomes from 10 to 15 inches long, half an inch in diameter, enveloped by membranaceous scales.

DESCRIPTIVE BOTANY.

Geography. — The maranta is a tropical and subtropical plant; it has been found growing in Florida about the 28th parallel.

Etymology. — The name *maranta* is derived from Bartholomew Maranti, a physician of Venice; and the specific name, *arundinacea*, reed-like, is from the Latin *arundo*, a reed. The popular name, *arrow-root*, is said to have been applied to this plant because of its efficacy in curing wounds inflicted by poisoned arrows.

History. — When this plant was introduced by civilized man into the *materia medica*, or as a food plant, is not known, — evidently since the discovery of the New World, as it is a native of tropical America.

Chemistry. — 100 parts of arrow-root yield to the chemist the following substances: —

Starch	26.
Woody fiber	6.
Albumen	1.50
Gummy extract, volatile oil, and salts	1.
Water	65.50
	100.00

Its starch-grains are convex-elliptical, approaching triangular shape, nearly uniform in size. In commerce it is frequently adulterated with rice, flour, wheat-starch, or potato-starch. These foreign substances are readily detected by the microscope.

Use. — The rhizomas when a year old are dug with instruments made of German silver, and knives made of the same metal are employed for removing the scales. They are then reduced to a pulp in wooden mortars, or by the use of a cylindrical rasp. The mass is placed in water, which holds it temporarily in suspension. After straining to remove the woody fiber, it is repeatedly stirred, and allowed to settle, the water being afterward carefully poured off. The starchy substance which remains as a sediment is then dried, either in the shade or sunlight. These manipulations are all conducted with great care. After drying, it is reduced to powder or flour, and is used for puddings, custards, and other culinary preparations. It is a favorite food for the sick-room. It constitutes the principal food for the people of many tropical and subtropical countries of the New World. In the West Indies it is administered to counteract the effects of poisons.

MARANTA ARUNDINACEA (Arrowroot).

Tacca integrifolia, Presl, and other species of this genus furnish starchy products which are used as substitutes for arrow-root in the East Indies.

ZINGIBERACEÆ.

There are also other plants whose roots yield starches, among which the most prominent are: several cannas, as C. echinus, C. glauca, Rose., C. edulis, and C. flaccida, Rose. These cannas are now under cultivation in Australia. The C. glauca and C. coccinia yield the starch known in commerce as *tous les mois*. The English have also introduced the cultivation of arrow-root-yielding plants into India, New South Wales, and Queensland. But no plant yields so much and of so good quality as the *maranta arundinacea*.

ELETTARIA, Maton. Calyx membranous, tubular, short, and 3-lobed. Corolla cylindrical, as long as the calyx; anterior lobe oblong, erect, and concave; side lobes narrower, curved backwards, slightly 3-lobed or toothed; lip obovoid; anthers inserted on the corolla, sessile and linear. Ovary 3-celled, containing many ovules. Style thread-like; stigma rising a little above the anthers. Fruit globular or ovoid; seed-vessel tough, and not opening when ripe; seed obovoid and rugose. Rootstock thick, growing just under the surface of the ground. Perennial.

E. cardamomum, Maton. Stems numerous, smooth, 6 to 12 feet high. Leaves sheathing, alternate, 9 to 12 inches long, and 1 to 5 inches wide, elliptical-lanceolate pointed, entire, smooth, and dark-green above, pale beneath, with strong midribs; foot-stalks short; flower-stalk starts from the base of the stem, prostrate. Flowers in a panicle; calyx tubular, toothed at the margin; corolla funnel-shaped, border lipped, 3-lobed, and spurred at the base. Fruit a 3-celled capsule, many-seeded.

There are five species of this genus, all natives of the tropical regions of Asia; but the cardamoms of commerce are from this species.

Geography. — The plant grows and fruits well at an elevation of 3,000 to 5,000 feet along the southern coast of India; but the seeds of commerce are shipped from Madras, Allepy, and Ceylon.

The plant yields fruit both in a wild state and under cultivation. It requires a temperature that does not fall below 70° Fahrenheit. It flourishes best in the mountains of Malabar, at an elevation of about 4,000 feet, under an annual rainfall of ten feet.

Etymology. — *Elettaria* is the name of the *cardamom* plant in the Malabar tongue; its meaning is not known. *Cardamom* is from the Greek καρδάμωμον, a spice, believed to have been applied first to the elettaria by Pliny, on account of the pungent spice of its seeds. The cardamoms of commerce are known under names derived from the places where they are grown: the Malabar, Madras, Allepy, and Ceylon.

Cultivation. — The cultivation consists in clearing the forest in spots where specimens of the plant are found growing spontaneously, and then keeping the ground free from weeds and underbrush. The rhizome or underground stem throws up from fifteen to twenty leafy stems or branches. The stems each throw off four flowering stalks near the ground, upon which the fruit finally appears. The plant fruits when four years old. An acre yields about twenty-eight pounds.

Use. — The medicinal properties of the seeds as well as their aromatic character are due to the presence of an essential oil, of which they yield three to five per cent. Their effect upon the human system is stimulant, carminative, and stomachic. They are used for flavoring sauces and for disguising unpleasant tastes in medicinal mixtures, and in tinctures, in confectionery and cordials, and they are a well-known masticatory.

ZINGIBER, Adans. Calyx membranous, tubular, and short, 3-lobed. Corolla tubular, cylindrical, dilated above; lobes narrow, inner ones incurved, concave, lateral ones spreading; lip small, middle lobe longer, entire or bifid. Anther on the erect filament oblong; connective linear, or awl-shaped, extending beyond the anther cells. Ovary 3-celled, containing many ovules. Style thread-like; stigma projecting beyond the anther cells, sometimes very small and globose. Capsule oblong or globular; seed-vessel hard, opening irregularly; seed large and oblong. It flowers in September, the stem withers away by the first of the following January, and the root is harvested in February.

Z. officinalis, Rose. (Ginger.) The stem is subterranean, of the rhizome character; the branches or aerial stems are 2 to 3 feet high, solid, erect, with imbricated membranous sheaths. Leaves lanceolate, acute, smooth, 5 to 6 inches long and 1 inch wide, alternate, with ovate, acuminate sheaths. The scapes are terminated with spikes of whitish or dirty yellowish flowers, whose lips are streaked with purple, and spotted; spikes bracteate, oval, and obtuse; bracts yellowish-green, with membranous edges.

ZINGIBER OFFICINALIS (Ginger).

The roots or underground stems are 3 to 4 inches long, made up of a number of short lobe-like shoots or knobs, whose tops are marked each with a scar, showing the spot where the stem grew. There are other species of this genus, about twenty in all; but the *officinalis* furnishes the ginger of commerce.

Geography. — The ginger plant is tropical, or, strictly speaking, subtropical. It is cultivated in southern Asia, and on the southern slopes of the Himalayas. It is an important crop in the West Indies, whither it has been carried by Europeans from southern Asia. In the same manner it has found its way to the western coast of Africa, where it is also largely cultivated.

Etymology. — *Zingiber*, the botanic name, is claimed by some to be geographical, from the Island of Zanzibar, where it first became known to Europeans. By others it is believed to be the Latinized form of the Sanscrit word *cringa*, a horn, due to the fancied resemblance of the root to a stag's horn. *Ginger*, the popular name, is a corruption of the word *zingiber*. *Officinalis* is the Latin for "useful," or "serviceable."

History. — Ginger was known to the Greeks and Romans in the first century of the Christian era, and was in common use in England before the Norman conquest, v. p. 1066. Where or when it was first cultivated is not known. It must have been brought under cultivation very early in the history of the Asiatic peoples, for it is not now known to be growing in a wild state. There is good reason to believe that it is a native of southern Asia, whence it has been carried to the western coast of Africa and to the West Indies.

Chemistry. — It yields to the chemist many distinct substances, among which starch is the largest in quantity. The pungency is due to an oily, resinous principle, called *gingerine*. The perfume and delicate flavor reside in an essential oil, which is entirely free from pungency.

Preparation. — There are three varieties of the root in American commerce: the *Jamaica*, *Cochin*, and *African*. The Jamaica and the Cochin are prepared before they are brought to market, by the removal of the epidermis, and are of a dull-yellow or buff color. The African root, or black ginger, is sent to market without preparation, and bears a dark, earthy hue. The Jamaica ginger of the shops is white, having been subjected to a bleaching process, and then whitewashed. This variety is preferred on account of its fine flavor and its light color. The Cochin ginger is valuable on account of its strength, and is much used by the manufacturers of the extract. The ground ginger of the grocers is made from the African root, which produces a dull, brownish-yellow powder.

Use. — As a medicine, ginger is stimulant, carminative, and anti-dyspeptic. It is used as a tonic for flatulency, a carminative in colic, and enters into the preparation of many medicines to assist their action. It is an ever-present domestic remedy. It is an important culinary spice and condiment for flavoring pies, preserves, cake, and aerated waters, beers, and cordials. In China, the root is largely preserved by cooking it in syrup. It reaches us in stone jars holding from a pint to a half-gallon. The Chinese also prepare an excellent candy by slicing the root and cooking it in syrup.

MUSA, Plum. Perianth attached to the ovary, 2-lipped or 2-parted, the lower lip cut at the extremity in five divisions. Stamens 5, ovary inferior, 3-celled; ovules numerous. Styles thick, shorter than the stamens, 3-6-lobed. Fruit oblong, angled, and 3-celled, 3-9 inches long, curved, $1\frac{1}{2}$-2 inches in diameter. Seeds numerous, in a soft, fleshy pulp; covering of the seed black. Flowers in an axillary spadix.

1. **M. paradisiaca**, L. (Plantain, or Yellow Banana.) Stem herbaceous, about 18 feet high, stout, formed by a succession of large sheathing. Leaves 4 to 8 feet long, 15 to 20 inches broad, oblong, entire. Flowers terminal, in a dense spike, 2 to 4 feet in length, nodding. Fruit 3 to 9 inches long, 1 to 2 inches in diameter, scimitar-shaped, pale-yellow when ripe, skin thick and coriaceous, pulp sweet and delicate to the taste, seeds distributed along the axis of growth in the pulp. Under cultivation the seeds seldom ripen, but the plant is propagated by offshoots at the base. If left in a state of nature, the seed matures, and many varieties are obtained. There are about thirty varieties under cultivation in Asia and the Asiatic islands, of which the M. paradisiaca is the parent or a fair type of the numerous species and varieties of the plantain.

2. **M. sapientum**, L. (Banana, or Red Banana.) The description of No. 1 covers the characteristics of the M. sapientum exactly, except that the stalk

is marked with purple spots, and the fruit is shorter, more cylindrical, of a dark greenish-red color, and 2 to 2½ inches in diameter, and has a more delicious taste.

There are many varieties of this species, differing in size and quality of the fruit. It is held by some botanists that the species M. troglodytarum, a native of the Molucca Islands, is the parent of all the cultivated species and varieties. But it is also held that the M. sapientum answers all the conditions of a parent.

MUSA PARADISIACA (Yellow Banana).

Geography. — The geographical distribution of the banana and plantain is very wide, extending all around the globe, between 38° north latitude and 35° south latitude. A mean temperature above 64° Fahrenheit is necessary to its existence, and it does not fruit freely where the temperature falls below 40°. It is found in the Indian Archipelago, China, Cochin China, and Hindustan. It grows in Australia and the islands of the Pacific, in Madagascar, and on the western shores of Africa. It also adorns the gardens of the Morea, of Sicily, and the south of Spain. In the New World it is grown in Mexico, Central America, Colombia, Peru, northern Brazil, Guiana, and in the greater part of the West Indies. Recently it has been introduced into the gardens of southern Florida and Louisiana.

Etymology. — The name *musa* is said to come from the Arabic *mouz*. It is also said to have been given by Plumier, a French botanist, in honor of Musa, a celebrated physician of Rome, who cured Augustus Cæsar of a disease which had been pronounced by the imperial practitioners incurable; for which service the emperor knighted him. The specific name *sapientum* is due to the circumstance that in India the leaves grow to a great height, and its groves were a favorite resort for sages (*sapientes*), who were accustomed to repose under their shade while partaking of the fruit; hence, *the fruit of the wise*. The name *paradisiaca* arises from the story that the banana of this species was the forbidden fruit of Paradise. The English or common name, banana, is very obscure in its origin as well as its signification, but is supposed to be a Portuguese corruption of some native name. The word *plantain* is from the Latin *plantago*, applied on account of its broad, spreading leaf.

History.— Many things point to Asia as the home of the banana, yet it is claimed that it had reached the western coast of South America before the arrival of the Europeans. It has also been suggested that it is indigenous to America, but the weight of history makes it an introduced plant; hence we are constrained to reckon the banana among the food plants of Asia. It was unknown to the ancient Egyptians, but is said to have been made known to the Greeks by Alexander's expedition into India.

Cultivation.— The banana has been very properly pointed out as an illustration of the wonderful fertility of tropical regions. In the temperate zones a fruit-tree requires years to mature fruit; but the dweller between the tropics takes a sucker from a banana-plant, places it in the soil on the river bank, and at the end of a few months gathers a crop of delicious fruit.

Use.— The banana is highly prized as a dessert. The natives of the tropics slice and fry it; it is baked in ovens, and is dried and ground into flour, of which bread and cakes are made. Two intoxicating drinks known as *banana wine* and *banana beer* are made from the juice by the Africans.

As to the economic value of the plantain and banana, they stand next to wheat and rice as food plants. Simmons, a recent English writer on tropical food plants, says : "The banana is to thousands of the dwellers of the tropics what rice is to the Hindoo, rye to the Muscovite, and wheat to the Englishman; it is their main dependance, in more senses than one, their staff of life, grown everywhere in small quantities throughout the tropics." Again he says : "Among the splendid, varied, and profuse vegetation with which tropical countries abound, . . . the magnificent herbaceous plant, the plantain, attracts particular notice."

The individuals of this family rank high among endogenous herbs. Their gigantic size, the magnificence of their foliage, the abundance and character of their fruit, the grandeur of their flowers, give them the very highest place among stemless endogens. The banana is the queen among ornamental herbs, and the household god of the laborer's cottage.

The structure of the stem furnishes a fiber, of which cordage, mats, and a coarse cloth and paper are manufactured. See *Musa textilis.*

The Chinese use the top of the stem for its juice, which also forms an important ingredient in the manufacture of ink. The pith of the stem and the top of the spike and the young shoots are edible. Humboldt estimated that an acre will yield 134,900 pounds of food. This far exceeds the product of any of the tuber-bearing plants of the temperate zones.

3. **M. textilis.** Nees. (Manilla.) This species of musa is treated separately because its characteristics are so different from the others. First, it is larger than any other, rising to the height of 30 to 50 feet. All the musas produce fiber, but the M. sapientum, M violacea, and **M. textilis** furnish the best quality, and most of it.

Geography.—The musa textilis is, like the other species, a tropical or sub tropical plant. It grows best on the slopes of volcanic mountains, among the larger trees that usually cover such declivities. The fiber of commerce, obtained from the M. textilis, thus far has been grown in the southern Philippines. Attempts have been made to grow it in other localities without much success. It is now cultivated in India and other parts of the south of Asia.

Etymology. — *Textilis*, the specific name, is from the Latin *texo*, weave, alluding to the use made of the fiber obtained from it. *Manilla*, the popular name, is due to the seaport of that name, from which most of the fiber and its products are exported.

History. — When this plant was first introduced into domestic economy, or when it assumed a commercial importance, is not known. The first recorded export was in 1850, and in 1856, 600,000 pounds were sent from three ports in the Philippines, three fourths of which came to the United States. In 1880 45,584,000 pounds reached England, a large part of which came thence to America. The amount used is rapidly increasing, and the uses to which it is applied are multiplying.

Preparation and Use. — Manilla is the prepared fiber of the *musa*. The plant is cut when eighteen months old. Just before it flowers, the leaves are removed; the stem is then opened lengthwise, and the flower-stalk, which forms a central column, is removed. From this, three successive coats of fiber are taken. The outer coat is the coarsest, and furnishes the material for making ropes, cordage, coarse bags, and mats. The next coat is of finer texture, and is used for manufacturing cloths and paper. The third and inner fiber is much more delicate than the outer coatings, and of it are made the finest fabrics, elegant shawls, and material for soft and delicate underclothing. Ropes and cordage made of the fiber of the outer coating are one fourth stronger than when made of the best hemp.

The fiber is prepared by bruising it between rollers in a mill; it is then boiled in large coppers, — potash, soda, or lime having been added to the water to remove the mucilage. It is taken from the coppers and hung up over manilla ropes on bamboo poles to dry. It is then beaten or combed, when it is fit for market or the factory. The three different sorts are kept separate. Formerly the preparation was all done by hand, but machinery is now employed, which does the work much better. Mixed with silk or cotton, it makes a very fine fabric.

Order LVI. **BROMELIACEÆ**.

Flowers perfect, usually regular, occasionally irregular, spiked, racemed, or panicled. Perianth inferior or superior, 6-parted, in two series, the outer series or calyx leaf-like, one sometimes shorter; inner whorl or corolla petioled, more or less coherent, twisted in the bud. Stamens 6, variously arranged; filaments subulate, dilated at the base, free or connate; anthers introrse. Ovary inferior, sometimes superior, 3-celled; style simple, 3-cornered, occasionally 3-parted; stigmas 3. Fruit a 3-celled berry, or capsule: seeds numerous. Leaves mostly all at the base of the stem, or scape; sheathing stiff, channelled; margin armed with spines, or toothed; epidermis clothed with scale-like hairs. Herbaceous, and occasionally woody plants, mostly without stems, sometimes epiphytic. Root perennial.

No. of principal genera, 27; species, 350; warm parts of America.

BROMELIACEÆ.

ANANASSA, Lind. (Pineapple.) Inflorescence densely strobiliform. Calyx 3-parted; petals 3. Stamens 6; style 1, 3-parted. Fruit a spike of densely packed berries, and fleshy bracts, forming a conical-shaped body from 6 to 12 inches long and 3 to 6 inches in diameter, crowned with a tuft of leaves.

A. sativa, Schult. (Pineapple.) Stem wanting, the flower-stalk springing from the midst of a tuft of radical leaves which are larger than the leaves that crown the fruit, armed with sharp, hooked teeth, curving upwards and tipped with a strong, sharp spine, lanceolate and fleshy.

There are numerous native sports, in a wild state. The plant, when not under cultivation, ripens seed, from which varieties are produced, which when found to be good are brought under cultivation.

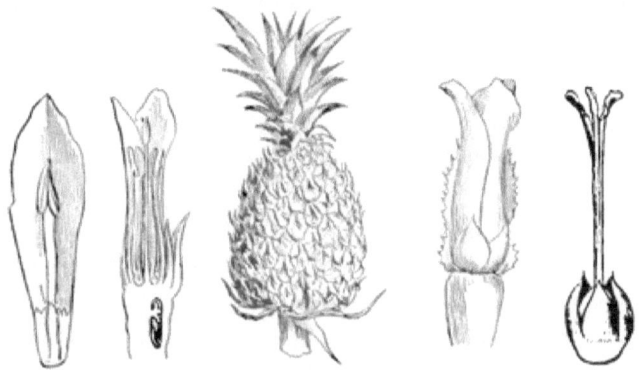

ANANASSA SATIVA (Pineapple).

The principal varieties under cultivation are as follows, varying in size and quality of the fruit. Ripley, Enville, Prickly Cayenne, Smooth Cayenne, Providence, Charlotte Rothschild.

These are the favorite varieties grown in the British West Indies.

Geography. — The cultivation of the pineapple was at one time confined to one of the islands of the Bahama group (Eleuthera), and neighboring islets; but it is now spread to all the tropical regions where civilized man has taken up his abode. It is the only important fruit which cannot be traced to Asia as its home. Now the southern shores of the Eastern Continent, the fields of eastern Africa, the isles of the Pacific, as well as the tropical regions of America, all give place to this interesting and delightful product. In India it has escaped from cultivation, grows and propagates itself in the jungles, and has become thoroughly naturalized.

Etymology. — Linnæus named the pineapple *Bromelia*, in honor of Olaf Bromel, a Swedish botanist. The name *ananassa* was given by Thunberg, from the Peruvian name *nana*. *Sativa*, the specific name, signifies "planted," or "sown." The names of varieties are for the originators or their friends, or from some quality of the plant. The common name, *pineapple*, arose from the resemblance of the fruit to pine cones.

History. — The home of the pineapple is tropical America. It is related that the Spaniards found it in Peru and took it to the West Indies, whence it was carried by the Portuguese to the East Indies. About the middle of the seventeenth century it was taken to Holland by Mr. Le Count, a Dutch merchant, who cultivated it under glass at his country seat near Leyden. It was thence carried to England, where it was successfully fruited under what is known in England as stove culture.

Use. — The pineapple is a most fragrant fruit, and is used for a dessert. It is also preserved in sugar and in brandy, and forms an important article of commerce, both raw and preserved. At Nassau, about two million cans are filled annually and sent to the United States.

Propagation. — The pineapple is propagated by suckers or by the tuft from the top of the fruit.

Marts. — New York is the great mart for this fruit; but it is now carried to Europe, in a crude state, by the fast steamers. It is matured in England under glass, and is sold for ten times more than the imported article.

Order LVII. **IRIDACEÆ.**

Flowers perfect, regular or irregular, terminal in a spike, corymb, or loose panicle, or solitary, each with 2 spathe-like bracts. Inflorescence with a double subfoliaceous bract. Perianth superior, petaloid, tubular, 6-parted, regular or bilabiate; segments in two series, equal, or the inner whorl smaller, dissimilar, usually falling early. Stamens 3; filaments equal, free; anthers extrorse. Ovary inferior, 3-celled, usually many-ovuled. Stigmas 3, opposite the stamens, or alternate, often dilated or petaloid. Capsule 3-cornered or lobed, 3-celled. Seeds numerous, subglobose or compressed, sometimes winged; testa membranous or papery, sometimes leathery or fleshy. Leaves usually radical, equitant, ensiform or linear, angular, entire, flat or longitudinally folded, those on the stem alternate, sheathing. Perennial herbs, with tuberous or bulbous rhizome.

Crocus sativus (Saffron).

No. of genera, 57; species, 700; warm parts of both hemispheres.

CROCUS, L. Flowers nearly sessile, among leaves, tube long and slender, the limb bell-shaped, divided into 6 nearly equal segments. Stigmas dilated, and colored at the top, often fringed; capsule buried among the radical leaves. Rootstock bulbous, coating fibrous and netted.

C. sativus, L. (Saffron.) Scape 1 to 3 inches high. Leaves radical, linear; margins revolute, with white furrow above. Flower-tube long, white, crowned with purple elliptical segments. Stigmas long, emarginate, red or deep orange, extending beyond the tube.

There are many other species, but the sativus is the most important.

Geography. — The geographical range is throughout the southern parts of the north temperate zone.

Etymology. — *Crocus*, the generic name, was given to this plant by Theophrastus, but the meaning is obscure. *Sativus* is Latin for "sown" or "cultivated."

History. — This plant was known to the ancients; it is mentioned by Dioscorides, Theophrastus, and Pliny. The Romans cultivated it for its perfume; and in later times it was an important crop in England. The parts producing the dye are the style and stigma; and it has been estimated that a single pound of saffron is composed of the stigmas of 200,000 flowers.

Use. — The medicinal properties are not important. It was formerly administered in infusions in a multitude of disorders, but especially to bring eruptive diseases to the surface. Modern practice has brought in remedies which have crowded it out.

There are two active principles contained in saffron flowers: an oil to which its flavoring and medicinal properties are due; and a coloring substance which furnishes the dye.

As a coloring agent saffron is now used to color creams and biscuits, and to color and flavor confectionery and cordials. As a dye, it is an important substance. The dye is produced by the presence of a principle which the chemists call polychroite; this substance, when brought into contact with different chemical reagents, produces a variety of beautiful colors and tints.

Order LVIII. **DIOSCOREACEÆ.**

Flowers diœcious, small, inconspicuous, regular, in axillary racemes or spikes. Perianth herbaceous, petal-like, in 6 parts, 2 seriate, equal, and persistent. Stamens 6, inserted at the base of the segments of the perianth; filaments short, free. Ovary inferior, 3-celled; styles 3, short; stigma blunt or 2-lobed; ovules 1-2 in each cell. Fruit capsular, 3-angled or winged, 3-celled; seeds compressed, winged. Leaves alternate or opposite, petioled, simple, prominently palminerved, entire or lobed. Perennial herbs, with twining stems; tuberous, fleshy roots.

No. of genera, 8; species, 160; chiefly tropical.

DIOSCOREA, L. Flowers small; stamens 6, at the base of the perianth; filaments awl-shaped. Styles of the fertile flowers 3; cells of the capsule usually 2-seeded. Seeds margined by thin membrane; pods 3-angled.

1. **D. sativa**, L. (Yam.) Stem long, slender, terete, smooth, twining or clambering over shrubbery. Leaves alternate, broad, ovate-cordate, glabrous, nerved, outer nerves forked, margin sinuate. Flowers in dense, axillary spikes, greenish-white. Root gives rise to large edible tubers.

2. **D. batatas**, Decaisne. (Chinese Yam.) Stem very long, resembling that of D. sativa. Leaves opposite, smooth, heart-shaped or halbert-shaped, in the

axils of which bulblets appear. The roots are large, tuberous processes, largest at the bottom, sometimes 6 feet long, tapering from the bottom (where they are sometimes 8 inches in diameter) to the top, the crown being an inch in diameter, the whole tuber sometimes weighing 40 pounds; edible.

There are about 150 species known to botanists. The most important edible species are the D. sativa and D. batatas, D. japonica and D. alata.

Geography. — The yam is tropical and subtropical, though it arrives at maturity in higher latitudes; but the tubers do not reach perfection except under a high temperature. The geographical distribution of this genus is a belt all around the earth, between the parallels of 30° on both sides of the equator. Some of the species are indigenous to America, some to Africa, but more to Asia. It is but recently that it has been brought under cultivation. Its cultivation is confined principally to Japan, the East India Islands, and Siam.

DIOSCOREA SATIVA (Yam).

Etymology. — *Dioscorea*, the generic name, was given by Linnæus in honor of Dioscorides, the Greek botanist. The specific name, *sativa*, signifies the "sown," or "cultivated," plant. The name *batatas* is supposed to signify "a club," referring to the shape of the tuber, which is small at the upper end and large at the other. The common name, *yam*, is supposed to be of African origin, the meaning being unknown.

Use. — The yam is used for a table vegetable, and is prepared by boiling, baking, frying, or roasting in hot embers. It is also beaten into a paste and made into cakes; also boiled, mashed, and made into puddings and custards. It constitutes a large portion of the food of the savage and half-civilized tribes of Africa and Malaysia.

ORDER LIX. **LILIACEÆ.**

Flowers perfect, mostly terminal, solitary, racemose or spiked. Perianth tubular, inferior; limb 6-lobed or parted, or perianth leaves distinct. Stamens 6, inserted on the receptacle or perianth-tube. Ovary free, usually 3-celled; cells few to many-seeded; style simple or 3-cleft. Fruit capsular. Leaves simple, entire, sheathing fascicled at the base, and sessile on the stem, flat or channelled. Herbaceous perennials, sometimes tree-like and woody. Root bulbous, tuberous, or with a creeping rhizome.

No. of genera, 187; species, 2000; cosmopolitan; mostly in damp places, but not in the water.

LILIACEÆ.

SMILAX, Tourn. Flowers diœcious, small, greenish, axillary; perianth deciduous, segments 6, in two series, outer ones broader, spreading, sepaloid; stamens equal to the number of divisions, shorter than the segments and inserted on their bases. Anthers adnate, 1-celled, with a cross partition. Ovary superior, usually 3-celled; stigmas 3, spreading, 2 ovules in a cell. Fruit a berry, globose, 6 seeds, sometimes 1-celled, with 1 seed only. Leaves entire, petioled, alternate, palmately 3-5 veined. Woody perennials.

1. **S. officinalis**, H. & Bonpl. (Sarsaparilla.) Stem woody, twining, nearly square, smooth, with scattered spines or prickles; slender, long, young shoots, without prickles. Leaves ovate-oblong, acute, cordate, 5 to 7 palmately nerved, thick and leathery, very large, 8 to 12 inches long, and half as wide. Foot-stalks an inch long, smooth, with tendrils. Root with long, creeping rootstocks.

2. **S. sarsaparilla**, L. (Sarsaparilla.) Stem prickly, nearly square, climbing by tendrils. Leaves 6 to 8 inches long, and 4 to 5 inches wide, oblong-ovate, deep-green cuspidate, subcordate at base, and 5-nerved. Flowers greenish, on long, flat, axillary peduncles. Berries large, globose, 1-seeded, pinkish-red when ripe, persistent. Root with long, creeping rhizomes.

This species has found its way into the southern United States, or is native there.

3. **S. medica**, Schlech et Cham. As above, except that the leaves are very smooth, prominently heart-shaped on the lower part of the stem, and ovate above.

This species furnishes the Mexican root.

4. **S. papyraceæ**, Poir. Leaves membranaceous, oblong-oval, blunt, otherwise as S. officinalis.

SMILAX SARSAPARILLA (Sarsaparilla).

This species yields the root known as the Para or Rio Negro Sarsaparilla.

Geography.—The species of smilax producing sarsaparilla are found in Mexico, and in central and northern South America.

Etymology.—*Smilax* is from the Greek word σμῖλαξ, a grater, alluding to the armed stems. *Officinalis*, the specific name, is from the Latin, signifying "of the shops." *Sarsaparilla*, the common name, is from the Spanish *zarza*, a bramble, and *parilla*, a little vine; hence a prickly vine, or prickly little vine. The specific name *medica* is due to the medical properties of this species; and *papyraceæ* indicates the membranous or paper-like character of the leaf.

History.—Sarsaparilla was introduced into medical practice about the middle of the sixteenth century, and attained a great reputation, which it has not been able to sustain.

Use.—When sarsaparilla first became known to the healing art, it claimed to be specific in all diseases which affected the blood.

Its effects upon the system are alterative, sudorific, and secretive; it is used for chronic rheumatism, skin diseases, and is considered especially efficacious as a restorative after an excessive mercurial course. At the present day it is largely used in the manufacture of patent medicines, — the regular physicians having found substitutes which they consider more efficacious.

Marts. — It is believed that the S. sarsaparilla has the same medicinal properties as the S officinalis; but this is disputed, and it is claimed that the root which supplies the market is from the S. officinalis only. Jamaica sarsaparilla is from Central America, and takes its name from the fact that it is shipped from Jamaica to Europe. The cortex of the root is red, and clothed with short rootlets; it is sent to market in bundles 18 inches long and 5 inches in diameter; these are made up into bales for shipment.

The North American market is supplied with Honduras sarsaparilla, procured from a species which is not known to botanists, or is not well determined; it is shipped from Balize in large bales, made up of small bundles, 30 inches long and 6 in diameter, weighing 3 to 5 pounds each; the ends of the bales being covered with green raw-hides.

The market of the United States is partly supplied by what is known as Mexican sarsaparilla; it is shipped from Vera Cruz, and is the root of the S. medica. It is a slender, shrivelled root, done up in large bundles, of 300 pounds each, fastened together with ropes.

Another sort, of very good quality, is known as Rio Nigro sarsaparilla, which is the root of S. papyraceæ. It is shipped at Para in neat rolls 3 feet long and 1 foot in diameter, bound together by a vine. Nearly all of this variety goes to Spain and Portugal.

ASPARAGUS, L. Flowers perfect; calyx with 6 equal, narrow, oblong, petal-like sepals, barely connected at the base, spreading. Stamens united near the base; anthers peltate; style short; stigmas 3. Ovary 3-cornered, top-shaped, and 3-celled; cells 2-seeded. Leaves reduced to minute scales. Branches thread-like. Fruit, globular berry; seed angular, outer covering black and leathery; embryo curved and eccentric. Root a mass of long fleshy fibers, about an eighth of an inch in diameter. Perennial herb.

A. officinalis. L. (Asparagus or Sparrow Grass.) Stem paniculately branched, 3 to 4 feet high, starting from the root in a stout shoot. Branchlets in fascicles, thread-like, bristly, and flexible, from half an inch to an inch and a half in length, pale pea-green. Flowers very small, axillary on the branches, solitary or in twos, yellowish-green. Berry with 6 seeds.

There are about 100 species, but the *A. officinalis* is the only one cultivated for food.

The asparagus is propagated from the seed, and sports. There are many varieties; but there are two known to the market gardeners as favorites, under the names of *red top* and *green top*. These change in form and size by high culture.

Geography. — The asparagus grows in the middle latitudes of the north temperate zone, is largely cultivated in England, Holland, throughout central Europe, in the countries of the Mediterranean, and on the sandy plains of Poland and southern Russia, about the Caspian Sea. It was introduced into Hindustan by the English, and is extensively cultivated there; it is also found

LILIACEÆ.

in Japan, and has a native name, which seems to point to its being indigenous to that region. It grows outside of cultivation in Great Britain. It is indigenous to the countries of the Levant; was brought to North America by

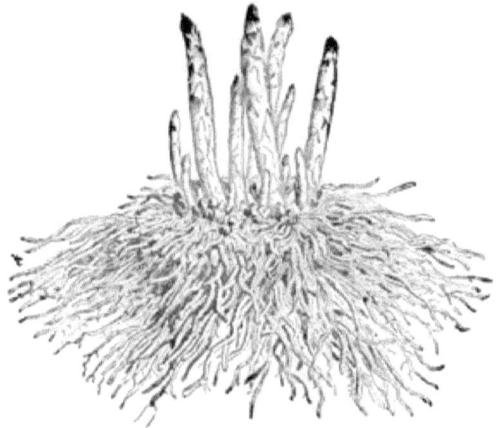

ASPARAGUS OFFICINALIS (Asparagus).

European colonists, where it is under cultivation, and is a favorite table vegetable. It has escaped from the garden, and is growing wild, especially along the seacoast.

ASPARAGUS OFFICINALIS (Asparagus).

Etymology. — The name *asparagus* is derived from the Greek word σπαράσσω, tear, alluding to the thorny character of some of the species. *Officinalis*, the specific name, is from the Latin *officina*, belonging to trade, or the market. The common name, *sparrow grass*, is a corruption of *asparagus*.

History. — Asparagus was known to the Greeks and Romans at least two thousand years ago.

Use. — Asparagus is a very delicate vegetable, and most highly prized as a pot-herb. It is greatly improved by cultivation, both in size and quality. Its tender, fleshy shoots are the parts eaten. When boiled in a small quantity of water with a little salt, it retains its natural delicate flavor, which to the epicure is not improved by disguising sauces. It is also cut into short pieces a quarter of an inch in length, boiled, and served as green peas are. Its medicinal properties are diuretic, and it is recommended for complaints of the kidneys. Its seeds have been used in Europe as a substitute for coffee.

PHORMIUM, Forst. Perianth incurved, segments connate at the base, in a short, top-shaped tube, the three exterior segments lanceolate, the three interior sometimes a little longer and narrower at the apex; stamens 6, attached to the base of the tube; filaments thread-like; anthers oblong or oblong-linear, erect, attached along their backs to the filaments; filaments intruding into the anther-cells. Ovary sessile, oblong, 3-cornered, and 3-valved; style filiform; stigma short, head-shaped; ovules numerous; capsule somewhat leathery, 3-sided, cylindrical, often strongly twisted. Perianth cylindrical, funnel-formed, united below. Leaves radical, keeled, not fleshy. Seeds oblong, flattened; seed-vessel black, membranaceous, sharply dilated at the margin, but not properly winged. Embryo long.

P. tenax, L. (New Zealand Flax.) Rhizome short, thick, radical fibers densely matted. Leaves radical, 3 to 9 feet long, narrow, linear, sword-shaped, leathery, tenacious, arranged a little way up the stem in two rows on opposite sides for a short distance from the root. Scape leafless and branching above; bracts under the branches, which sometimes fall off. Flowers pedicelled, in terminal panicles, or fascicled with the bracts along the branches, varied in color; flower-stalk jointed under the flower. The seed matures in the third year, the stalk rising to the height of 16 feet, when the whole plant dies down, and renews itself from its roots.

Geography. — Its home is New Zealand, where it was applied by the natives to the manufacture of clothing, cordage, and mats before the arrival of Europeans. It has been introduced into southern Europe and northern Africa. It grows in the open air in the southern parts of England and of Ireland; and is grown in the United States in greenhouses.

PHORMIUM TENAX (New Zealand Flax).

Etymology. — *Phormium* is from the Greek φορμός, wicker-work, or a basket, in reference to the use made of the fiber of the leaves. *Tenax* is Latin, signifying "tenacious," due to the strength of the fiber. *New Zealand flax*, the common name, indicates the country where it is found native.

History. — The plant was discovered in New Zealand, and carried to Europe by Captain Cook in 1773, and described by Linnæus. It is found wild and also under cultivation in its native country.

Preparation. — The leaves are cut into strips, and the outer green part scraped off; the fibers are then separated and kept straight, and are washed, rubbed, and bleached in the sun; it is then creamy white, and has a silky luster. While growing, it resembles the American agave or century-plant.

Use. — It is used in New Zealand for fabrics for garments, and for cordage, and enters into commerce as a material for cordage and coarse bagging. It is also used to adulterate the fiber of the musa textilis in the manufacture of manilla cordage.

ALOE, L. Perianth tubular, contracted above, straight or slightly curved, 6-lobed at the summit, nearly closed. Stamens 6, inserted below the ovary, exserted; filaments awl-shaped, as long as the perianth; anthers linear-oblong, introrse, 3-celled, many-ovuled; style thread-like; stigma small; capsule leathery, ovoid or oblong; seeds numerous, 3-angled or flattened, 3-winged; testa membranaceous and black. Leaves thick, fleshy, in two or three series, crowded near the base of the stem, lanceolate, ends curved down; margins dentate; teeth armed with spines; scape terminal. Flowers yellowish.

1. **A. vulgaris**, Lamarck. (Aloes.) Stem perennial, 2 to 3 feet high, 6 to 8 inches in diameter, crowned with a dense cluster of leaves, bases wide, lower ones spreading, upper ones more erect, lanceolate, thick, fleshy, concave above, convex beneath; margins toothed; teeth armed with hooked prickles; surface of the leaf smooth, dark green or mottled; scape either simple or branched, rising from the crown of leaves, terminated with a slender spike of flowers; bracts triangular, acute, longer than the pedicels, veined, and persistent. Stamens 6, a little longer than the perianth; anthers small, oblong. Ovary oblong-oval, bluntly triangular, 3-celled, double row of ovules in each cell; style as long as stamens; stigma terminal. Fruit oblong ovoid, blunt capsule, an inch long, 3-celled; pericarp thin, brown, smooth; seeds numerous, flattened; testa thin and membranaceous. Flowers yellow, tinged with green.

1. **A. socotrina**, Lam. This species differs from the last in a taller stem, which, as it grows old, becomes forkedly branched, with tufts of leaves at the summits of the branches, prickles on the leaves. Perianth an inch and a half long, red or pinkish, with greenish-white tips, part of the stamens extending beyond the perianth.

3. **A. spicata**, L. f. Stem about 3 feet high. Leaves enveloping the base of the stem, curved, and then deflexed, about 2 feet long, flat near the base, becoming nearly cylindrical near the end, armed with strong, sharp, triangular prickles, with several small ones at the summit. Flowers bright orange-yellow. Stamens longer than perianth; style longer than stamens.

There are in all about 80 species of this genus, natives of the hot regions of the Old World. But the aloe of commerce is the product of the species above described, and of others whose botanical characters have not been determined, and whose home is not known.

Geography. — The aloe is a tropical and subtropical plant, and may be cultivated in all regions of no frost. It is found in southern Asia, Arabia, southern Europe, and northern Africa; but it abounds in south Africa, near the Cape of Good Hope. It is largely cultivated in the British West Indies.

Etymology. — *Aloe*, the generic name, is said to be derived from *alloch*, an Arabic word. The Greek for the same is ἀλόη; and the Latin, *aloe*, whose figurative meaning, "bitterness," seems to suit the case, for the name could not be more appropriate. *Vulgaris*, the specific name, is Latin for "common." *Socotrina* is from the Island Socotra, where it was formerly cultivated. *Spicata*, Latin for "spike-bearing," alludes to the form of inflorescence. *Ferox*, Latin for "wild," in this case signifies "coarse" and "strong."

History. — The aloe was known to the Greeks and Romans; it was mentioned during the first century by both Dioscorides and Pliny. Its home is Asia and Africa; and it has been carried to the West Indies by Europeans, and is cultivated there.

Preparation. — There are three kinds of aloe known in commerce, namely: *Cape aloes*, *Barbadoes aloes*, and the *aloes of Socotra*. The Cape aloes is obtained from the leaves of A. spicata and A. ferox, and perhaps other species. The leaves are cut from the plant and set into vessels to drain; the sap, or juice, is evaporated in iron vessels over a fire; when reduced to a thick syrup, it is poured into vessels to cool, in which condition it is sent to market in large cases. At a temperature below 40° Fahrenheit, it is hard and brittle; at a higher temperature, it runs like pitch. The Barbadoes aloes is the product of the A. vulgaris, which is under cultivation in that island, and in Jamaica.

The juice, or sap, having been drained from the leaves, is stored in casks until the collecting is over, when it is evaporated in copper kettles to a syrup, then poured into gourds holding from three to seventy pounds, and thus sent to market. This kind was first sold in London in 1693.

Socotrine aloes is manufactured somewhere in the interior of Africa, taken to the ports on the Red Sea, and thence to Zanzibar, from which place it is sent to Bombay, where it is purified and shipped to England. Where the plants grow that yield it, or how it is prepared, is not very well known, but it is very certain that the drug does not come from Socotra.

Use. — Its medicinal properties are, when administered in small doses, stimulant, tonic, stomachic, and anti-bilious. When taken in large doses, it acts violently upon the bowels, producing drastic purgation. It was formerly extensively used in complaints of the liver.

In Africa cords and nets are made of its leaf-fiber. Its juice has been used in embalming, and as a varnish to protect against insects.

Note. — The aloe of the Bible is a different plant. The reference made in the Bible to aloe is always to the wood, and not to the plant; and it is believed that the resinous wood known to the Egyptians and western Asiatics was the product of a tree found in eastern Asia, *Aquilaria Agallocha*, Roxb., of the order Thymeleaceæ. The wood is very valuable, mostly for its fragrance when burned. It sells in Bombay as high as $3.00 per pound for fine specimens.

Order LX. PALMÆ.

Flowers small, diœcious or monœcious, seldom perfect, sessile or on short pedicels; perianth double, persistent, leathery, formed of a calyx and calyx-like corolla; sepals 3, distinct or coherent; petals 3, usually distinct, valvate in the bud of the male flower, imbricate in the female; stamens hypogynous, or perigynous, 6 in two whorls, opposite the sepals and petals. Inflorescence axillary. Fruit a berry or drupe, with smooth or scaly epicarp; sarcocarp fleshy, oily or fibrous; endocarp membranous, fibrous, woody, or bony. Seed oblong, ovoid, or spherical. Leaves springing from the terminal bud, alternate, base of petiole sheathing the stem, petiole convex below; blade pinnate, fan-shaped or simply split. Perennial woody plants, mostly beautiful and majestic trees. Trunks often tall and slender.

No. of genera, 132; species, 1100; tropical and subtropical.

ARECA, L. (Betel Nut. Feather Palms.) Flowers monœcious, small, numerous, sessile, bractless, spadix branched. Staminate flowers very numerous; calyx with 3 small ovate segments; petals 3, much longer than the calyx, broadly ovate, smooth, thick, and yellow; stamens 3-6; filaments short, red, and attached to the backs of the arrow-shaped anthers. Pistillate flowers with a calyx of 3 ovate rigid sepals; petals 3, like the sepals, but thinner. Ovary large, 1-celled, one ovule in each cell; stigmas 3. Fruit two and a half inches long, containing a single seed an inch long and two thirds of an inch in diameter, somewhat in shape of a nutmeg, distinctly marked by a network of red veins, which penetrate the mass and give a marbled appearance to the internal structure. A tree produces about 300 nuts. Leaves pinnately divided.

ARECA CATECHU
(Betel-nut).

A. catechu, L. (Betel Nut.) Stem slender, 40 to 60 feet high, 18 to 20 inches in diameter. Fronds 3 to 4 feet long, all terminal. Leaflets numerous and opposite.

Geography. — The geography of the areca is tropical and subtropical. It has spread through the Sunda Isles, the Philippines, Cochin China, Sumatra, and southern India, and has names in each of these countries which point to the probability that it is native to all these localities; but there seems to be insufficient evidence to locate its home in any one of them. It grows best on plains and terraces.

Etymology. — *Areca* is from *arec,* the name applied to the young tree in Malabar. *Catechu* was applied to this species because it was erroneously supposed to yield the gum catechu. The fruit is called *betel nut,* because the leaf of the piper betel is used in connection with it. *Feather Palm* alludes to the feathery appearance of the leaves, and is applied to several of the genera of this order.

History. — When the betel nut was first used as a masticatory is not known; it was in use in the East Indies when the country became known to Europeans, and its use upon state occasions dates back to the fourth century.

Use. — When the nut is in an unripe state it is cut into slices and wrapped in the leaves of the piper betel, with a little lime, and in this condition it is chewed. It affects the nervous system, somewhat in the same manner as tea, coffee, and cocoa; it is stimulant and astringent, and is said to check perspiration and alleviate fatigue. When used moderately it is supposed to preserve the teeth, and to fasten them when loose in the gums; but used to excess it destroys the teeth, causing them to crumble and waste away and eventually it causes the death of its victim. It is of very general use among the inhabitants of southeastern Asia, extending to all classes, and it is offered to guests on state and other ceremonial occasions. It is carried in a box, frequently of gold or silver, or ornamented with precious stones, corresponding to the snuffbox used for tobacco snuff.

A liniment made of the juice from the leaves, mixed with oil, is considered a specific in lumbago; the nut is also a reputed remedy for tape-worm.

PHŒNIX, L. (Feather Palm.) Flowers diœcious, very small, in large clusters, emerging from a spathe or fleshy shield; number of flowers in a spadix sometimes reaching 12,000. Staminate flowers oblong or ovoid; calyx cup-shaped, 3-toothed; petals 3, oblique, oblong, or ovoid, slightly attached at base; stamens 6, sometimes 3–9, inserted on the base of the corolla; filaments awl-shaped; anthers linear, oblong, erect, and attached to the filaments along their backs. Pistillate flowers with a rotund calyx; stigma sessile. Fruit oblong, terete, 1-seeded. Pericarp fleshy; seed hard, somewhat cylindrical, plane on one side and convex on the other, with a groove extending the whole length of the plane side, from half an inch to an inch in length, having sweet, nutritious pulp. Leaves all terminal, pinnate.

P. dactylifera, L. (Date Palm.) Stem 30 to 60 feet high, 12 to 18 inches in diameter, scarred with the marks of fallen leaves, which are pinnate, glaucous, 8 to 10 feet long; leaflets lanceolate, acuminate, pinnæ close together. Flowers in branching spadices, the main flower stem long, the heavy mass of flowers causing it to bend downward, each cluster of the female tree bearing from 150 to 200 dates; each flower produces three, two of which are usually abortive. The fruit has a vinous, gummy, sugary taste.

There are twelve species of the phœnix, but the only one of importance as a food plant is the P. dactylifera. The varieties of this species are very great; it is propagated by seed and sports freely, producing varieties which differ from each other in the size and quality of the fruit.

Geography. — The geographical distribution of the date is throughout the hot and arid region between 15° and 30° north latitude, from the Atlantic coast along the fringes of the great deserts to the river Indus, and on the oases of the Sahara. It can grow in a higher latitude, but refuses to ripen its fruit except in a very high temperature. The tree itself will live just in the edge of the region of no frost, but will not fruit, nor even flower. Somewhere in the above-named region it had its origin, but no naturalist has been able to name the spot. According to Loudon, the date derives its generic name, *Phœnix*, from the fact that the best dates were brought from Phœnicia. But if fine dates came from Phœnicia, they must have been carried there from further south, for the temperature of the most southern part of Phœnicia is too low to ripen the date. They may have been taken from Arabia or from more southern parts of Syria, by caravans, to the seaports of Syria in early times, and shipped there for southern and western Europe.

The date-tree is becoming rare in Syria. It is common near Acre, skirting the lagoons and crowning the sandhills which abound there. The city of Palmyra got its name from this tree, which is still abundant in its vicinity. It was the symbol of Palestine in the time of the Jews. It adorned the medals of Vespasian and Titus, as emblems of the country. It is now carefully cultivated near Jaffa.

Etymology. — The derivation of *Phœnix*, the generic name, has been given. *Dactylifera* is from the Greek δάκτυλος, finger, from a fancied resemblance of the fruit to the portion of the finger between the joints. *Date*, the common name, is said to be a corruption of *dactyl*, and is also claimed to be a Sanscrit word signifying "fruit," or "finger fruit."

History. — The date has been cultivated as a food plant since prehistoric times. The ancient Egyptians understood the sexes of the trees, and learned to assist nature by taking the pollen from the male to the female tree. Cakes of dates have been found in the tombs of Thebes, one of which is preserved in the British museum. The ancient Hebrews must have been acquainted with the date, though it could not have been among the fruits of Palestine, for it is only indirectly referred to in the Scriptures. The tree grows in Judæa, and it is said that the fruit ripens as far north as Jerusalem. Bethany, which is near Jerusalem, signifies "the home of dates." It has been represented as a tree delighting in dry and arid regions, but recent investigations reveal the fact that it requires an abundant supply of water; and wherever it is found, even in the midst of the desert, it indicates the presence of surface springs.

PHŒNIX DACTYLIFERA (Date Palm).

Cultivation. — It is artificially fertilized by shaking a branch of staminate flowers over the pistillate inflorescence.

Use. — The date is eaten raw in all cases, and furnishes the only food for thousands of the inhabitants of the region where it grows. It is conserved with sugar, and packed in bags made of the leaf. It is said that the native Arab will exist for days without inconvenience on a few handfuls of this fruit, while his camels are content with date-stones ; this may serve to indicate the remarkable amount of nutriment which the date contains.

The uses of the leaf and stem are somewhat similar to the uses to which the cocos is applied. Date trees are employed as articles of trade and dower.

COCOS, Linnæus. (Feather Palm.) Flowers monœcious, on the same spadix. Spathe simple, woody, spindle-shaped; staminate flowers on the upper part of the spadix, and the pistillate ones below. Calyx of the staminate flower 3-leaved; sepals lanceolate, keeled, and frequently united at the base; corolla 3-parted, membranous or fleshy, usually erect. Stamens 6; filaments awl-shaped, nearly equal in length; anthers linear and erect; calyx of the pistillate flower ovate or sub-orbicular, 3-leaved; sepals convolute; corolla 3-parted; petals membranaceous, imbricated; ovary ovate or flattened-globose, 1-celled; style short or wanting; stigmas 3. Fruit ovate-elliptical or 3-sided, outer coating thick and fibrous, 6 to 12 inches long, and 4 to 8 inches in diameter. Seed nearly globular; testa horny, marked with three spots at the top. Large palms; trees with a crown of feathery leaves.

C. nucifera, Linnæus. (Cocoanut Palm.) Trunk 50 to 100 feet high, and 1 to 2 feet in diameter. Leaves springing from the terminal bud, alternate, pinnate, sheathing at base; pinules with decurved margins. Flowers enveloped in a spathe. Fruit in clusters of 10 to 20, 3-sided, tapering to a blunt point, 6 to 9 inches long and 3 to 5 in diameter; outer husk fibrous, enveloped in a smooth, glossy cuticle; the whole an inch or more in thickness. Seed a prolate spheroid, 4 to 5 inches long and 3 to 4 in diameter, inclosed in a horny shell, rough on the outside, and one fourth of an inch thick, inside of which and adhering to it is a coating or wall half an inch thick, hard, white, crisp, oily, and edible; the cavity within is filled with a rich, limpid liquor, which is very grateful to the palate. A single tree sometimes has upon it 120 of these gigantic nuts at one time.

Cocos NUCIFERA (Cocoanut Palm).

There are about thirty species to this genus, all tropical or strictly subtropical.

Geography. — The geographical zone of the cocoanut is equatorial, extending twenty-five degrees both sides of the equator in seacoast regions. It is found in higher latitudes, but does not fruit well above 25°, except in localities especially favored by a high temperature, — a temperature which is seldom below 75° Fahrenheit. It abounds in the coast regions of Brazil, on the West India Islands, the western and eastern coast, and islands of Central America;

it is extensively grown in Ceylon, and in all the coast countries of southern Asia, especially in Farther India and southwestern China, and delights in a sea exposure.

Etymology. — *Cocos*, the generic name, is supposed to be derived from the Portuguese word *coco*, a monkey, due to the three spots on the end of the nut, which make it resemble the face of that animal. It is also believed to come from the Greek word κόκκος, a fruit or berry. *Nucifera*, the specific name, is compounded of the two Latin words, *nux*, a nut, and *fero*, bear, and signifies "nut-bearing." *Cocoanut*, the common name, is made up of *coco* and *nut*.

History. — It is on record that this fruit, if not the tree, was known to the ancient Egyptians centuries before the beginning of the Christian era. Columbus found it growing in Central America on his fourth voyage to the New World. The Arabs were acquainted with it, and their physicians wrote of it during the Middle Ages. It has Indian, Arabic, Persian, and Malay names, showing it to be native in those countries four thousand years ago. The bulk of evidence is that the home of the cocoanut is the East Indian Archipelago, whence it has sown itself, by means of the ocean currents, on the coasts of eastern Africa and the far off shores of western America. The character of its fruit is highly favorable to its distribution by sea; and its preference for the seacoast leads to the inference that it has drifted across the ocean and been thrown upon the isles and low shores, where it has become naturalized, — the seed being covered by a strong shell, and the whole inclosed in a light, porous husk, which is impervious to water, on account of a smooth, glossy cuticle which envelops it.

Use. — There is not another individual of the whole vegetable kingdom which furnishes so many useful substances to man, no single tree of the vast flora of the world is so completely utilized, as the cocoanut palm. Its products are: sugar, milk, solid cream, wine, vinegar, oil, cordage, cloth, cups, timber for building, and materials for implements of husbandry, furniture, baskets, mats, and culinary utensils. The inhabitants of large districts subsist almost exclusively upon its fruit alone. The shells are used for goblets, ornamented with carving.

In Ceylon a man's credit and commercial standing is measured by the number of cocoanut trees he has in his plantation.

The fruit is largely used in America and Europe for the preparation of dessert dishes, as pies, custards, cakes, and confections; and the desiccating of the cocoanut has become an important industry in the eastern United States. The oil of the nut is treated to extract the stearine which is employed for the manufacture of candles; the more fluid parts are used for salad dressings, for illuminating purposes, and for the manufacture of soap. Soap made with this oil is so soluble as to make it capable of use in sea water. Among fibrous plants the cocos holds high rank. The fiber of the husk (coir) is employed in the manufacture of floor cloths, door mats, strings, bags, brushes, brooms, and many other articles for domestic use. It does not decay in water. The leaves serve for thatch, and their strong midribs are divided into splints, and woven into baskets.

METROXYLON, Rottb. (Feather Palm.) Flowers polygamous. Staminate flowers with funnel-shaped calyx, 3-toothed; corolla 3-parted, lapping each other; stamens 6, inserted on the base of the corolla: filaments united at the base; anthers linear and dorsifixed. Pistillate flowers much like the staminate ones; ovary oblong, 3-celled, conical, with 3 tooth-like stigmas; ovules 3. Fruit ellipsoidal or subglobular, 1-celled, 1-seeded; pericarp clothed with scales, imbricated. Leaves terminal, nearly erect, pinnatisect; segments usually opposite, linear-lanceolate, acuminate; petiole convex underneath, furrowed above. Spadix large, branched, loose; spathe leathery. Flowers immersed in a woolly covering, small.

M. sagu, Rottb. (Sago Palm.) Trunk 30 to 50 feet high, and 6 to 15 inches in diameter, straight, cylindrical, gray, while young armed with strong, sharp spines to protect it against the attacks of the wild hog; these spines fall off when the rind has become hard. Outer coat of the trunk hard, inner part spongy. Leaves few, crowning the stem; entire leaf, including petiole, about 20 feet long, erect, slightly curved; petiole 6 to 8 inches in diameter near the base, clasping; rachis square below, and keeled or triangular above. Leaflets many and opposite, middle ones longer, straight, stiff, narrow, linear-lanceolate, acute, entire, leathery and smooth, 3-veined, bright-green above, pale beneath. Flowers unisexual or perfect, sometimes mixed in the same spadix, numerous, small, each with three small bracts, smooth within, covered on the outside with a yellowish wool, and in the axil of a strong smooth bract. Inflorescence in a cylindrical, dense spike, 4 inches long. The whole inflorescence consists of numerous spikes arranged along on 6 to 9 main stalks, which crown the tree, making a center-piece for the leaves. Calyx rigid, smooth, and 3-lobed; corolla much larger than the calyx, cut into three obtuse, thin segments; stamens 6, as long as the corolla, and inserted on it at its base; ovary short-stalked, imperfectly 3-celled, 1 ovule in a cell; style conical, tapering, pointed. Fruit one and a half inches in diameter, spherical, mucronate at summit, clothed with greenish-red scales. Seed solitary, enveloped in a dark-brown testa. The tree matures in from 15 to 20 years, when it flowers and fruits and dies. The seeds seldom mature, and the tree is propagated by stolons.

METROXYLON SAGU (Sago Palm).

There are six species of this palm, but the sago of commerce is obtained from the following species:—

M. sagu, M. spinosa, and M. lævis.

Geography.— This tree is found only in the tropics, in the hottest and wettest parts of Asia. It flourishes throughout the Eastern Archipelago, extending about 10 degrees both sides of the equator, abounding in swampy localities.

Etymology. — *Metroxylon* is from the Greek μήτρα, the heart or pith; ξύλον, wood or tree; hence *pith-tree*. *Sagu* is the Malay name for "bread" or "food." *Sago* is a corruption of *sagu*.

History. — There is no reason to believe that the ancients were acquainted with the product of this tree. When the East Indies were first visited by

Europeans the sago constituted the principal food of the natives of Malacca, southern China, and the adjacent isles.

Preparation. — The tree grows best in a muddy bog. An acre yields about 300,000 pounds, and a single tree produces about 500 pounds, — a very large tree, 900 pounds. It is cut at the foot, just as it is about to fruit, which occurs when the tree is from fifteen to twenty years old. The top is taken off, and a strip of the shell or outer coating is removed the whole length of the trunk, and with an instrument constructed by fastening a sharp stone to a stick the pith is removed, leaving a thin shell not more than an inch thick. The pith is kneaded in water, in large troughs, by which process the starch or sago is washed out and sinks to the bottom; the water is then run off. This starch, or sago, is then made into balls or rolls, which weigh thirty pounds each, when it is ready for use, as crude or raw sago. The natives make it into cakes, and bake it, in which condition it will keep for years, and when used it merely needs soaking. One tree will produce sufficient food to keep a man for a year.

There are several varieties of the sago, due to different methods of preparation. Pearl sago, the variety which reaches the American market, is prepared in Singapore by the Chinese; the raw sago is made into a paste, forced through sieves and dried in pans over fire, and constantly stirred while undergoing the process.

Use. — Sago furnishes the inhabitants of Malaysia and southern Asia a food material equal in importance to rice in other parts of Asia. Many millions know no other food but fish and the crude sago.

In Europe and the United States it is a popular material for puddings and custards. It is pure starch, free from all irritating character, and hence constitutes an excellent food for infants, old people, and convalescent invalids.

The leaves are used for thatching; the leaf-stalks, immense in size, furnish building material, and the fiber is used for cordage; in fact, all parts of the plant are applied to some use.

Order LXI. **GRAMINEÆ.**

Flowers perfect, occasionally monœcious or diœcious, sometimes polygamous, each mostly with 2 opposite bracts or paleæ, the lower of which is larger. Inflorescence in spikelets, the spikelets variously collected into panicles or spikes. Perianth imperfect, seldom wanting, composed of whorled hypogynous, membranous, or fleshy, irregular scales, free or connate, 3, 2, or 1 in number, the outer alternate with the paleæ. Stamens hypogynous, 3 or 6 in number, seldom 4-2 or 1; ovary free, 1-celled, 1-ovuled; styles 2, very rarely 3, free or connate at base, or united; stigmas with simple or branched hairs. Leaves alternate, springing from the nodes; petiole dilated, convolute, sheathing; sheath split in front, blade entire, mostly linear; stipule axillary at the top of the leaf-sheath. Stem cylindrical, rarely flattened, fistular or solid, mostly jointed at the insertion of the leaves; nodes annular, solid, and swollen.

Annuals or perennials, with fibrous or creeping rhizome, frequently stoloniferous at the lower node.

Some of the cultivated species of this order are not known in a wild state. Number of genera, 1,298; species, 3,200; found in all parts of the world. Our most useful plants belong to this order. The stems and culms of most of species are largely composed of silicates.

ZEA, L. Flowers monœcious: staminate flowers in terminal panicles; pistillate flowers in 1-3 axillary spikes; staminate spikelets 2-flowered, with 2 concave glumes, the lower one 3-nerved, the upper one 2-nerved. Paleæ membranous and without arms, the two collateral and fleshy scales glabrous. Stamens 3 in number, and the linear anthers 4-sided. Pistillate flowers with very short glumes on an axillary spike, which is inclosed by many spathe-like bracts; these form the husk by which the fruit is protected. Style simple, very long, thread-like, far exserted, and hanging.

Z. mays, L. (Indian Corn, Maize.) Stem or culm from 3 to 15 feet high, 1 to 2 inches in diameter, composed of bundles of woody fiber imbedded in pith; the whole inclosed in a smooth, flinty cortex, terete, grooved on one side with a smooth, semi-circular channel, and divided into nodes whose intervals are from 5 to 9 inches long, crowned at the top with a compound panicle of staminate flowers called "the tassel." Root fibrous; the stem throws off aerial roots from the nodes next to the ground, which are called "brace roots."

The pistillate flowers are borne on a close axillary spike, inclosed in a mass of spathe-like bracts, called "the husks," and are characterized by the elongated, filiform styles, which extend far beyond the orifice of the bracts and hang in tresses over the ear or head, like bundles of silk threads; they are called the "silk." The leaf appears at the nodes, clasping the stalk by striate sheaths, which are eared, from 2 to 4 feet long, and from 2 to 4 inches broad, linear-lanceolate. There are no radical leaves.

Fruit a flat, reniform or cuneate-shaped seed, arranged in rows on the rachis, which is from 5 to 12 inches long, and called the "cob"; the number of rows on a cob varies from 8 to 14, with from 20 to 50 seeds in a row. The cob ranges from three fourths of an inch to two inches in diameter.

Flowers in July or August, and ripens its fruit in September and October.

In common with other plants grown from seed, it sports and forms varieties which depart from the specific character in color, size, and shape of the grain and quality of the meal made therefrom. The gourd-seed varieties are in every way larger, and the seeds not so hard and flinty. The flour-corn has reniform seed, but is softer, and is used largely in the Southern States for bread. The gourd-seed varieties are raised in the South and West. The eight-rowed flint, both white and yellow, are grown in New York and the eastern Atlantic States north of New York.

The number of forms resulting from free sporting is very great; about 30 of these are sufficiently characteristic and constant to make species, and would be so regarded if outside of cultivation.

Geography. — Indian corn is now cultivated and is an important crop in all the United States, Upper Canada, Mexico, South America, southern Europe, Africa, and western Asia. It does not grow well above the forty-fifth parallel, but flourishes best below the fortieth, and delights in a hot, sunny clime.

Tropical and subtropical countries seem to be the lands of its birth. As it becomes acclimated farther north, the ears diminish in size, and the whole plant becomes dwarfed. In the warm regions it reaches the height of 12 to 15 feet; in Maine and southern Canada it ranges from 3 to 4 feet. The ears in Canada seldom exceed 8 inches in length, while in the southern United States the length is from 8 to 15 inches.

Etymology. — *Zea* is from the Greek ζάω, live, alluding to the capacity this grain has to sustain life. *Mays* and *maize* are derived from *mahiz*, the name by which the American aborigines called this plant, — the meaning of which is obscure. *Indian corn* is a name given by Europeans, on account of the use of the plant by the aborigines of America.

History. — This cereal is undoubtedly of American origin. Attempts have been made to show that in an old Chinese book found in the national library in Paris there is a figure of a plant identical with corn; and hence the inference is drawn that it is also indigenous to Asia; but there is good reason to doubt that it was known in the Old World before the discovery of America, or before Columbus introduced it into Spain in 1520, twenty-eight years after the discovery of the New World. Humboldt and other good authorities do not hesitate to say that it originated solely in America. In a marvelously short time it spread over southern Europe, northern Africa, and western Asia, showing conclusively that had it been known in Asia it would have reached Europe before the discovery of America. It is found in the tombs of the ancient Peruvians, and in the mounds of the Mississippi. C. Darwin found it buried with shells fifty-five feet above tidewater. It is nowhere found wild, propagating itself.

Cultivation. — Maize delights in a light loam, which cannot be made too rich by fertilizers. In the well-worn lands of the Atlantic States large supplies of fertilizers are necessary to abundant crops; but in the rich alluvial bottoms of the Mississippi Valley nothing is needed to insure an abundant harvest but to plough and plant, and stir the soil. It is planted in the intersections of cross-drills, four feet apart each way, by dropping four to five grains in a place and covering with a hoe. In the Atlantic States from thirty to forty bushels to the acre are a satisfactory yield; fifty and seventy-five bushels are frequently reached; in the bottom lands of the Mississippi Valley, sixty to eighty bushels are not uncommon.

In the Atlantic States north of New Jersey the yellow and white flint varieties are grown, but in the lowlands of New Jersey and further south the

ZEA MAYS (Indian Corn).

gourd-seed varieties are planted, which require a longer season than the climate north of central New Jersey affords.

Use. — As an article of food, Indian corn is used for bread, mush, griddle-cakes, puddings, dumplings, etc. When merely cracked, it is called "samp," and is prepared for the table by boiling and serving like a vegetable, or by boiling it with salt meats. When coarsely ground it is called "hominy," and is cooked and served as oatmeal, and eaten with milk or sauce.

Large quantities of starch are manufactured from it, both for the table and the laundry. It is also used in large quantities for distilling, and is the chief material for manufacturing the celebrated Bourbon whiskey.

But the most important use made of this grain is as feed for cattle, sheep, horses, and swine; it excels all other feed for its fattening properties.

The leaves and upper parts of the stalks are good substitutes for hay, and the cobs are excellent fuel. The quantity of corn raised in the United States exceeds the amount of all the other cereals combined. It yields the largest returns of all the cereals.

Marts. — The great marts of North America are Chicago, Buffalo, St. Louis, New Orleans, New York, and San Francisco, in the United States, and Toronto in Canada.

TRITICUM, L. Spikelets from 2- to many-flowered, in a stout spike. Florets distichous; rachis zigzag; glumes nearly opposite, not quite equal, sometimes with awns; inner paleæ herbaceous, lower one concave and sometimes awned or mucronate, upper one with 2 aculeate and ciliate keels. Scales 2, usually entire and ciliate. Stamens 3. Ovary sessile, crowned with 2 plumose stigmas.

T. vulgare, L. (Wheat.) Stem or culm 2 to 5 feet high, tapering from the root to the base of the head or the ear, divided by nodes into several internodes, or lengths, from 4 to 7 inches long. At each node is a single, clasping, lance-shaped leaf, strongly veined and rough on the upper side. Flowers appear at the top of the culm in a close panicle.

The grains, or seeds, are oval in shape, a quarter of an inch in length, flat, and marked on the side next the rachis by a groove the whole length, outside convex. It is an annual, and when planted in early spring, it flowers and fruits the same season; when thus cultivated it is known as "summer wheat" or "spring wheat." The best wheat is biennial; it is planted in early autumn, in time to take root and form root or radical leaves before winter sets in; it ripens in July of the following year, and is called "winter wheat," because it remains in the ground during the winter.

As wheat is grown from the seed it sports or produces new forms; growers have taken advantage of this circumstance to obtain improved varieties, and very many such varieties have from time to time been recommended by wheat growers, especially in Europe and America. The varieties under which the forms may be classified in America are: —

Var. **hybernum**. Winter Wheat. **T. hybernum**, L.
Var. **æstivum**. Summer Wheat. **T. æstivum**, L.
Var. **nudum**. No-bearded Wheat.
Var. **album**. White Wheat.
Var. **rubrum**. Red Wheat, or Mediterranean Wheat.

There are many other forms, all referable to the above, which are sufficiently constant to be considered varieties.

T. compositum deserves to be treated as a species, and in all respects resembles T. vulgare, except that the stalk is thicker, and the head branching. It is grown in northern Africa, and in southern Italy, and was no doubt the wheat of the ancient Hebrews and the Romans.

Geography. — Wheat does not grow well north of the fiftieth degree of north latitude in North America, nor south of the thirtieth degree. In Europe it grows well in southern Russia below 51°, and is cultivated with success throughout central and western Europe, and as far south as southern Italy. It is also cultivated successfully in Turkey, Syria, northern and southern Africa, and in the south temperate zone in South America, — in Brazil, Chile and Buenos Ayres; also in Australia, where it constitutes the most important object of agriculture. The great wheat-growing regions of the world are the southwestern plains of Russia, the great central plain of North America, the southern plains of California, and recently, northern India and England.

Etymology. — *Triticum* is from the Latin verb *tero*, whose participle is *tritus*, rubbed, — alluding either to the practice of rubbing it to separate the grain from the chaff, or to the mode of grinding it into flour. The specific name *vulgare* is the Latin for "common." The word *wheat*, the common name, is supposed to be derived from the Sanscrit *sereta*, meaning "white," and arises from the circumstance that the flour made from this grain is white.

History. — No form of wheat, nor any species closely resembling it, has ever been seen wild. It must, therefore, either have been very much altered from the original wild grass, which tradition and probability would lead one to consider a native of some part of central Asia; or else, by reason of changes of climate in the country of its origin, it has become extinct as a wild plant. In favor of the latter supposition in preference to the former is the fact that, like other annual cereals, the wheat shows very little tendency to vary. The forms cultivated in ancient Egypt, in China, and in Palestine, appear to be identical in all respects with those we are now familiar with. The home of the wheat is generally believed to be western Asia, in the countries watered by the Tigris and Euphrates, whence it has found its way into every favorable clime where agriculture is practiced. Among the ancient Egyptians, and the inhabitants of Palestine, Mesopotamia, and northern Syria,

TRITICUM VULGARE (Wheat).

it was the most important crop. In the western hemisphere wheat was not known until the sixteenth century. Humboldt mentions that it was accidentally introduced into Mexico with rice brought from Spain by a negro slave belonging to Cortes, and the same writer saw at Quito the earthen vase in which a Flemish monk had introduced from Ghent the first wheat grown in South America.

Cultivation. — Wheat is a true patrician; it will not thrive upon scanty fare, nor flourish without attention. It is what agriculturists call a gross feeder; it not only demands a deep, heavy soil, but the soil must be well tilled and highly fertilized in order to satisfy this prince of the cereals. Without these conditions it refuses to yield largely, but responds with liberal harvests to generous cultivation.

In the rich bottom lands of the Mississippi valley, the southern plains of California, and the wheat-growing lands of southern Russia, nothing is needed but to prepare the ground and sow the seed to insure a large yield; but in old and long cultivated districts the most careful attention to suitable fertilizers is necessary to secure even moderate crops. The yield per acre varies from ten to sixty bushels. In the well-worn fields of the Atlantic states the yield is frequently not more than twelve bushels, and thirty bushels is a very satisfactory crop. In the rich alluvial soil of the central states the yield frequently reaches fifty bushels, and sometimes sixty. The quantities vary with soil, climate, and mode of cultivation.

A notable case of high farming was brought before the court in ancient Rome. A farmer was accused of sorcery for raising better crops of wheat than his neighbors. When the accused appeared before his judges, who sat in the open air, he brought with him and exhibited his agricultural implements, superior in construction, his well-fed oxen, and his callous hands. Pointing to his cattle and implements, he exclaimed: "Here, O Romans, are my tools of witchcraft, which I employ to make my crops." His judges pronounced him innocent, reprimanded his accusers, and advised them to follow his example.

Use. — A bushel of wheat weighs 60 pounds and will make 47 pounds of flour, leaving 13 pounds of middlings, bran, and waste.

This grain now constitutes the staple food of most of the civilized peoples of the earth. The flour is made into bread, cakes, puddings, pastry, crackers, biscuit, etc., and is so well known as to need no further description. The Jews were acquainted with the making of leavened bread, which they no doubt learned while in Egyptian bondage. When leavened bread was first used, and who first made it, is nowhere recorded. Homer speaks of leavened bread at the time of the Trojan War. Pliny states that there were no public bakers in Rome until about 200 years before Christ.

The straw of wheat is used for the manufacture of hats for both men and women. The fine Leghorn straws are manufactured from the stalks of the wheat collected while green, and bleached in the sun. Wheat is sometimes used as a forage crop. A variety has been introduced from Japan which seems to be very useful for this purpose.

Marts. — The great wheat markets of the world are: Odessa, on the Black Sea; Riga, on the Baltic; the North German ports; Constantinople; London and Liverpool, in England; Chicago, San Francisco, and New York, in the United States; and Toronto, in the Dominion of Canada.

ORYZA, L. (Rice.) Spikelets 1-flowered, in compound panicles; flowers perfect, with 2 very small, bristle-formed glumes: paleæ 2 in number, boat-shaped and flattened, the lower one broader, and tipped with a straight awn; stamens 6 in number; stigmas clothed with hairs.

1. **O. sativa**, L. (Common Rice.) Stem from 2 to 5 feet high, somewhat like the culm of wheat, with shorter internodes. Leaf linear, elongated, and rough. Flowers in close panicles. Branches erect, from 5 to 10 inches long, outer pale, strongly veined and keeled; hispid ciliate, terminating in an awn. Grain white, somewhat fusiform, compressed, slight grooves and ridges extending lengthwise, one fourth of an inch long, and about an eighth of an inch in diameter. Flowers in July; fruits in August and September.

Besides the sativa, we have the following species:—

2. **O. præcox**, Early Rice.
3. **O. mutica**, The mountain Rice.
4. **O. glutinosa**, Clammy Rice.

These last three and several others are said to be well-marked and constant species; but as rice is grown from the seed, it sports freely, and many varieties have arisen.

Geography. — Rice grows well in all the low lands of the tropics where alternate flooding and drying can be effected, and in the temperate zones as high as the thirty-sixth parallel. In India there is a species, O. coarctata, that grows upon the uplands, and as high above the sea as 4,000 feet. It is extensively cultivated in China, Japan, the East Indies, especially in the southern parts, in Japan and the islands of the Indian Ocean, and in Africa, having been introduced into Egypt in the days of the caliphs. It has in later years been raised sparingly in southern Europe. The mountain or upland rice has been successfully grown in Hungary. It is planted and successfully grown in South America and in the southern United States.

Etymology. — The word *oryza* is Latinized from *eruz*, an old Arabic or Sanscrit word, which signifies "grow," and is supposed to have been applied to this plant on account of its prolific character. *Sativa*, the specific name, means "sown," or "cultivated." The common name *rice* is supposed to be a corruption of the botanic name.

ORYZA SATIVA (Rice).

History. — There is no record that reveals the time when rice first became a food of the human family. It is one of the cereals yearly sown by the Emperor of China; the first record of such sowing was made in the year 2800 B. C. The little that is known about its early history points to southern Asia as the land of its nativity. Alexander the Great brought it to the notice of the Greeks on his return from his expedition to India 330 B. C. Its cultivation in Italy dates from 1468 of the present era. Since anything has been known of Asia, rice has constituted the principal food of all classes in that country not only on the continent, but also on the adjacent islands.

The growth of rice in America dates from about the year 1700. It is related that a vessel from Madagascar entered one of the ports of South Carolina, believed to be Charleston, and the captain of the vessel presented Mr. Woodward, a settler, with a small quantity of seed rice, which he planted. Soon after this occurrence Mr. Dubois, the treasurer of the East India Co., sent to Carolina a bag of rice. From these two small quantities of seed, coming from different countries, sprang the three varieties of rice grown in America, one of which is now the favorite in the markets of America, as well as in Europe, and is pronounced the best in the world.

Rice of excellent quality is raised in the Sandwich Islands, most of which is brought to the United States *via* San Francisco.

Cultivation. — The mode of culture is to prepare the ground, plant the seed in drills a foot apart in the row, with the rows far enough apart to work between them in keeping the ground free from weeds. After planting, the field is flooded for some days, and then the water is drained off. When the plants make their appearance above the ground the water is again let on to kill the young weeds. After two or three weeks, in the month of April, the water is again withdrawn, and the ground kept free from weeds with the hoe. When the plants are some eight inches to a foot in height, late in August or early in September, the water is let on and left till the grain is ripe. Then the water is finally withdrawn, and as soon as the ground is sufficiently dry, the crop is reaped, bound in sheaves, and taken to the high land to cure and to be threshed.

A continuation of rainy weather about harvest time, which frequently occurs in southern India, renders the rice crop uncertain there. In 1770 the crop failed, or was destroyed, and ten million persons died of starvation; in 1860 one and a half million persons perished from the same cause.

Use. — Rice, in the United States, is used in many ways: it is prepared by boiling in water and eaten as a vegetable; cooked with milk and eaten as a porridge for dessert; baked with milk and eggs for dessert puddings; ground into flour and used to thicken soups and gravies; and also made into griddle-cakes. It is said that the modes of cooking rice in the East are very numerous; but the masses of Asia boil it in the most simple manner, and eat it without any sort of dressing.

Rice, in India, China, Japan, Egypt, and the islands off the coast of Asia, forms the principal article of food for more than five hundred million persons. The greater part of the teeming millions of Japan, China, and southern India seldom taste any other food. It furnishes food for a far greater number of people than any other plant. Rice does not possess the nutritive qualities of the other food grains, being constituted largely of starch; it should not be eaten until six or eight months after harvesting.

SACCHARUM, L. (Sugar Cane.) Spikelets panicled, in pairs, one pedicellate, and the other sessile, spikelets made up of 2 flowers each, at the base of which is a tuft of long silky hairs; lower floret without stamens or pistils, a single bract at the base; upper floret perfect; glumes 2, equal, and without awns; stamens 1 to 3; ovary sessile, glabrous; styles 2, elongated; stigma plumose; hairs simple and toothed. Fruit free, perennial.

S. officinarum, L. (Sugar Cane.) Stem or culm 10 to 20 feet high, composed of a strong cortex filled with a pith, charged with a sugary substance.

GRAMINEÆ.

Internodes short. Leaves flat, linear-lanceolate, clasping, like the leaves of maize. Flowers in a panicle, 2 feet long; racemes thread-like, **erect, and spreading, clothed with silky hairs.**

Sugar cane grown from the seed sports freely, hence there are many species. — The following have been found growing without cultivation: **S. contractum, S. polystachyum, S. dubium, S. rubicundum, S. atrorubens, S. fragile.**

Geography. — The sugar cane is a tropical and subtropical plant. The geographical range is a belt extending around the earth, including the torrid zone and some twelve additional degrees both north and south of the tropics. The West Indies, Brazil, Mexico, and the southern United States, in America, the Isle of Mauritius, southern India, the islands of the Pacific, and northern Australia produce most of the sugar of commerce. In the United States, Louisiana is the sugar-producing region. A small part of Mississippi, and also of Missouri, produce cane.

Etymology. — The name *saccharum* is from the Arabic name *sakkar*, *sukkar*, corrupted into *sugar*. The specific names are all due to some characteristic of the plant; as, *officinarum*, of the shops; *contractum*, smallness of the whole plant, etc.

SACCHARUM OFFICINARUM
Sugar Cane.

History. — The sugar cane is a native of Cochin China; but where it was first brought under cultivation is not known. Circumstances point to India. It is known that the Venetians imported it thence by way of the Red Sea as early as the middle of the twelfth century; and previous to the discovery of America it was grown upon the islands of the eastern Mediterranean, — having been introduced by the Saracens, and carried to southern Spain by the same enterprising people. Soon after the discovery of the West Indies the Dutch manufactured sugar in the Isle of St. Thomas, in 1610, and the English, in 1643, commenced its manufacture in Barbadoes and Jamaica. Pliny and Galen both speak of sugar as "sweet salt," which was used at that time as a medical remedy. It was first substituted for honey in compounding medicines by Actuarius, a physician of the tenth century, and was then called "Indian salt," which points to India as the country of its origin.

Five hundred years ago sugar as an article of food was not known in Europe, now it is one of the necessaries of life throughout the civilized world, and has largely superseded honey, whose sweetness was so much extolled by the ancients. About the end of the sixteenth century the Portuguese began to import it from Brazil; it was then used in medicine or as a great delicacy.

Up to 1872 the variety known as the Creole had been cultivated in the United States to the exclusion of all others. It was found to have greatly degenerated, and this led to efforts to introduce new varieties. Mr Lapice, one of the largest and most experienced sugar planters, visited the East Indies and the islands of the Pacific to examine the character and condition of the cane. Selecting what seemed to him the best, he sent home a ship loaded with 11,000 cuttings of new varieties. No report has thus far been made of the comparative productiveness of the kinds introduced.

A few years ago thirty-two varieties were sent from Mauritius to the British West Indies; about one half of these proved to be constant, and worthy of the attention of sugar-growers. Besides these forms there are many others, which are either sports or modifications, due to the effects of soil, climate, and mode of culture.

Chemistry. — Saccharose or cane-sugar yields to the chemist carbon, hydrogen, and oxygen, $C_{12}H_{22}O_{11}$, and has a specific gravity of 1.60. It crystallizes in prisms, which are phosphorescent when broken or electrified. Its solutions turn polarized light to the right, and are hence called dextrose. It fuses at 310° F.; it is soluble in water, and slightly so in alcohol.

Preparation. — Sugar is obtained by crushing the cane between grooved rollers, whose grooves are armed with iron. The expressed juice is then evaporated in pans at a low temperature and in a partial vacuum. By the use of lime, charcoal, etc., it is refined. The drainings and uncrystallized parts are molasses and syrup.

The cane, after passing through the press, is soaked in water, and the strainings and rougher parts of the molasses are mixed with the water; then it is allowed to ferment. It is then distilled, and produces rum. Large quantities of sugar are obtained from the beet in Europe, and from the maple tree in America. (See Beet, and Sugar Maple.) The sugar of China is obtained from sorghum.

Use. — Sugar is so well known that a description of its use would seem superfluous. It is found on our tables in some form at every meal. It is a perfect preservative for fruits of every description, and the principal ingredient in all confectionery. In fact, few articles of food have attained so wide a usefulness.

Propagation. — To preserve the constancy of species and varieties, the cane is propagated from cuttings either from the upper nodes of the culm or from the rootstock. The new varieties are seedlings, though the plant seldom matures seed in a state of cultivation.

SORGHUM, L. Spikelets panicled in twos or threes on the spreading branches; the middle spikelet 2-flowered, perfect, lower flower abortive; side spikelets sterile, without awns; pedicels usually smooth. Glumes leathery. Stamens 3. Annual.

S. saccharatum, L. (Broom Corn.) **(Andropogon saccharatum, Pers.)** Culm 6 to 9 feet high, solid, with pith intermingled with woody fiber like Indian corn, about ¾ of an inch in diameter. Leaves 1½ to 2½ feet long, 2½ inches wide, lanceolate, acuminate, smooth, pubescent at the base; panicle 2 feet long, branches simple or nearly so; flexuous, rough, with short hairs.

This plant is propagated from seed, hence it sports freely, producing many varieties.

Geography. — Sorghum grows best in the warmer parts of the temperate zone, but will mature its seed in eastern Massachusetts. It is grown in southern India, northern Africa, southern and middle Europe, and throughout the United States of North America.

Etymology. — The word *sorghum* is derived from *sorghi*, the Indian name of this plant, the meaning of which is obscure. The common name, *broom corn*, arises from the use to which the rigid branches of the panicle is applied, i. e., the making of brooms. *Saccharatum* owes its name to the fact that this species is used for making sugar.

History. — Sorghum is native in the middle of Africa, and was taken to England in the latter part of the eighteenth century, whence it was brought by colonists to eastern North America, where it has for many years been cultivated. It has also been cultivated in Egypt, Abyssinia, and the Deccan.

Use. — Sorghum was grown formerly in the eastern United States for the manufacture of brooms, and in the South for feed for cattle. A coarse meal is made of the seed, which is fed to poultry. During recent years it has been used for making syrup and sugar.

The government of the United States has given much attention to this plant, in order to ascertain its value as a sugar producer, in comparison with that of the cane. Mr. Leonard Wray claims that varieties grown in Natal compare with the sugar cane in the ratio of five to six; that is, where the cane yields thirty, sorghum yields twenty-five. It is cultivated in France and French Africa for the production of alcohol, and in Italy for a syrup used in doctoring wine.

SECALE. L. Spikelets 2-flowered, crowded into a cylindrical spike: florets sessile, distichous, perfect, with a linear rudiment of a third terminal floret. Glumes subopposite, nearly equal, keeled, and sometimes awned. Paleæ herbaceous, the lower one awned, and keeled with unequal sides, outer side broader and thicker, the upper palea shorter, 2-keeled; scales 2 in number, entire, ciliate; stamens 3; ovary sessile, hairy; stigmas 2, subsessile, terminal, and plumose; hairs lengthened, simple, and sharply denticulate; grains hairy at the top. Spike simple, compressed, and linear.

S. **cereale,** L. (Rye.) Stem hairy near the head, and ranging from 3 to 5 feet in height. Leaves lance-linear, edges and upper side rough, glaucous. Heads about 5 inches long, linear, flattened. Paleæ lower, ciliate on the keel and margin. Awns rough and ciliate, long, straight, erect. Annual and biennial.

There is but one species under cultivation, S. cereale; but, like all plants grown from the seed, it sports, and the varieties are numerous, though far less attention has been paid to its cultivation in that direction than to wheat.

Geography. — The geographical range of rye is the colder parts of the temperate regions of the world, between 48° and 69° north latitude all around the globe, where the cereals are cultivated. In northeastern United States it is an important crop for bread; in the central states it is largely raised for distilling. It grows well, and is the great cereal of northern Europe, and especially of the sandy districts of the Baltic provinces, and the shores of the Gulf of Finland.

SECALE CEREALE (Rye).

Rye of an excellent quality for bread making is grown upon the great plain on Long Island. William Cobbet, who in his day owned a large tract on the northern edge, near the west end, boasted that his rye bread was better than, and nearly as white as, the Englishman's wheaten loaf. Cobbet named his place Hyde Park, a name it still retains.

Etymology. — The word *secale* is supposed to be derived from the Celtic word *sega*, a sickle, or from the Latin *seco*, cut, in allusion to the sharp, rough edges of the leaves. The specific name, *cereale*, means "bread-corn" or "bread-material," from *Ceres*, the goddess of food plants. The common name, *rye*, is from the Anglo-Saxon *ryge*.

History. — We have no positive knowledge when rye was first cultivated. It was spoken of by writers in the first century of our era. It is native to southern Russia, and the regions north of the Black and Caspian seas. To the North German, the Pole, the Norwegian, the Swede, and the Russian, rye is what wheat is to the inhabitant of southern Europe, the Briton, and the American.

It was used largely in England in early times, probably having been introduced by the Danes and Saxons. The wheat introduced earlier by the Romans was regarded as a delicacy, and its use was confined to a few. History relates that among the upper classes in Great Britain hospitality was a prominent feature, and when visitors came, the most lavish profusion was exercised in their entertainment. Among the delicacies proffered on such occasions was wheat bread; but when the guest prolonged his stay, he began to be treated as a member of the household, and the rye bread was returned to the table. At first this was taken as a compliment, but it finally came to be understood as a hint that the visit had been sufficiently long. Hence the proverb: "Do not prolong your visit till the rye loaf comes on."

Use. — Bread, cakes, biscuit, and puddings are made of rye; in fact it is applied to most of the purposes for which wheat is used. In the central states rye is extensively used in the manufacture of whiskey. In Holland it is mixed with both barley and buckwheat for distilling; the liquors thus produced are called "Hollands," and when flavored with juniper berries, they form gin.

Rye is an excellent feed for cattle, and especially for cows when giving milk.

HORDEUM, L. (Barley.) Three spikelets at each point of the rachis, each 1-flowered, the side florets sometimes abortive; glumes linear-lanceolate, flat, stiff, awns awl-shaped; paleæ herbaceous, lower one concave, terminating in an awn, upper one 2-keeled; scales 2, sometimes 2-lobed, ciliate, sometimes smooth; stamens 3; ovary sessile, hairy at the top; stigmas 2, nearly terminal, and sessile. Caryopsis terminating in a hairy summit. It is an annual, flowering and fruiting the same season it is sown.

1. **H. vulgare.** L. (Barley.) Culm or stem 2 to 3 feet high, smooth. Leaves linear-lanceolate, keeled, and striate, smoothish, eared at the throat; heads 3 inches long, stout, 4-sided, sometimes somewhat 6-sided, lower paleæ crowned with long awns, serrulate on the margin; upper paleæ obtuse or emarginate. Flowers in May. Fruits in July.

2. **H. distichum,** L. Stem 2 to 3 feet high. Leaves like the last. Heads about 4 inches long, flattened, and 2-ranked. Husk attached to the ripe grain. Flowers in June, and ripens in last of July to August.

Like other plants raised from the seed, it sports freely, and varieties are numerous; but the two species here described are pretty constant, and little attention has been paid to improving or perpetuating varieties, especially in America. The distichum ripens later, and is in some places preferred for that reason.

In Europe the following species are also grown: —

3. **H. hexastichon**, Six-rowed Barley.
4. **H. zeocriton**, Battledoor Barley.

Barley stands next to rye in importance as a food plant. Its characteristics as to cultivation so resemble those of wheat and rye that little needs to be said about them. It is, however, a more gross feeder than rye, and will not yield heavily without high tillage.

Geography. — Barley grows and ripens over a larger geographical range than either wheat or rye. It ripens and yields generous crops in latitudes where no more than two months in the year are free from frost. It grows well in northern Russia and Siberia, where the ground thaws out only to the depth of two feet, and even less. It will ripen also in warm climates, even in the regions of no frost, but delights in a short, hot summer, such as characterizes the higher regions of the temperate zones, and like wheat and rye is to be found an emigrant to all the cereal-growing abodes of civilized man.

Etymology. — The name *hordeum* is derived from the Latin *hordus*, heavy, because bread made from it is usually heavy. The specific name *vulgare* signifies "common," and *distichum*, "two-ranked." The common name, *barley*, is supposed to mean "bearded grain."

History. — The ancient home of the grain is not known. A traditional history of barley among the Egyptians makes it the first grain used by man. They hold that their goddess Isis taught men its use. It was among the food plants used by man as far back as we have any history of human customs.

HORDEUM VULGARE (Barley).

A six-ranked barley, H. hexastichon, cultivated by the ancients, has been found in Egyptian monuments and in the Lake-dwellings of Switzerland, in deposits belonging to the Stone Age. A species known to the ancient Greeks, and denominated the "sacred barley," was used to decorate the hair of the goddess Ceres. Some make its native country Tartary, while others claim that it is indigenous to Siberia. History bears out the belief that its home is in the middle parts of the temperate zone in western Asia. It does not fruit without cultivation; when it escapes cultivation it ceases in a year or two to ripen its seed, and is lost. In fact this is the case with other cereals, and is looked upon as a great mystery; it does not favor or bear out the doctrine of development, for the other species under this genus refuse under the most careful cultivation to be anything more than ordinary forage grasses. Barley among the Romans was used for feed for horses and cattle, and it also constituted the bread-grain of the plebeian classes. Pliny informs us that the gladiators were called "hordearii" (barley-eaters), from the circumstance that they subsisted on barley.

Use. — Barley at the present day is not largely used as a bread plant. It is regarded as a valuable fattening feed for cattle. In England and Germany it forms the beer-making grain. In Holland and north Germany it is used for distilling, and is the principal grain from which whiskey and Hollands are distilled.

It is used to thicken soups, and sparingly for porridge and cakes. Pearl barley is prepared by removing the hull, and is cooked as rice is. In northern Scotland and adjacent isles it is an important bread grain, and with oats constitutes a large part of bread material.

Marts. — For Russia the principal markets are Odessa on the Black Sea, and Riga on the Baltic; for Turkey, Constantinople and Rodosto on the Sea of Marmora; for France, Marseilles on the Mediterranean, and Havre on the English Channel; for Germany, Hamburg on the Elbe, and Bremen on the Weser. In the United States the markets are local, being confined to brewing centers, as Chicago, Cincinnati, Milwaukee, etc.

AVENA, L. Spikelets panicled, each having 2 to 5 flowers; glumes 2, loose, membranous, and without terminal awns, about as long as the paleæ, the lower one usually toothed at the top, with a twisted awn on the back; the upper one awnless, with two keels; scales forked and large: stamens 3; stigmas 2, sessile. Fruit subterete, sulcate on the upper side, summit hairy. Annual with pendulous spikelets, or perennial with erect spikelets.

A. sativa, L. (Oats.) Stem 2 to 4 feet high, smooth. Leaves about a foot long, linear or linear-lanceolate, nerved and rough; sheaths striate and loose, ligulæ cut; panicle loose and nodding; spikelets all with peduncles, and hanging; lower floret usually with an awn on the back, upper one awnless. Annual. Flowers in July. Fruits in August.

Oats, like wheat and the other cereals, have a tapering stem and numerous root leaves, and possess the same habit of tillering; but the plant is wholly different in its inflorescence and the form of its head, which in wheat, rye, and barley is a compressed spike, or compound compressed spike, whose spikelets are sessile. In the oat the head is a loose panicle; the branches near the base of the head in some cases are four inches long, decreasing towards the top, forming a pyramidal or conical-shaped head. Some varieties have the branching all on one side, and on that account are called one-sided or secund oats.

As this grain is raised from the seed, it departs from the specific form, producing varieties.

The A. sativa, however, is very constant, and little or no attention has been given to the perpetuation or improvement of its varieties. A large-grained secund form and a black-seed variety have in turn attracted the notice of cultivators, but neither of these has become constant enough to gain importance.

Geography. — The geographical range of oats is not so great as that of any of the cereals before described. It endures a colder climate than any other, but does not fill well south of the fortieth parallel in the north temperate zone; in the regions of no frost it does not fruit, except in elevations far above the sea.

Etymology. — *Avena,* the botanic name of the oat, was given by Linnæus, and is supposed to be derived from the Celtic word *aten,* eat. The specific name *sativa* means "sown," or "cultivated." *Oats*, the common name, is supposed to come from the Anglo-Saxon word *ata*, food. These derivations are not entirely clear, but are the most probable.

History. — The native country of this grain is supposed to be west-central Asia and east-central Europe. It was known to the ancient Greeks and Romans, and was used by them to feed horses and cattle. It also constituted the food of the slaves and plebeians. It was found in the Swiss Lake-dwellings and in the ancient tombs in Germany.

Though it possesses less nutritive material than either wheat or rye, it has held and still holds an important place as a food plant. It is found upon the tables of the rich, as well as of the poor in Great Britain, northern Europe, and in the United States and Canada. As a feed for live stock it is as highly valued in the British Isles as is maize in the United States.

In attestation of its value as a bread plant as well as for feed, the following anecdote is in point. Dr. Johnson, the English lexicographer, had a deep-rooted dislike for the Scotch, and lost no opportunity to make it manifest. At one time, in conversation with a Scotch gentleman, Dr. Johnson remarked that oats were a grain that Englishmen fed to their horses, but that Scotchmen ate it themselves. To which the gentleman with characteristic Scotch readiness replied: "Indeed it is true; but see what horses you have in England, and what superior men we have in Scotland."

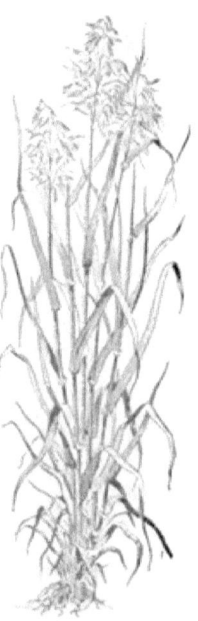

AVENA SATIVA (Oats)

Cultivation. — Oats need a generous soil to yield large crops. The ground is prepared by the plough and harrow; the seed is sown broadcast in most countries, though in Scotland it is sown in drills and worked with a hoe. It is sown as soon as the frost leaves the ground in the early spring, and ripens about the first of August. It fills best where the days are sunny and the nights cool. The market value of oats varies according to quality; northern oats, so-called,— that is, oats grown in a cool climate, — are preferred. In good soil and a cool climate an average yield is forty-five bushels to the acre, but sixty bushels is not an uncommon yield.

Use. — In Scotland, Ireland, and the north of Europe, especially in Norway, oats constitute a large portion of the bread material. They are used as a porridge, cooked with milk, or made into a thick pudding and eaten with milk; they are also eaten in the form of griddle-cakes or "scons." In most countries where horses are used, oats are the staple feed. Ordinarily they are best when ground; but the trainers of race-horses prefer to feed the oats to them in the grain, since, when thus fed, the horse needs no hay, or not so much.

The grain of oats is largely composed of starch; it contains also sugar, gum, and oil. About 12 per cent of its substance is a proteid, known to chemists as avenine, a substance resembling casein. It is mixed with barley in the manufacture of whiskey. Its chaff is used for filling beds.

SETARIA, Beauv. (Millet. Bengal Grass.) Inflorescence a spike-like panicle; spikelets 2-flowered, surrounded by persistent bristles; the upper floret perfect, lower one staminate or neutral or wanting; lower glume usually short. Stamens in the perfect flowers 3, the collateral scales truncate, fleshy, and smooth. Number of styles 2, terminal, elongated. Stigmas plumose; inflorescence a spike. Seeds free and compressed.

S. Italica, Beauv., Var. **Germanica,** Kunth. Stem or culm 2 to 5 feet high, terete, leafy. Leaves from 6 to 15 inches long, linear-lanceolate, broad, flat, and finely serrate on the edges, clasping the stem in a striate sheath; spike compound, yellowish or purplish, oblong, ovoid, or somewhat cylindrical; rachis hirsute, hairs long. Flowers in July. Fruit ripens in August.

Geography. — It grows well where oats and rye can be successfully cultivated; it has escaped in America, and is ranked among the weeds when without cultivation.

Etymology. — *Setaria* is from the Latin word *seta*, a bristle, on account of the bristly character of the spikelet involucres. The specific name *Italica* denotes the plant's home; and the name of the variety indicates that it arose in Germany. Millet is from the French *millet*, diminutive of *mil*, the old French name of the plant. *Bengal grass* is named from Bengal, where this plant is native.

History. — Setaria is one of the grains which the Emperor of China sows at a public ceremony which has occurred annually since 2700 B. C., and it is believed to be native to China, Japan, and India. Its cultivation spread toward the west through Russia. It is found in the Swiss Lake-dwellings of the Stone Age. It is either completely naturalized or native to western Asia, southern Europe, and Egypt. Though cultivated for forage, it is so inferior in every way to oats that little attention is paid to its improvement. There are a number of species, but the var. Germanica is altogether superior to the others.

Use. — This plant is grown in Italy for seed, used to feed caged birds.

ORDER LXII. **CONIFERÆ.** (CONE FAMILY.)

Flowers in catkin-like spikes, monœcious or diœcious, naked, or without floral envelopes. Staminate flowers composed of anther-bearing scales; catkins longer than broad. Pistillate catkin more or less conical, subglobular or cylindrical, with few or many flowers, occasionally 1-flowered; ripened catkin becoming a strobile or conical subglobular body, formed of woody, imbricated scales, bearing 1 — many naked ovules, with 2–15 cotyledons. Seed in most cases furnished with a membranous wing, sometimes solitary, in a fleshy cup. Leaves simple, needle-shaped, alternate, linear or lanceolate, in some cases in groups, inclosed at the base in a membranous sheath, resinous. Trees or shrubs.

No. of genera, 32; species, 300; temperate zones and mountains.

PINUS, L. Evergreen trees, ranging from 30 to 100 feet in height. Leaves linear, grouped, 2 to 5 in a group, very rarely solitary, inclosed at the base in a scaly sheath. Male flowers grouped in catkins.

Female flowers in cones mostly ovate, their scales dry, thickened at the tips and spiny, inclosing 2 ovules. Cotyledons 3 or none. Flowers appear in spring; seeds mature in the autumn of the following year.

Travellers and naturalists speak of the pine forests of Carolina and Mississippi as possessing an unsurpassing romantic beauty. The tall trunks rising 50 to 70 feet without a branch, with no undergrowth to break the view, the branches festooned with the gray tillandsia mingling and contrasting with the deep-green foliage, give the appearance of a vast decorated roof supported by a countless number of graceful columns, which, fading away in the dark distance, present a scene whose beauty is exceeded only by its grandeur. These valuable pine forests are undergoing rapid destruction, and unless some mode of economic forestry be at once adopted, this valuable timber will soon be exhausted.

1. **P. monophylla**, Torr. and Fremont. (Single-leaved Nut-pine.) Small tree, branching irregularly, forming an unsymmetrical head. Bark pale, and falling off in plates. Leaves solitary, tapering, 1½ inches long, terminating in a spine; sheaths one third to half an inch in length. Male flowers inclosed by 6 bracts; cones nearly globular, about 2 inches in diameter; seeds long egg-shaped, half an inch in length, shell thick, yellowish-brown; cotyledons 10 or less.

Geography. — The nut-pine is an American tree; the zone of growth is between the parallels of 30° and 45° north latitude. It grows along the eastern slopes of the Sierra Nevada mountains, at great elevations, in arid localities.

Etymology. — *Pinus* is derived from the Latin *pinus*, a pine-tree. *Monophylla* is from the Greek μονος, one, and φύλλον, leaf, one-leaved, referring to the solitary leaf.

Use. — The tree is of no value in planted grounds on account of its small size and irregular branching. The wood is hard, and makes excellent fuel; it is too small for use as lumber. The fruit is sweet and edible; it constitutes an important article of food for the wandering aborigines of the Pacific slope. It is collected and sold to passengers on the line of the Northern Pacific Railway.

2. **P. Austriaca**, Hoffm. (Austrian Pine.) Synonym, P. nigra, Link. Stem from 80 to 120 feet in height. When growing in open grounds it forms a subcylindrical head, branching regularly, though the branches are crooked. Bark grayish. Leaves long, rigid, slender, mucronate in twos, incurved, and inclosed in short, dark-green sheaths; cones 2 to 3 inches long, curved, light-brown; scales smooth, shining, armed with a blunt spine.

Geography. — Its geographical range is Lower Austria, Carinthia, Styria, and the neighboring regions.

Etymology. — The specific name, as well as the common name of this tree, is from the country where it is found native, Austria.

History. — It is not long since it was brought to the notice of nurserymen and planters. Its fine head, hardy character, and deep dark-green foliage have brought it into favor.

Use. — The tree is valuable for ornamental purposes, and is found in most planted grounds. The wood is good for lumber and for fuel.

3. **P. strobus**, L. (White Pine. Weymouth Pine. Apple Pine. Sapling Pine. New England Pine. Pumpkin Pine.) Trunk 50 to 160 feet in height, from 3 to 7 feet in diameter. With other timber the stem frequently rises to the height of 80 feet without a branch. Branches whorled in the young trees. Bark light gray. Leaves 3 to 4 inches long, bluish-green, in fives, and glaucous, forming a soft, beautiful foliage; sheaths deciduous; cones 5 to 6 inches in length; scales 1 to 2 inches long, one half to three fourths of an inch wide; seed less than a quarter of an inch in length, obovate, tapering to a point below with a wing; cotyledons 6 to 5.

Geography. — It is native to all parts of North America, from the Atlantic west, reaching as far south as Virginia.

Pinus strobus (White Pine).

Etymology and History. — *Strobus* is the Latin for an odoriferous gum. It was successfully grown in the grounds of Lord Weymouth (England), and was hence called *Weymouth pine*. It was introduced into England in 1705.

Use. — This tree is a symmetrical grower; it forms a beautiful object, and produces a pleasant contrast with other evergreens in planted grounds.

The wood is fine-grained, white, and free from resin, strong, easily worked, takes a good polish, and does not warp easily. It is highly prized in carpentry and in joiner's work. The woodwork of the insides of dwellings, doors, sashes, floors, wainscoting, etc., are largely composed of white pine. It is the most valuable lumber tree in the United States.

4. **P. mitis**, Mx. (Yellow Pine. Spruce Pine.) Synonym: **P. variabilis**, Pursh. Trunk 30 to 80 feet high, 18 inches to 3 feet in diameter, branching regularly; bark dark and rough. Leaves channelled, slender, 3 to 5 inches long, in twos or threes, inclosed in lengthened sheaths, bluish-green, scattered over the branches. Cones lateral, conical, or egg-shaped, 2 inches long, solitary; seeds small, with reddish wings.

CONIFERÆ. 303

Geography. — The P. mitis is found in the eastern United States, from New England to the shores of the Gulf of Mexico. It attains its full size south of Virginia, and flourishes in the hot sands of the Carolinas and the Gulf states.

Etymology. — *Mitis*, the specific name of this pine, is derived from the Latin *mitis*, soft, or tender, and is due to the delicate feel of the soft young leaves. *Variabilis*, Latin for variable, refers to the inconstant number of leaves in a cluster, — usually 2, but occasionally 3. *Yellow pine* refers to the color of the wood. The origin of the name *spruce pine* is not apparent.

PINUS SYLVESTRIS (Scotch Pine).

Use. — The P. mitis is sparingly used in planted grounds; it forms a symmetrical pyramidal head, and its bluish-green foliage makes it a desirable ornamental tree. The wood is rich in resin, fine-grained, and takes a good polish; it is used for flooring, and of late years has been largely used in naval architecture, taking the place of oak, especially for decks.

5. **P. sylvestris**, Ait. (Scotch Pine. Scotch Fir.) Trunk 60 to 100 feet high, sometimes reaching the diameter of 6 feet. Branches somewhat straggling, forming a picturesque head. Grows rapidly in planted grounds. The new growth is frequently crooked by its own weight before it becomes hardened, by reason of which the trunk is frequently very crooked. Leaves rigid, in

pairs, 1½ to 3 inches long, twisted, bluish-green; sheaths short, and torn; cones conical, ovate, 2 to 3 inches long, color grayish-brown; point 4-sided, recurved. Seeds small, with a long reddish wing. Cotyledons 5 to 7.

There are about a dozen varieties, for as it is propagated from the seed it sports freely.

Geography. — Its home is in the upper and northern exposures of the Pyrenees, and the Tyrolese, Swiss, and Vosgian mountains It forms exclusive forests throughout Europe and northern Asia, and its trunks are found in great abundance buried in the peat bogs of Great Britain; hence the name *bog fir*. It grows with great rapidity in almost any soil. On account of its straggling, irregular branching, it forms a picturesque rather than a beautiful object; yet it is a favorite with nurserymen and amateurs.

Etymology. — *Pinus sylvestris* may be translated "the pine of the woods." The common name, *Scotch pine*, would seem to indicate that its home is Scotland, but there is no reason to believe it is indigenous there.

Use. — The pinus sylvestris, besides being a favorite ornamental tree, is for Europe what the pinus strobus is for America; it is the red and the yellow deal of England; it enters into the structure of buildings, and is largely used in naval architecture all over Europe.

6. **P. resinosa**, Ait. (Red Pine.) Synonyms: **P. rubra**, Mx. **P. Canadensis bifolia**, Du Hamel. Trunk varying in size from 50 to 80 feet in height, and 2 feet in diameter, branching regularly; when standing alone forming a symmetrical head. Bark smooth and red. Leaves in pairs, channelled, and semi-cylindrical, dark-green, 4 to 6 inches long, appearing near the ends of the branches; sheaths long. Cones egg-shaped, 2 inches long, in clusters, on short peduncles; scales unarmed; wood fine-grained, moderately charged with resin, heavy, strong, and very durable.

Geography. — Its home is northeastern North America, from Canada to southern Pennsylvania.

Etymology. — *Resinosa* is from the Latin *resina*, resin, due to the resin-bearing character of the wood, although it is one of the least resinous of the pines. *Rubra* is from Latin *ruber*, red, referring to the color of the bark. *Canadensis bifolia* is Latin, meaning Canadian two-leaved pine.

Use. — The wood of the red pine is hard and strong, valuable for frames of buildings, and for boards and planks. It is used for flooring and wainscoting.

7. **P. rigida**, Miller. (Pitch Pine.) Trunk 40 to 70 feet in height, branched irregularly. Bark dark, deeply furrowed. Leaves in threes, rigid, 6 inches long, sheaths short. Cones ovoid, pyramidal, in clusters; scales tipped with short, stout, reflexed spines; seed nearly smooth; cotyledons 5; wood heavy, with resin.

Geography. — The P. rigida is native throughout eastern North America, but north of Virginia it does not attain to its full size. In Georgia, the Carolinas, Alabama, and Mississippi it is a tree 60 to 80 feet high.

Etymology. — *Rigida* is from the Latin *rigidus*, stiff, referring to the leaf. *Pitch Pine* owes its name to the sap which exudes from the bark.

Use. — The pitch pine is used in the South for the manufacture of turpentine, resin, and tar. The wood is very hard, takes a fine polish, is much used for floors and ship-building, and is excellent fuel.

8. **P. cembra**, L. (Swiss Pine. Stone Pine.) Trunk 50 to 80 feet in height, branching in whorls, branches semi-upright. Bark smooth and dark. Leaves in fives, 2 to 3 inches long, triangular, slender, straight, crowded, dark-green, sheaths falling. Cones egg-shaped, 3 inches long; scales blunt, hooked; seeds large, wedge-shaped, shells hard; cotyledons about 13; seeds edible.

There are two varieties to be found in planted grounds.

Var. **Siberica**, Loudon. Leaves shorter, lighter green, and longer cones; said to be indigenous to eastern Siberia.

Var. **pygmæa**, Fischer. Dwarf, 2 to 4 feet in height. Leaves short; cones globular. Found in dry, sterile grounds and rocky hills.

Geography. — The pinus cembra may be found in a range of territory extending from the mountains of Switzerland, through Austria, northeastward to Siberia. It adapts itself to almost any soil or climate.

Etymology. — The specific name of this tree signifies "pine;" so that *pinus cembra* may be translated "pine pine." The common name, *stone pine*, is due to the hard shell of the seed.

History. — The stone pine was introduced into planted grounds about the middle of the eighteenth century, and was brought from elevations in the Alps, ranging from four thousand to six thousand feet.

Use. — The Swiss stone pine is a favorite evergreen with amateurs, on account of its symmetry of growth, its compact branches, and its deep green foliage. Its slow growth adapts it to the wants of those who have small grounds. It bears the knife well, and can be kept back, and almost dwarfed. The wood is soft, has a fine grain, takes a good polish, and is a favorite wood for joiners, turners, and carvers. Large quantities are used in Switzerland in the manufacture of toys. It also furnishes a fragrant resin. The seeds are eaten in Siberia.

9. **P. palustris**, L. (Southern Pine. Green Pine. Long-leaved Pine. Broom Pine. Yellow Pine. Pitch Pine. Georgia Pine. Red Pine. Fat Wood.) Trunk 60 to 80 feet high, and 2 to 3 feet in diameter. Bark furrowed. Leaves in threes, sometimes reaching the length of 4-8 inches, dark-green, crowded at the ends of the branches; sheaths long, lanceolate, light-colored; cones cylindrical, tapering at each end, 10 inches long; seeds in a thin white covering or testa.

Var. **excelsa**, Loudon. Whole tree larger. Found in the north of Europe; said also to have been seen on the northern Pacific coast.

Geography. — This pine has a narrow geographical limit, which extends from southeastern Virginia south to middle Florida, thence along the Gulf coast to Louisiana and Texas, in a strip less than 200 miles wide.

Etymology. — The name *palustris*, the Latin for "swampy," does not seem very appropriate, as this tree does not grow in swamps. The number of popular or common names applied to this species is very unusual, due no doubt to the great value of its products.

Use. — The pinus palustris is one of the most important of all our timber trees; no other yields so many valuable products. The wood is hard, takes a fine polish, and is very heavy, weighing 45.62 pounds to the cubic foot, and is highly valuable for building purposes, furnishing timber for the frame of a building, boards for the covering, beams and planks for the floors, and material for the joiner's work. It is also excellent fuel.

Turpentine is the product of this pine. It is obtained by cutting a pocket-like cavity in the side of a tree. The wound thus made discharges the fluid into the pocket or box, which holds about three pints. These cavities fill in about ten days; the contents are then removed, two grooves are cut in the bark above the pocket leading to the cavity, another filling takes place, and the process is continued. The liquid is put into casks made on the spot, and hardens or becomes a semi-fluid, and forms the turpentine of commerce. There are in the markets of the world several varieties obtained from other coniferous trees.

In America a single person attends to the emptying of 4,000 pockets, which yield in a season sixteen barrels, weighing net 320 pounds to the barrel. The crude turpentine has a fixed resin dissolved in oil, with succinic acid.

Spirits of turpentine is procured by distilling crude turpentine and water; the water and spirits go over together, and are allowed to cool in a vat. The mass will arrange itself into two layers, when the spirits may be drawn off into vessels for market.

Spirits of turpentine is largely used in the arts for dissolving gum resins, for varnishes, and for mixing paints, and before the discovery of petroleum was used for illuminating purposes; mixed with alcohol it formed the burning fluid previously used for illuminating. As a medicine it is diaphoretic and anthelmintic, acting directly upon the kidneys, and in large doses it is powerfully cathartic. It is used by veterinary practitioners as a liniment. The annual product in the United States is about 17,500,000 gallons. Rosin (or colophony) is a residuum of distillation, after the volatile oil of turpentine has gone over.

Tar is obtained from the dead branches and trunks of trees that have been exhausted by six or more years' tapping. The wood is cut into suitable lengths and placed on end in a pyramidal stack upon a floor made of clay, well pounded and concave, outside of which is a well, also lined with clay. In arranging the wood a cavity is left in the center, and filled with combustibles, such as dry cones, shavings, etc.; the whole is then well covered with earth, a few openings being left around the base. When all is ready fire is thrown down among the combustibles in the center, and when well lighted the top is closed. Great heat is produced, by which the tar is liquefied; it then passes down into the dish-like floor, whence it flows through an opening made for the purpose into the well outside, from which it is ladled into barrels. It is interesting to note that the Greeks obtained tar by a process precisely similar, centuries prior to the beginning of the Christian era. The entire annual yield of tar in America is about 100,000 barrels. It is also obtained in large quantities from Norway and Sweden.

Tar is largely used in the manufacture of naval cordage, as a paint, and otherwise, in naval architecture. As a medicine it was formerly used as a diaphoretic, and as an ointment for scald head, and the vapor was inhaled for lung affections. Tar water, that is, water which has for a time rested upon tar, is also used for stomach disorders.

Oil of tar is procured by distilling tar.

Pitch is a solid black substance, the residuum of distilling tar for tar oil, and is largely used in pavements, and for waterproof cements.

Resin oil is a viscid whitish opalescent liquid obtained by the distillation of resin, and was formerly used as a lubricating substance, and for the manufacture of illuminating gas.

CONIFERÆ.

PICEA, Don. (Spruce.) Leaves scattered, articulated to the persistent base of the petiole, partly tetragonal, rigid. Staminate flowers solitary in the axils, the connective not produced as a scale-like appendage. Scales of the reflexed cone persistent, mostly concealing the small subtending bract.

1. **P. nigra**, Link. (Black Spruce.) Trunk 70 to 80 feet high. Branchlets spreading horizontally. Bark smooth and dark. Head pyramidal, symmetrical, and when young very graceful. Leaves half an inch long, erect, stiff, 4-sided, very dark-green. Cones egg-shaped, hanging, about 2 inches long, dark-purple, changing to reddish-brown; scales ellipsoid; margin unevenly toothed.

Geography. — This tree was introduced into England in 1700, and grows well there; but it nowhere rises to its native grandeur except in its forest home, in the highlands of southern Canada, and the northern United States. Its geographical range is from the Atlantic coast west to the head waters of the Mississippi, between 39° and 50° north latitude.

Etymology. — The generic name is from the Latin *piceus*, pitchy. *Nigra*, the specific name, is Latin, signifying "black," given to this tree on account of the heavy, dark-green color of its foliage and bark. It forms a fine conical head with a tapering trunk. The common name *spruce* is derived from the old English *Spruce* or *Pruse*, Prussia, the tree having been first known as a native of Prussia.

Use. — The black spruce is largely used in planted grounds. The wood is hard, light, strong, and elastic; it is extensively used for architectural purposes, for framework and flooring in the construction of dwellings. On account of its elastic and sonorous qualities it is much used for piano sounding-boards.

PICEA ALBA (White Spruce).

2. **P. alba**, Link. (White Spruce.) Trunk 50 feet high, 1 to 2 feet in diameter, tapering, forming a pyramidal head. Leaves less than an inch long, sprinkled over the branchlets, needle-shaped, curved upwards, light, glaucous, green. Cones about 2 inches long, subcylindrical; scales entire; the winged seeds very small.

Geography. — When young this tree is very beautiful, and rivals the celebrated Norway spruce. Its geographical range is between 42° and 67° north latitude. It abounds in the forests of southern Canada and the northern United States.

Etymology. — *Alba*, the specific name, is Latin, meaning "white," and refers to the light color of the foliage; the popular name is due to the same characteristic.

Use. — The white spruce is a favorite with nurserymen and amateurs, for planted grounds. The wood is strong, and makes excellent lumber for building purposes, and is largely used for floor planks.

3. **P. excelsa**, Link. (Norway Spruce.) Trunk 80 to 150 feet high, branched profusely. As the branches lengthen, the weight of their ends causes them more and more to assume a horizontal position, and finally to droop. Leaves elongated, and though scattered assuming a semi two-ranked arrangement, quadrangular. Cones cylindrical, terminal, and pendent, sometimes 8 inches long and 2 in diameter; scales broad, apex projecting and notched; seed small, one eighth of an inch in length, and half as broad, with a wing. Cotyledons 7 to 9. Flowers in May, and the cones ripen the following year in the spring.

This tree sports freely, and there are about a dozen well-marked varieties, by the use of which the nurserymen and amateurs are enabled to produce varied effects in planted grounds.

Geography. — The Norway spruce is indigenous throughout northern and middle Europe, and on the northern declivities of the mountains of southern France and Spain. Vast forests on the Alps, at an elevation of nearly 7,000 feet, are wholly composed of the Norway spruce; but it is said to attain its perfection in the forests of Norway, its home and the country from which it derives its common name, where it is the grand monarch of the woods. It is propagated in America from seeds brought from Europe, formerly from Norway. It grows to its full size in deep, damp soils.

Etymology. — The specific name of this tree, *excelsa*, is the Latin for lofty, and usually relates to excellence in rank or character, not to size or height; but in its application to this fine tree, it seems to have been used with the latter signification.

Use. — On account of its hardness and the symmetrical head it forms, its beautiful deep green, and the patience with which it bears the knife, the Norway spruce has become the most popular and the standard evergreen in our nurseries and in planted grounds, for ornamental purposes. In Europe, though not in America, its wood is used for architectural purposes. The wood is light, elastic, durable, and of a yellowish-white color. It is charged with resin, which is the base of Burgundy pitch. The young trees are cut when six to ten inches in diameter, and used by builders for scaffolding. The larger trees are sawed into planks for flooring. On account of its elasticity and sonorousness it enters into the structure of musical instruments, — especially of the backs of violins and of piano sounding-boards. Cabinet makers line furniture with it, and it is largely used for boxes for packing merchandise. It is fine-grained, takes a good polish, receives a black stain well, and is especially suitable for picture-frames and other articles that are gilded. It is also highly prized by carvers for their purposes.

ABIES, Link. (Fir.) Flowers monoecious; aments in terminal or nearly terminal clusters; scales of the cones thin and flat; seeds winged. Leaves solitary, and without sheaths at the base, scattered, bearing a slight scar, linear, flat above. Staminate flowers solitary in the axils, the connective being barely prominent above the anthers. Scales of the erect cone deciduous with the seeds, the subtending bract conspicuous, but not thickened nor prickly tipped, often equalling or exceeding the ovuliferous scale.

A. balsamea, Miller. **(Balsam Fir. Balm of Gilead.** Called in England American Balsam Fir.) **Synonyms: Pinus balsamifera,** L.; Picea balsamifera, Marshall. Trunk 20 to **50 feet in height, 1 to 2 feet in diameter.** Bark dark **gray,** smooth or **blistered, with resinous** vesicles. **Branches nearly** horizontal, numerous and **slender,** drooping when old. **Leaves about** an inch long, nar**row,** linear, spreading, **and** slightly **turned back,** green above, **silvery** underneath. Cones cylindrical, 4 inches long, **violet** colored, scales **thin,** smooth, **obovate or** subspatulate, and slightly **mucronate.** Seeds small, **angular.**

Var. longifolia, Booth. Leaves longer, branches more upright, than A. balsamea.

Var. variegata, Knight. Some of the leaves have a yellowish cast, contrasting with the silvery sheen of the others, and forming a beautiful object for the lawn. This feature is made the most of by nurserymen and dealers in trees.

Geography. — The home of the balsam fir is northeastern North America. Lower Canada, especially Quebec, New Brunswick, and Nova Scotia abound in this tree. It is found in the mountainous parts of the middle states, especially in the Adirondack regions, and west to Wisconsin. It loves a cold, damp soil.

Etymology. — *Abies* is the Latin for fir tree. *Balsamea,* Latin for " of balsam," is due to the resinous character of the bark. *Fir* is from the Anglo-Saxon *furh*, a kind of oak. *Balm* is a contraction of *balsam*.

History. — The balsam fir is a native of North America, and was introduced into England by Bishop Compton about the end of the seventeenth century. It is a beautiful object while young, but on account of the dense ramification, the branchlets and leaves of the lower branches near the trunk die for want of light, and the tree becomes unsightly; on this account it has been for many years discarded by planters. In its native woods, in Nova Scotia particularly, the snow lodging upon its branches causes them to droop, and thus partially conceal the silvery sheen of the under surface of the leaves. In the hilly regions of Nova Scotia the sides of the wood-roads are walled forty to fifty feet in height with the silvery green of this beautiful tree.

Use. — The wood of the balsam fir is resinous, yellow, soft, and easily worked, but is not large enough to be valuable for building purposes. The gum, or resin, known as Canada Balsam, is obtained from the bark by puncturing it. From these wounds the resin flows out in a viscid fluid, about the consistency of honey, which hardens after exposure to the air.

It enters into the *materia medica*, and is administered in the form of pills for stomach troubles, and also for bronchial affections. It is transparent, and used to incase insects and other perishable objects, for the microscope, and for setting the glasses of microscopic lenses, and is an ingredient in the manufacture of varnish.

LARIX. Mx. **(Larch.** Tamarack. Hackmatac.) Aments scattered over the branches, resembling buds; anthers 2-celled; cones reflexed, subglobular; scales persistent, the subtending scale conspicuous; seeds winged. Leaves deciduous, **soft,** thread-like, in fascicles or scattered on this year's shoots.

1. **L. laricina,** Du Roi. **(American Larch.** Black Larch.) Trunk 80 to 100 **feet in height. Bark dark and rough.** Branches horizontal or **drooping;**

branchlets pendent, regular, forming a graceful pyramidal head. Leaves thread-like, slender, and soft to the touch, in fascicles of 10 to 20. Cones ovoid when young, when open subglobular and purplish, about an inch long, and nearly the same in diameter; scales thin; margins turned in. Seeds small, with short wings.

2. **L. Europæa**, DC. (European Larch.) Trunk same as No. 1. Branchlets more pendulous. Leaves an inch long, obtuse, and flat, bright-green. Cones much larger than No. 1, long egg-shaped, 1-1½ inches in length. Scales orbicular, reflexed, bracts extending beyond the scales. Seeds small, ovate, winged. Cotyledons about 7.

Of this species there are several varieties; the most interesting one is a dwarf, remarkable for its pendent or weeping branches.

Geography.—The American larch is found native as far south as southern New York, and north to the fiftieth parallel, in a belt quite across the continent to the Pacific coast. The European larch is found in the mountainous regions of middle Europe.

Etymology.—*Larix*, the generic name, and *laricina*, the specific name of this tree, are derived from the Celtic *lar*, fat, due to the resinous wood. *Tamarack* and *Huckmatac* are Indian names of obscure meaning. *Europæa* indicates that this species is a native of Europe. The common name, *larch*, seems to be merely a corruption of the word *larix*.

History.—The larch was much used in Venice, in the period of its prosperity, for frames and other parts of buildings; and it is said that buildings framed of that material show no signs of decay even at the present day. The paintings of some of the great masters were executed upon larch panels, and their excellent preservation is said to be due to the hardness and perfect condition of the wood upon which the work was executed.

LARIX LARICINA (Larch).

Use.—The larch is a favorite in planted grounds; no collection of trees would be considered complete without it. The wood is hard, heavy, strong and durable; it is used for fencing, for agricultural instruments, bridges, and for heavy and strong carriages for transporting stone, hardware, coal, lime, and other coarse and hard articles. It is prized for dock logs and frames for canal gates. When the larch log is sawed into planks it is necessary to season them in close piles to prevent warping. The bark is highly charged with tannin, and is used in the manufacture of leather. The resin of this tree yields Venice turpentine, which is used in medicine.

JUNIPERUS, L. (Juniper. Cedar.) Flowers diœcious, occasionally monœcious, on separate branches; aments usually axillary, sometimes lateral, small, ovoid; stamens many, inserted on all sides of the axis. Fertile aments imbricately bracted at the base. Involucre composed of 3 to 6 scales, united at the base, a concave ovule at the base of each scale. Fruit formed of the fleshy scales, subglobose, berry-like, containing 1 to 3 hard seeds. Cotyledons oblong, 2 in number. Leaves scale-like, subulate, lanceolate, evergreen. Trees and shrubs.

1. **J. communis**, L. (Common Juniper.) Trunk 5 to 10 feet in height (in cultivation 15 to 20 feet high), branches numerous, erect. Leaves in whorls, from half to three quarters of an inch long, sharply lanceolate, channelled, keeled below, and bristly pointed, green underneath, and glaucous above. Sterile flowers in little axillary cones. Fertile flowers, on a separate plant, axillary and sessile. Fruit dark-blue, subglobose; berries ripening the next year after the flower appears; sweet, with a taste of turpentine; they contain sugar, and on fermentation yield a beverage resembling gin.

Var. **pyramidalis** of the nurserymen is a seedling, and is a very compact, graceful form in planted grounds.

Var. **prostrata** (synonyms: Var. **alpina**, L., **J. nana**, Willd.) is a prostrate, straggling form, covering sometimes an area of fifty square feet; branchlets assurgent.

2 **J. Virginiana**, L. (Red Cedar.) Trunk 25 to 50 feet in height, branches extended, making a broad, pyramidal head. Leaves very small and scale-like on the old branches, longer and sharper on the young wood, closely imbricated, and very dark-green; the fertile and sterile florets on separate trees, inconspicuous in longish terminal aments. Fruit, a blue berry, covered with a whitish bloom, size of a small pea; sap wood white; heart wood reddish, light, close-grained, and very durable. It sows itself and sports freely, producing several forms as to foliage and ramification, one of which is sufficiently conspicuous to be entitled to the dignity of a variety.

Var. **forma cylindrica**. Stem 10 to 30 feet high, branching profusely, branches growing nearly parallel with the trunk, forming a compact, cylindrical head, making a beautiful object in the landscape.

JUNIPERUS VIRGINIANA (Red Cedar).

3 **J. Bermudiana**, L. (Pencil Cedar.) This species is a beautiful, long-leaved variety found in the West Indies. The wood is soft and close-grained, and used for making lead pencils.

There are other cedars in Europe and Asia.

Geography.—The J. communis is a native of middle and western Europe, northern Asia, and North America. J. Virginiana is a native of North America and the West India islands; it is found all along the eastern coast of the Atlantic, and as far west as the foothills of the Rocky Mountains.

Etymology.—*Juniperus* is the Latin for juniper, from *juvenis*, young, and *parere*, produce, referring to the presence of the old fruit after the new has appeared. Cedar is from the Greek κέδρος, a cedar-tree.

Use.—The J. communis is a favorite in planted grounds; it bears the knife, and may be pruned into any degree of dwarfage. The fruit is used to flavor gin, and in medicine is administered for kidney complaints; it is considered one of the most active diuretics known. The J. Virginiana or red cedar of North America is used sparingly as an ornamental tree, but the wood is very durable, fine grained, and takes a good polish. Its durability makes it valuable for fencing; its fine grain adapts it to fine cabinet work and for fancy boxes. Trunks are also made of it for storing furs and woollens to protect them against the depredations of the moths, to which its odors are destructive. The wood of this species, as well as that of Bermudiana, is largely used in the manufacture of lead pencils.

THUJA, Tourn. (Arbor Vitæ.) Flowers monœcious, on the ends of separate branches; sterile flowers in an egg-shaped ament. Anther-cells 4 in number, on a scale-like connective or filament. Fertile aments or cones rough or angular, subglobose; scales few. Seeds winged, 2 under each scale, covering membranaceous. Cotyledons 2. Leaves evergreen, imbricated; scales lying close to the flattish branchlets.

T. occidentalis, L. (American Arbor Vitæ.) Trunk 30 to 50 feet high, and 1 to 2 feet in diameter, branching profusely. Branches upright and compact, forming a pyramidal or oblong cylindrical head. Leaves small, scale-like, imbricated, in 4 rows on the 2-edged branchlets. Cones egg-shaped; scales spreading; seeds winged. The frond-like branchlets are densely ramified, and spread in a lateral direction. The leaves when bruised emit an aromatic odor.

THUJA OCCIDENTALIS
(Arbor Vitæ).

Geography. — The T. occidentalis is a native of North America, north of 40° north latitude. It has been introduced into England as an ornamental tree.

Etymology. — *Thuja* is derived from the Greek word θυία, an African tree with sweet-smelling wood. *Occidentalis* is Latin for western, and refers to the western world, the home of this species. The popular name *arbor vitæ* is Latin for "tree of life," and is supposed to arise from the circumstance that the fruit of some of the species is used for medicine. In the East the cypress is called the *tree of life*, for the reason that the berries are supposed to be a remedy for all diseases.

Use. — The Thuja occidentalis is largely used for ornamental purposes in planted grounds. It bears the knife well, and is on that account well adapted for hedging purposes.

The wood is light, durable when exposed to the weather, and furnishes excellent material for fencing. It is also extensively used in the manufacture of casks for packing

TSUGA, Endl. (Hemlock.) Flowers monœcious; aments in terminal or nearly terminal clusters. Scales of the cones thin and flat, reflexed and persistent, nearly hiding the subtending bract. Leaves linear, flat, and somewhat 2-ranked.

T. Canadensis, Carr. (Hemlock Spruce.) Trunk 50 to 100 feet in height, branching freely. Bark gray, smooth on young trees, but very rough and furrowed on old trees. Leaves solitary, flat, slightly toothed, blunt at the apex, in 2 ranks, half an inch long, and less than an eighth of an inch wide. Cones three fourths of an inch in length, and less than half an inch in diameter; scales suborbicular, half an inch long; wing less than half an inch broad. Flowers in June; seed matures in the following year in June.

Geography. — The geographical range of the hemlock is confined to a belt on both sides of the forty-fifth parallel, in the Northern Hemisphere, reaching down to Pennsylvania in mountainous regions, and even to North Carolina, and as far north as Oregon and Hudson Bay.

TSUGA CANADENSIS (Hemlock Spruce).

Etymology. — *Tsuga* is Japanese for yew-leaved or evergreen. The name *Canadensis* comes from Canada, the home of the tree. The origin of *hemlock*, the common name, is not so easily determined; it is suggested that it comes from *hem*, the edge or border, and *loc*, fasten, inclose, alluding to the use of the tree in hedging. Again, *hem* means "injure" or "cripple," and may allude to the poisonous properties of the cicuta, called *hemlock*. These inferences are not to be relied upon, and the origin and meaning of the name must be left in obscurity.

History. — The hemlock is an American tree; it was taken to Europe soon after the settlement of northeastern America, and grows well in the northern parts of England.

Use. — As an ornamental tree the hemlock is a favorite in large grounds. When not crowded it rises to the height of 40 to 80 feet, a perfect pyramid, its lower branches resting on the ground. Its foliage is the most delicate of all the coniferæ. It bears the knife well, and makes a compact and beautiful hedge. The wood is soft, easily split, and has a very coarse grain; yet it is strong, holds a nail well, requires a great force to produce a cross fracture, and is very durable. It is largely used in the frames of edifices, for joists and for sheathing, being the cheapest of all the soft-wood lumber. The lumber is obtained from the stripped trunks, which are sawed into 13-feet lengths, which during the following winter are drawn to the frozen streams and left till the spring thaw, when they are floated down to the saw-mills, where they are sawed into boards, scantling, and ceiling laths, and thence sent to market.

The bark is highly charged with tannin, and is used in immense quantities for manufacturing leather. It is obtained by felling the tree in the early summer when the sap is in its greatest activity; girdles are cut around the trunk and large branches by means of an axe, and with a wedge-shaped bar the bark is stripped; it is then piled to dry. Its value is estimated by the cord.

CHAMÆCYPARIS, Spach. (Cypress.) Flowers monœcious, on different branches of the same tree, in terminal catkins. Staminate flowers in ovoid aments, 4 anthers under the scales. Pistillate flowers in a globular cone; ovules bottle-shaped; scales thick and woody, peltately dilated, bossed in the middle; cones globose; seeds few, with narrow wings attached to the base; cotyledons 2, or 3. Trees with closely appressed evergreen leaves.

1. **C. thyoides, L. (Cupressus thyoides, L.)** (White Cedar.) Trunk 40 to 80 feet in height, and 1 to 3 feet in diameter. Branches somewhat spreading, and pendent at the extremities. Bark brown, ragged, soft, exfoliating in strips. Leaves imbricated in four rows, short, scale-like, with a small tubercle on the back of each. Cones in groups; very small, globular scales, shield-shaped, blunt-pointed; seeds small, subglobular.

There are many species, but the thyoides is the most important in eastern North America.

2. **C. sempervirens, L. (Cupressus sempervirens, L.)**, is a native of the countries of the Levant. It is there the gloomy sentinel of the graveyard. It is pyramidal, and hence the emblem of death.

3. **C. pendula (Cupressus pendula)**, of China, has pendulous branches; in habit like the weeping willow.

Geography. — The geographical zone of the chamæcyparis is from 30° to 42° north latitude. In America it extends across the continent, and it occupies about the same zone in the Old World. It is found sparingly in the middle Atlantic States, especially in New Jersey, and west to the Great Lakes, but it reaches perfection in the swamps of the Carolinas, Georgia, and Florida.

The cedar swamps of New Jersey, made famous by the botanical excursions of Pursh, Nuttall, Michaux, Bartram, and Gray, have since their day been the Mecca to which every young botanist longs to make a pilgrimage. In these swamps a very important industry is carried on, consisting of mining sunken logs of the cypress, and working them into shingles and barrel staves.

Etymology and History. — *Chamæcyparis* is derived from the Greek χαμαί, on the ground, and κυπάρισσος, cypress, from κύω, produce or contain, and πάρισος, equal, alluding to the regularity of the branches. According to some authors, the name is derived from *Cyparissus*, son of Telephus, who for killing the stag of the gods was transformed into a cypress tree. The common name was derived from the isle of Cyprus, where a tree of this genus abounded.

Use. — The wood of chamæcyparis thyoides is light, soft, and easily worked, has a fine grain, and takes a good polish. It is used in the manufacture of trunks, boxes for preserving linen and woollen goods, shingles, and staves for casks. Cabinet-makers use it for drawers in fine cabinet ware; and large use is made of the small trees by builders for scaffolding poles, — the poles being light, slender, and strong. It is also used for masts for small vessels. It is remarkable for its durability. The doors of St. Peter's church at Rome were of cypress wood from the Levant, and were found to be quite sound after a service of 1,100 years. Its peculiar bitterness preserves it from the attacks of insects, and in part explains its durability.

GLOSSARY.

A.

ab nôr'mal, *contrary to the usual or natural structure.*
ab o rĭg'i nal, *original in the strictest sense.*
a bŏr'tion, *non-development of a part.*
ab rŭpt', *terminating suddenly.*
ab sôrp'tion, *the act of taking in or sucking up.*
ac au lĕs'çent, *apparently stemless.*
ac çĕs'so ry, *something added.*
ac crĕs'çent, *growing after flowering.*
ac crēte', *grown to.*
ac cŭm'bent, *lying against.*
ă çĕph'a lous, *without head.*
ăç'er ose or ăç'er ous, *needle-shaped.*
a chē'ni um (*pl.* achenia), *a small, dry, hard-shelled, one-celled, one-seeded, indehiscent fruit.*
ăch la mȳd'e ous, *without floral envelopes.*
a çĭc'u lar, *finely needle-shaped.*
ā'corn, *the fruit of the oak.*
a cot y lēd'o nous, *without cotyledons.*
ăc'ro gens, *summit growers.*
a cū'le ate, *armed with prickles.*
a cū'mi nate, *drawn out into a point.*
a cūte', *ending in a sharp angle.*
a dĕlph'ous, *having the stamens joined in a fraternity.*
ad e nŏph'o rous, *producing glands.*
ad hēr'ent, *growing to.*
ad hē'sion, *the union of organs of different kinds, as stamens to petals, etc.*
ăd'nate, *growing fast to.*
ad prĕssed', or ap prĕssed', *brought into contact with, but not united.*
ad ven tĭ'tious, *out of the usual order; accidental.*
ad vĕn'tive, *applied to foreign plants sparingly introduced into a country, but not naturalized.*
ā ĕr ā'tion, *same as respiration.*
æ ru'gi nous, *verdigris-colored.*
æs ti vā'tion, *arrangement of the parts of a flower in the bud.*
af fĭn'i ty, *resemblance in essential organs.*
ăg'a mous, *sexless.*
ăg'gre gate, *assembled close together.*

a glu mā'ceous, *without glumes; same as petaloid.*
a grĕs'tis, *growing in fields.*
air' blăd'der, *a sac filled with air.*
air' plants, *plants whose roots are in the air.*
ā'kene, or a kē'ni um, *an indehiscent seed-vessel; a nutlet.*
ā'la (*pl.* alæ), *a wing.*
al a băs'trum, *a flower-bud.*
ā'late, *winged.*
al bĕs'çent, *whitish, or turning white.*
al bū'men, *a deposit of nutritive material within the seed-coats.*
al bū'mi nous, *like albumen.*
al bŭr'num, *sap-wood.*
ăl'gæ, *sea weeds.*
al li ā'ceous, *having the odor of garlic.*
al lŏg'a mous, *having cross-fertilization.*
ăl'pīne, *belonging to high mountains above the limit of forests.*
al tĕr'nate, *distributed singly at different heights of the stem.*
ăl've o late, *with pits like the honey-comb.*
ăm'ent, *a deciduous spike.*
am en tā'ceous, *catkin-like, or catkin-bearing.*
a môr'phous, *without definite form.*
am phi căr'pous, *producing two kinds of fruit.*
am phĭt'ro pous, *turned both ways.*
ăm'pho ra, *a pitcher-shaped organ.*
am plĕc'tant, *embracing.*
am plĕx'i caul, *clasping the stem, as the base of some leaves.*
am pul lā'ceous, *swelling out like a bottle or bladder.*
am ȳl ā'ceous, ăm'y loid, *composed of starch; starch-like.*
a nǎl'y sis (botanical), *the process of classifying and finding the names of plants.*
an ăn'drous, *without stamens.*
an ăn'ther ous, an ăn'thous, *without anthers.*
a nas to mō'sis, *union of vessels or veins.*
a nǎt'ro pous, *having the ovule inverted at an early period in its development, so that the chalaza is at the apparent apex.*
an çĭp'i tal, *two-edged.*
an dro̅'ci um, *the stamens of a flower taken together.*

317

318 GLOSSARY.

an drŏg′y nous, *having stamens and pistils on the same peduncle.*

ăn′dro phore, *a column of united stamens, as in the Mallow.*

an e mŏph′i lous, *wind-loving, said of wind-fertilized flowers.*

an frăc′tu ose, *bent hither and thither, as the anthers of a Squash.*

ăn′gi o sperms, *plants whose seeds are inclosed in a vessel.*

ăy′gu lar, *a kind of divergence of leaves.*

an i sŏm′er ous, *having the parts unequal in number.*

ăn i so pet′a lous, *with unequal petals.*

an i soph′yl lous, *having the leaves of a pair unequal.*

an nŏt′i nous, *yearly, or in yearly growths.*

ăn′nu al, *yearly.*

ăn′nu lar cells, *cells with ring-like markings.*

ăn′nu late, *marked by rings, or furnished with a ring like that of the spore-case of most ferns.*

an tē′ri or, *adjacent to the bract.*

an thē′la, *an open paniculate cyme.*

an thel mĭn′tic, *expelling or killing worms.*

ăn′ther, *the part of the stamen that contains the pollen.*

an ther Id′i um, *the organ in cryptogams corresponding to the anther in flowering plants.*

an ther If′er ous, *anther-bearing.*

an thē′sis, *the opening of the flower; flowering.*

an tho căr′pous, *having the fruit and flower united.*

ăn′tho phore, *a stipe between the calyx and the corolla.*

an trōrse′, *facing toward the anterior.*

a pĕt′a læ, *plants without petals.*

a pĕt′a lous, *without petals.*

ā′pex, *the top or point, especially of a leaf.*

ăph′yl lous, *without leaves.*

ăp′ic al, *belonging to the apex or point.*

a pīc′u late, *tipped with a small, distinct point.*

ap o căr′pous, *having the several pistils of the same flower separate.*

ap o pĕt′a lous, *having the petals entirely disconnected.*

a pŏph′y sis, *a swelling.*

ap o sĕp′a lous, *having separate sepals.*

ap o thē′ci um, *the fructification of lichens forming masses of various shapes.*

ap pĕnd′age, *any superinduced part.*

ap pen dĭc′u lar, *having appendages.*

ap prĕssed′, *see adpressed.*

ăp′ri cate, *to grow in dry and sunny places.*

ăp′ter ous, *without wings.*

a quăt′ic, *living in water.*

a răch′noid, *resembling cobwebs.*

ar bo rē′tum, *a collection of trees.*

ăr′bor ous, *tree-like.*

ar che gō′ni um, *the organ in mosses analogous to the pistil of flowering plants.*

ărc′u ate, *arched or curved like a bow.*

ăr′e nose, *growing in sand.*

a rē′o late, *having the surface divided into little spaces or areas.*

ar gĕn′te ous, or ăr′gen tate, *silvery.*

ar gĭl′lose, or ar gĭl′lous, *growing in clay.*

ar gū′tus, *acutely dentate.*

ăr′il, *an extra seed-covering.*

a rĭs′tate, *with an arista or awn, as the Barley.*

a rĭs′tu late, *short-awned.*

ärmed, *bearing prickles, spines, etc.*

ar rĕct′, *upright in position.*

as cĕnd′ing or as cĕnd′ent, *arising obliquely; assurgent.*

as cĭd′i um, *a tubular, horn-shaped, or pitcher-like formation.*

ăs′cus, *a sac; the spore-case of lichens and some fungi.*

as per gĭl′li form, *shaped like the brush used to sprinkle holy water, as the stigmas of many grasses.*

ăs′per ous, *rough to the touch.*

as sim i lā′tion, *the function of producing starch or other plant food.*

as sŭr′gent, *same as ascending.*

ăt′ro pous or ăt′ro pal, *not inverted; orthotropous.*

at tĕn′u ate, *becoming slender or thin.*

au răn′ti ā′ceous, *orange-colored.*

au′re ous, *golden.*

au tŏg′a my, *self-fertilization.*

awl-shaped, *sharp-pointed from a cylindrical base.*

awn, *the bristle or beard of Barley and like plants.*

ăx′i al root, *the main root growing downward; tap-root.*

ăx′il, *the angle between the petiole and the branch on the upper side.*

ăx′il la ry, *growing out of the axils.*

ăx′is, *the stem.*

B.

băc′cate, *berry-like; covered with pulp.*

bac tē′ri um, *the smallest organism known: micro-organisms, destitute of chlorophyll, which multiply with marvelous rapidity and cause putrefaction and disease.*

băn′ner, *the upper petal of a papilionaceous flower.*

bär′bate, *bearded.*

bärbed, *furnished with a barb or double hook.*

bär′bel late, *beset with stiff, short hairs, as bristles.*

bar bĕl′lu late, *diminutive of* barbellate.

bärk, *the outer covering of an exogenous tree or shrub.*

bāse, *the extremity of any organ by which it is attached to its support.*

bā′si fixed, *attached by its base.*

GLOSSARY. 319

băs'i lar, *attached to the base; basal.*
bast-cells, *long cells of bark.*
beaked, *ending in an extended tip.*
beard'ed, *having tufts of long hairs.*
bi ar tic'u late, *twice-jointed; two-jointed.*
bi au ric'u late, *having two ears, as the fig-leaf.*
bi căl'lose, *having two hard spots.*
bi căr'i nate, *two-keeled.*
bi çĭp'ĭ tal, *two-headed; dividing into two parts.*
bi'col or, *two-colored.*
bi cŏn'ju gate, *twice-paired, as when a petiole forks twice.*
bi cŭs'pid ate, *with two points or cusps.*
bi děn'tate, *with two teeth.*
bi ěn'nĭ al, *of two years' **duration.***
bi'fid, *cleft into two parts.*
bi fō'li ate, *with two leaflets.*
bi fŭr'cate, *forked; twice forked.*
bij'u gate, *bearing two pair**s.***
bi lā'bi ate, *two-lipped.*
bi lăm'el late, *of two plates.*
bi'lobed, *two-lobed.*
bi lŏ çĕl'late, *divided into two secondary cells.*
bi lŏc'u lar, *divided into two cells.*
bi'nate, *two by two; in pairs.*
bi nō'dal, *having two nodes.*
bi nō'mi al, *having two names.*
bi pǎl'mate, *twice palmately divided.*
bĭp'a rous, *having two branches or axes (applied to a cyme).*
bi pĭn'nate, *twice pinnate.*
bi pin nǎt'ĭ fid, *twice pinnatifid.*
bi pin năt'ĭ sect, *twice pinnately **divided.***
bĭp'li cate, *twice folded together.*
bi sē'ri al, bi **sē'ri ate,** *occupying two rows, one within the other.*
bi sĕr'rate, *doubly serrate; as when the teeth of a leaf are themselves serrate.*
bi sĕx'u al, *having both stamens and pistils.*
bi tĕr'nate, *twice ternate.*
bi'vălved, *two-valved.*
blāde, *the expanded part of the **leaf.***
blanched, *whitened by lack of light.*
blōōm, *a whitish powder on fruits, leaves, etc.*
bōat'-shaped, *concave within and keeled without, like a small boat.*
bŏt'a ny, *the science which treats of **plants.***
brănch'e ate, *with opposite **spreading** branches.*
brăct, *the small **leaf** or scale **from the axil** of which a flower or its pedicel proceeds.*
brăc'te ate, *furnished with bracts.*
brăc'te o late, *furnished with bractlets.*
brăct'let, brăc'te ole, *a **bract on the pedicel** or flower-stalk.*
branch, *a shoot growing from the stem.*
brĭs'tles, *stiff, sharp hairs.*
brĭst'ly, *beset with bristles.*
brŭsh'-shaped, *aspergilliform; shaped like the brush used in sprinkling holy water.*
bry ŏl'o gy, *that part of botany which treats of mosses.*

bry ŏph'y ta, *moss-like plants.*
bŭd, *the growing point; an undeveloped plant or flower.*
bŭd'ding, *the process of forming buds.*
bŭd'-scales, *coverings of a bud.*
bŭlb, *an underground bud.*
bul bĭf'er ous, *bearing or producing bulbs.*
bŭlb'lets, *little bulbs.*
bŭl'late, *appearing as if blistered.*
bys sā'ceous, *composed of fine, flax-like threads.*

C.

ca dū'cous, *dropping off early.*
cæ s'pi tose, *turf-like; having many stems from one rootstock, or from many entangled roots.*
ca lăth'i **form,** *cup-shaped.*
căl'căr ate, *furnished with a spur.*
căl'çe o late, *slipper-shaped.*
căl'lous, *hardened.*
ca lyç i flō'rous, *having the petals and stamens adnate to the calyx.*
căl'y çĭne, *calyx-like.*
ca lȳc'u late, *having an outer calyx or calyx-like involucre.*
ca lȳp'tra, *the hood of the spore-case of a moss.*
ca lȳp'tri **form,** *shaped like a candle-extinguisher.*
cā'lyx, *the outer floral envelope.*
căm'bi um, *an old name for the sappy cells between the wood and bark; nascent structure.*
cam păn'u late, *bell-shaped.*
cam py lŏt'ro pous, *having the ovule curved, with the apex near the hilum.*
can a lĭc'u late, *channelled.*
căn'çel late, *latticed; resembling lattice-work.*
căn'di dus, *pure white.*
ca nĕs'çent, *grayish white.*
căp'ĭl la ry, *or* cap il lā'ceous, *resembling **hair**; long and slender.*
căp'i tate, *head-shaped, growing in close clusters or heads.*
ca pĭt'u lum, *a little head.*
căp're o late, *bearing tendrils.*
căp'sule, *a dry, dehiscent seed-vessel with more than one carpel.*
căr'bon di ŏx'ĭde, *a substance consisting of one atom of **carbon to two of** oxygen.*
ca rī'na, *a keel.*
cǎr'ĭ nate, *boat-shaped; having a sharp ridge beneath.*
căr i ŏp'sis, the *one-seeded fruit of grain or grasses.*
căr'ne ous, *flesh-colored; pale red.*
căr'pel, *a pistil.*
car pŏl'o gy, *that department of botany which relates to fruits.*
căr'po phore, *part of a receptacle prolonged between the carpels.*

GLOSSARY.

car ti lăg′i nous, *firm and tough in texture, like cartilage.*
căr′un cle, *an excrescence near the hilum of some seeds.*
căr y o phyl lā′ceous, *relating to the Pink family.*
car y ŏp′sis, *a grain; a thin, dry, one-seeded pericarp.*
cas sĭd′e ous, *helmet-shaped.*
căs′sus, *empty; sterile.*
căt′e nate, or **ca tĕn′u late,** *end to end, as in a chain.*
căt′kin, *an ament.*
cau′date, *tailed, or tail-pointed.*
cau′dex, *the trunk or stem of a plant.*
cau′di cle, *the stalk of a pollen-mass.*
cau lĕs′çent, *having a distinct stem.*
cau′li cle, *a little stem, or rudimentary stem of a seedling.*
cau′line, *relating to the stem.*
cau lo căr′pic (*stems*), *same as perennial.*
cau′lome, *the cauline parts of a plant.*
cĕll, *a sac or bag-like body containing protoplasm.*
cĕll-growth, *formation and enlargement of cells.*
cĕl′lu lar tĭs′sue, *tissue formed of cells.*
cĕl′lu lose, *the substance of which cell-walls are formed.*
cen trĭf′u gal in flo rĕs′cence, *a flowering from the center.*
cen trĭp′e tal in flo rĕs′cence, *a flowering outside toward the center.*
çĕph′a lous, *head-shaped; growing in close clusters; capitate.*
çē′re al, *relating to grains, corn, etc.*
çĕr′nu ous, *nodding, but less inclined than pendulous.*
chăff, *the husks of grasses and grains.*
chăff′y, *abounding in or resembling chaff.*
cha lā′za, *the part of an ovule where the covering and the nucleus join.*
chăn′nelled, *hollowed out like a gutter.*
chăr′ac ter, *a word expressing the essential marks of a species, genus, etc.*
char tā′ceous, *having the texture of paper.*
chlō′ro phyll, *the green substance of leaves and bark.*
chlo rō′sis, *a condition in which naturally colored parts turn green.*
cho ri pĕt′a lous, *having separate petals; polypetalous.*
chō′ri sis, *separation of an organ into two or more parts.*
chrō′mule, *coloring matter in plants.*
çĭc′a trix, *the scar left by the fall of a leaf or other organ.*
çĭl′i ate, *fringed with marginal hairs.*
çi nē′re ous, *ashy-gray, ash color.*
çi′on, *a young shoot.*
cir′çi nate, *rolled inward from the top.*
cir cu la′tion, *a moving around (as of the sap).*
cir cum sçĭs′sile, *opening by a transverse slit.*

cir cum scrip′tion, *general outline.*
çĭr′rhose, *furnished with a tendril.*
çĭt′re ous, *lemon-yellow.*
clā′dose, *branched or ramose.*
clăth′rate, *latticed; cancellate.*
clā′vate, *club-shaped.*
cla vĭc′u late, *having clavicular, or little tendrils or hooks.*
claw, *the narrow or stalk-like base of some petals, as of Pinks.*
clĕft, *cut into lobes.*
clei s tŏg′a mous, *fertilized in closed buds.*
clei s tŏg′a my, *fertilization in closed buds.*
climb′ing, *rising by clinging to other objects.*
clȳp′e ate, *buckler-shaped.*
co a lĕs′çent, *growing together.*
co ărc′tate, *contracted, drawn together.*
cōat′ed, *having an integument, or covered in layers.*
cŏb′web by, *bearing hairs like cobwebs or gossamer; arachnoid.*
coc çĭn′e ous, *scarlet-red.*
cŏc cus (pl. *cocci*), *a berry; a one-seeded carpel of separable fruits.*
coch le ăr′i form, *spoon-shaped.*
cŏch′le ate, *spiral, like a snail-shell.*
co hē′sion, *union of one organ with another of the same kind.*
cō′hort, *a division next above the Order.*
col lăt′er al, *placed side by side.*
cŏl′lum, *the part of the root where the stem meets it.*
cŏl′ored, *of any color except green, which in botany is not a color, while white is.*
col u mĕl′la, *the axis to which the carpels of a compound pistil are often attached, as in Geranium; or which is left when a pod opens, as in Azalea.*
cŏl′umn, *the combined stamens and styles.*
co lŭm′nar, *shaped like a column or pillar.*
cō′ma, *a tuft of hair.*
cŏm′u la sure, *the joining of the carpels of the cremocarp.*
cŏm′mon, *belonging alike to several.*
com plēte′ flow′er, *one that has all the organs, — calyx, corolla, stamens, and pistils.*
cŏm′pli cate, *folded upon itself.*
cŏm′pound flow′er, *one composed of a number of separate flowers crowded on the torus.*
cŏm′pound leaf, *one composed of separate leaflets, or little leaves.*
com prēssed′, *flattened on the sides.*
cŏn′chi form, *shell- or half-shell-shaped.*
cŏn′col or, *all of one color.*
con dū′pli cate, *folded on itself lengthwise.*
cōne, *a strobile; a multiple fruit having the shape of a cone.*
con fer rŭ′mi nate, *stuck together, as the cotyledons in a Horse-chestnut.*
cŏn′flu ent, *uniting; coherent.*
con formed′, *similar to something associated or compared; closely fitted, as the skin to the kernel of a seed.*

GLOSSARY. 321

con gès'ted, *conglomerate; crowded together.*
con glòm'er ate, *crowded together; densely clustered.*
còn'nate-per fō'li ate, *having the leaves connate, or united round a stem.*
con tin'u ous, *not jointed or articulated.*
con tòrt'ed, *twisted.*
con tor tū'pli cate, *twisted back upon itself.*
con tráct'ed, *either narrowed or shortened.*
còn'tra ry, *turned in an opposite direction to the ordinary way.*
còr'date, *heart-shaped.*
cor i ā'ceous, *leather-like.*
còrk'y, *of the texture of cork.*
còrm, *a sort of bulb, or fleshy stem.*
còr'ne ous, *horn-like in texture.*
cor nĭc'u late, *having a small horn or spur.*
còr'nute, *horned; bearing a horn-like projection or appendage.*
co rōl'la, *inner perianth, made up of petals.*
cor ol lā'ceous, *like, or belonging to, a corolla.*
còr'ol line, *pertaining to the corolla.*
co rō'na, *a crown.*
còr'o nate, *crowned; furnished with a crown.*
còr'ti cal bark, *outer bark.*
còr'ti cate, *coated with bark or bark-like covering.*
còr'ymb, *a flat-topped or convex cluster of flowers, each on its own foot-stalk, and arising from different points of a common axis.*
cor ym bif'er ous, *bearing corymbs.*
co rȳm'bose, *in corymbs; approaching the form of a corymb, or branched in that way.*
còs'tate, *ribbed; having rib-like ridges.*
cot y lē'dons, *lobes or seed-leaves, or first leaves of the embryo.*
cra tēr'i form, *of the form of a goblet.*
creep'er, *a plant that trails on the ground.*
crēm'o carp, *a dry fruit of two one-sided carpels, which separate when ripe.*
crē'nate, *bordered with rounded teeth.*
crēn'u late, *finely scalloped.*
crēst'ed or crĭs'tate, *having an elevated ridge.*
cre tā'ceous, *chalky, or chalk-like.*
crĭb'ri form, *pierced like a sieve with small apertures.*
crī'nite, *bearing long hairs.*
crō'ceous, *saffron-colored; of a deep reddish-brown.*
cròss'-breeds, *the progeny of interbred varieties.*
cròss-fer ti li zā'tion, *the fertilization of a plant by pollen from a different individual.*
cru'ci form, *in the form of a Roman cross.*
crūde sap, *sap before it has been exposed to the sun in the leaf.*
crus tā'ceous, *hard, thin, and brittle.*
cryp to gā'mi a, *name of the division of plants without flowers.*
cū'cul late, *rolled up into a hood shape.*
cūlm, *the straw of grasses.*

cūl'trate, *shaped like a trowel or broad knife.*
cū'ne ate, or cu nē'i form, *wedge-shaped.*
cūp'-shaped, *in the form of a drinking cup.*
cu'pu late, *provided with a cupule.*
cū'pule, *a little cup, as the cup of the acorn.*
cu pu lif'er ous, *cupule-bearing.*
cūr vi-sē'ri al, *in oblique or spiral ranks.*
cūr'vi-veined, *with curved ribs or veins.*
cush'ion, *the enlargement at the insertion or base of a petiole.*
cūs'pi date, *having a sharp, stiff point.*
cūt, *having sharp or deep divisions; incised.*
cū'ti cle, *outer lumina of wall of epidermis.*
cy ăn'ic, *blue, or of any color except yellow.*
cy ăth'i form, *cup-shaped.*
cy̆c'lic al, *rolled up circularly, or coiled into a complete circle.*
cy clō'sis, *the circulation of protoplasmic granules within a living vegetable cell.*
cy lin drā'ceous, *approaching to the cylindrical form; not tapering; columnar.*
cȳm'bi form, *boat-shaped.*
cȳme, *flower-cluster with the oldest flowers at the top or center.*
cy'mose, cy'mous, *having the nature of a cyme; bearing cymes.*
cy'mule, *a partial diminutive cyme.*
cy to blas tē'ma, *the viscous fluid in which vegetable cells are produced and held together.*
cy'tode, *a nucleated mass of protoplasm.*
cy'to plasm, *a vessel or chamber which surrounds or contains the protoplasm.*

D.

de căg'y nous, *having ten pistils or styles.*
de căm'er ous, *having ten parts.*
de căn'drous, *having ten stamens.*
de cĭd'u ous, *falling at the end of the season.*
dēc'li nate, *bent downward.*
de com pound', *much compounded or divided.*
de cŭm'bent, *reclining with the top ascending.*
de cūr'rent, *running down into, or upon.*
de cŭs'sate (leaves), *opposite and having the pairs at right angles.*
de du pli cā'tion, *separation of an organ into many parts.*
de flēxed', *bent downward.*
de flō'rate, *past the flowering state, as an anther after it has discharged its pollen.*
de fo li ā'tion, *the casting off of leaves.*
de hĭs'cence, *mode of the opening of seed-vessel or anther.*
de hĭs'cent, *opening by regular dehiscence.*
del i quēs'cent, *branching, so that the stem is lost in branches.*
dēl'toid, *like the Greek letter Δ in form.*
de mērsed', *growing below the surface of water.*
dĕn'droid, *tree-like in form.*

Pr. Fl. — 22

dĕn'tate, *toothed.*
den tĭc'u late, *toothed with fine or small teeth.*
de nū'ded, *become naked.*
de pau'per ate, *less developed than usual.*
de pĕnd'ent, *hanging down.*
de prĕssed', *flattened from above; low.*
de scĕnd'ing, *tending gradually downwards.*
de scĕnd'ing ăx'is, *the root.*
dĕx'trin, *a gummy substance produced by the action of diastase upon starch.*
dĕx'trorse, *twining; turning to the right.*
di a dĕl'phous, *having stamens grouped into two sets by united filaments.*
di ag nō'sis, *a brief statement of the distinctive character of a plant or group.*
di al y pĕt'a lous, *having separate petals; polypetalous.*
di ăn'drous, *with two stamens.*
di ăph'a nous, *transparent, or translucent.*
di'as tase, *a peculiar ferment in malt, altering starch into dextrine.*
di cär'pel la ry, *having two carpels.*
di chla mȳd'e ous, *having both calyx and corolla.*
di chŏg'a mous, *having stamens which ripen before the pistils, or vice versa.*
di chŏt'o mous, *forked, or two-forked.*
dĭc'li nous, *having flowers of separate sexes.*
di cŏc'cous, *splitting into two cocci or closed carpels.*
di cot y lĕd'on ous, *having two cotyledons or seed-lobes.*
di cŏt y lē'dons, *plants which have two seed-leaves in their embryos.*
dĭd'y mous, *double.*
di dȳn'a mous, *having the stamens of a four-androus flower in two pairs, one pair shorter than the other.*
dif füse', *much divided and spreading.*
dĭg'i tate, *having several distinct leaflets palmately arranged, as in the leaf of the Horse-chestnut.*
dĭg'y nous, *having two pistils or styles.*
dĭm'er ous, *made up of two parts, or having organs in two.*
di mĭd'i ate (*anther*), *halved.*
di mŏr'phous, *having two forms.*
di œ'cious, *having staminate and pistillate flowers borne on different plants.*
di pĕt'al ous, *having two petals.*
dĭph'yl lous, *two-leaved.*
dĭp'ter ous, *having two wings.*
dĭs'ci form, *or disc-shaped, flat and circular, like a disk or quoit.*
dĭs'coid, *having no rays.*
dis cŏl'or, *of two different colors or hues.*
dis crēte', *separate; opposite of concrete.*
di sĕp'al ous, *having two sepals.*
dĭsk, *face of an organ, or a circular spot on cells.*
dĭsk-flow'ers, *flowers of the disk in Compositæ.*
dis sĕct'ed, *cut into deep lobes.*

dis sĕp'i ment, *a partition, a separating tissue.*
dis sĭl'i ent, *bursting into pieces.*
dĭs'tich ous, *arranged in two rows.*
dis tinct', *separate; not united.*
di thē'cous, *having two thecæ, or anther-cells.*
di vâr'i cate, *widespread; straggling.*
di vēr'gent, *spreading with a smaller angle.*
di vid'ed (*leaves*), *cut into divisions, down to the base or midrib.*
do de căg'y nous, *having twelve pistils.*
do de căn'drous, *having twelve stamens.*
do lăb'ri form, *ax-shaped.*
dŏr'sal, *on, or relating to, the back.*
dŏt'ted cells, *cells with small spots.*
dŏt'ted ducts, *ducts with spots or dots.*
doū'ble flow'er, *a flower in which the stamens become petals.*
down'y, *clothed with short, weak hairs.*
dru pā'ceous, *like a drupe.*
drupe, *a stone fruit, as the Peach and Cherry.*
drȳ'ing-press, *an apparatus for drying botanical specimens.*
dŭcts, *elongated cells through which the fluids of a plant pass.*
du mōse', *bushy; like a bush.*
dū'pli cate, *in pairs; double.*
du rā'men, *heart-wood.*
dwarf'ing, *preventing a plant from growing to its full size.*

E.

ēared, *auriculate; having ears.*
e brăc'te ate, *without bracts.*
ĕch'i nate, *prickly; with rigid hairs.*
e dĕn'tate, *toothless.*
ef fēte', *sterile, exhausted.*
ef füse', *very loosely branched and spreading.*
e glăn'du lose, *destitute of glands.*
ĕl'a tera, *spiral, elastic threads accompanying certain spores.*
el lip soi'dal, *shaped like an ellipsoid.*
el lĭp'tic, el lĭp'tic al, *having the form of an ellipse.*
e lŏn'ga ted, *lengthened; extended.*
e mär'gi nate, *having a notch at the apex or top.*
ĕm'bry o, *the young plant in the seed.*
ĕm'bry o sắc, *the cell in the ovule in which the embryo is formed.*
ĕm'bry o nal, *belonging, or relating to, the ovary.*
e mersed', *raised out of water.*
en de căg'y nous, *having eleven pistils or styles.*
eu dĕm'ic, *peculiar to a country geographically.*
ĕn'do carp, *the inner layer of a seed-vessel.*
ĕn'do chrome, *the coloring matter of plants.*

GLOSSARY.

en dŏg'e nous struc'ture, *structure in which the pith and woody fiber are indiscriminately mingled.*

en'do gens, *plants whose structure is endogenous.*

en do phlœ'um, *the inner layer of bark.*

eu do pleū'ra, *the inner coating of a seed.*

en do rhi'zal, *radicle, or root, sheathed in germination.*

ĕn'dos mose, *a thrusting, causing liquids of different densities to pass through thin membranes and mingle.*

ĕn'do sperm, *the albumen of a seed.*

ĕn'do stome, *the orifice in the inner coat of an ovule.*

en ne ăg'y nous, *having nine pistils or styles.*

en ne ăn'drous, *having nine stamens.*

ĕn'si form, *sword-shaped.*

en tire', mär'gined, *having a continuous edge.*

en to môph'i lous (*flowers*), *frequented and fertilized by insects.*

e phĕm'er al, *enduring for one day.*

ep i cā'lyx, *an involucel like that of the Malvaceæ.*

ĕp'i carp, *the outer layer of a seed-vessel.*

ep i dĕr'mis, *outer layer of cells.*

ep i gē'ous, *growing on the earth, or close to the ground.*

e pĭg'y nous, *adnate upon the top of the ovary.*

ep ĭ pĕt'a lous, *growing upon the petals.*

e pĭph'yl lous, *borne on a leaf.*

ĕp'i phytes, *plants on other plants.*

e pĭp'te rous, *winged at the top.*

ĕp'i sperm, *the skin of the seed.*

ē'qual, *alike in number or length.*

ĕq'ui tant, *astraddle.*

e rōs'trate, *not beaked.*

e ryth ro cär'pous, *red-fruited.*

es sĕn'tial ōr'gans (*of a flower*), *stamens and pistils.*

ē'ti o la ted, *colorless for want of light.*

ex al bū'mi nous, *without albumen.*

ex cŭr'rent, *running through or beyond.*

ex ĭg'u ous, *puny; small.*

ĕx'o carp, *outer layer of a pericarp.*

ex ŏg'e n ĕ, ĕx'o gens, *plants which increase by the addition of new material outside of last year's growth.*

ex ŏg'e nous struc'ture, *structure like an exogen.*

ex o rhi'zal, *radicle in germination, not sheathed.*

ĕx'os mose, *flowing out.*

ĕx'o stome, *the orifice in the outer coat of the ovule.*

ĕx'pla nate, *spread or flattened out.*

ĕx sert', *projecting out of or beyond.*

ex stĭp'u late, *without stipules.*

ĕx'tine, *outer coat of a pollen grain.*

ĕx'tra-ăx'il la ry, *growing from without the axils.*

ex trōrse', *turned outward.*

F.

făl'cate, *scythe-shaped; curved.*

făm'i ly, *in botany, same as Order.*

fa ri'na, *meal or starchy matter.*

far i nā'ceous, *flour-like in texture.*

făr'i nose, *mealy on the surface.*

făs'ci ate, *banded; also applied to monstrous stems which grow flat.*

făs'ci cle, *a bundle.*

făs'ci cled, fas çic'u late, fas çic'u lā ted, *in the form of a fascicle; growing in a bundle or tuft, as the leaves of Larch and roots of Peony.*

fas tĭg'i ate, *close, parallel, and upright; as the branches of Lombardy Poplar.*

faux (*pl.* fau'ces), *the throat of a calyx, corolla, etc.*

fa vē'o late, *honeycombed; alveolate.*

feath'er-veined, *with all veins from the sides of the midrib or midvein.*

fē'male flow'er or plant, *one having pistil only.*

fe nĕs'trate, *pierced with one or more large holes, like windows.*

fer ru'gi nous, *of the color of iron-rust.*

fĕr'tile flow'er, *seed-producing flower.*

fer ti li zā'tion, *act of making fertile.*

fi brĭl'la (*pl.* fibrillæ), *fibril, minute thread.*

fi bro-văs'cu lar, *containing woody fibers and ducts.*

fĭd'dle-shaped, obovate, *with a deep recess on each side.*

fĭl'a ment, **the stalk** *of a stamen.*

fĭl'i form, slender; *like a thread.*

fĭm'bri ate, *fringed; having the edge bordered with slender processes.*

fĭs'sion, *a splitting into parts.*

fis sĭp'a rous, *multiplying by division of one body into two.*

fĭs'tu lar, *hollow, as the leaf of an Onion.*

fla bĕl'li form, *fan-shaped.*

fla gĕl'li form, *whip-shaped; long, tapering, and supple.*

fla vĕs'çent, *yellowish; turning yellow.*

flĕsh'y, *composed of firm pulp, or flesh.*

flĕx'u ous, *zigzag, or wavy.*

flōc'cose, *having tufts of soft hairs.*

flō'ra, *the spontaneous vegetation of a country; a written description of the same.*

flō'ral, *relating to flowers.*

flō'ral ĕn'vel ōpe, *the perianth of a flower.*

flō'ret, *one of the small flowers composing a compound flower.*

flō rĭf'er ous, *producing flowers.*

flōs'cule, *a floret.*

flow'er, *the organ which produces the seed.*

flow'er bud, *an undeveloped blossom; the dormant, or sleeping, flower.*

fo li ā'ceous, *leaf-like in texture or form.*

fō'li ate, *provided with leaves.*

fo li ā'tion, *the act of leafing.*

fō'li o late, *relating to or bearing leaflets.*

GLOSSARY.

fō'li ole, *a leaflet ; a distinct part of a compound leaf.*
fō'li ose, *bearing many leaves.*
fō'li um (*pl.* folia *or* foliums), *a leaf.*
fol'li cle, *a one-celled, many-seeded carpel, opening by the ventral suture.*
fol lic'u lar, *like, pertaining to, or consisting of follicles.*
fŏot'stalk, *a pedicel, petiole, or peduncle.*
fo rā'men, *a small opening or orifice.*
fo răm in if'er ous, *having small openings.*
fo răm'i nous, *full of holes.*
fōrked, *branched into two or three or more.*
fōr'ni cate, *having an arch or scale over the throat of the corolla.*
fōr'nix, *a small arched scale in the throat of a corolla, as in Comfrey.*
fō've o late, *having shallow pits.*
free, *not adherent or adnate.*
frĭnged, *edged with soft hairs.*
frŏnd, *an organ which is both stalk and leaf.*
frŏn'dent, *covered with leaves.*
frou dēs'cence, *the act of bursting into leaf.*
frŏn'dose, *frond-bearing, leafy.*
fruc ti fi cā'tion, *the act of producing fruit.*
frŭit, *a ripened pistil ; a seed-vessel with its contents.*
fru tēs'cent, *shrubby in character.*
fu gā'cious, *lasting but a short time.*
fŭl'cra, *accessory organs, such as tendrils, stipules, etc.*
fu lĭg'i nous, *of a sooty color ; smoky-brown ; blackish.*
fŭl'vous, *dull yellow ; tawny.*
fū'ni cle, fu nĭc'u lus, *the stalk of an ovule or seed.*
fŭr'cate, *forked ; forked-veined.*
fur fu rā'ceous, *scurfy.*
fŭr'rowed, *marked lengthwise with channels.*
fŭs'cous, *grayish or blackish brown.*
fū'si form, *spindle-shaped.*

G.

găl'bu lus, *the woody cone of the Juniper and Cypress.*
gā'le a, *the name applied to the upper lip of labiate corollas.*
gā'le ate, *shaped like a helmet.*
găm o pĕt'a læ, *plants whose petals are united.*
găm o pĕt'al ous, *having the petals united ; sympetalous.*
ga mŏph'yl lous, *having united perianth leaves.*
găm o sĕp'al ous, *with the sepals united.*
gĕm'i nate, *having two together.*
gem mā'tion, *formation of new individuals by budding.*
gĕm'mule, *a small bud ; the plumule.*
gĕn'e ra, *see* genus.
ge nĕr'ic, *pertaining to a genus.*
ge nĭc'u late, *bent at an angle.*

gē'nus (*pl.* genera), *a group of species alike in their fructifying organs.*
gĕrm, *the ovary.*
gĕr'men, *the old name for the ovary.*
ger mi nā'tion, *the development of the seed into a plant ; sprouting.*
gĭb'bous, *having a rounded protuberance at the side or base.*
glā'brate, *becoming glabrous, or almost glabrous, with age.*
glā'brous, *smooth ; not hairy.*
glăd'i ate, *sword-shaped.*
glănd, *an organ of a leaf or branch which secretes a fluid, as oil, resin, etc.*
glăn'du lar, *consisting of glands.*
glăns, *the acorn or mast of the Oak and similar fruits.*
glăr'e ose, *growing in gravel.*
glau cēs'cent, *slightly glaucous, or bluish gray.*
glau'cous, *with a bloom or whitish wavy powder, as on fresh plums.*
glo bōse', *nearly spherical in form.*
glo chĭd'i ate, *tipped with barbs, or with a double-hooked point.*
glŏm'er ate, *collected into close heads.*
glŏm'er ule, *a cyme grown in the form of a head.*
glu mā'ceous, *glume-like ; glume-bearing.*
glu mĕlles', *the inner husks of grasses.*
glūmes, *bracteal coverings of flowers or of the seeds of grains and grasses.*
gŏn'o phore, *a stipe below stamens.*
gŏs'sy pine, *cottony ; flocculent.*
grăft'ing, *the act of inserting a shoot or scion from one tree into the stem or some other part of another, so that they unite and produce fruit of the kind from which the shoot was taken.*
grain, *the gathered seeds of cereal plants ; a grain-like prominence or tubercle, as on the sepals of Dock.*
gra mĭn'e ous, *grass-like.*
grāu'u lar, *composed of grains.*
grĭs'e ous, *gray ; bluish gray.*
gru'mous, *or* gru mōse', *formed of coarse, clustered grains.*
gŭt'tate, *spotted as if by spots of something colored.*
gym no cär'pous, *having a naked fruit.*
gym no spĕr'mæ, *a class of exogenous plants characterized by naked seeds.*
gym no spĕrm ous, *having naked seeds.*
gy năn'drous, *bearing stamens on the pistils.*
gȳn'o base, *a process of the torus, an or around which the carpels are suspended.*
gy nœ'ci um, *the collective pistils of a flower.*
gȳn'o phore, *a produced torus bearing the ovary on its summit.*
gȳn o stē'gi um, *a sheath around pistils.*
gȳn o stē'mi um, *the column in Orchids consisting of style and stigma with stamens combined.*
gy'rāte, *curved inward ; circinate.*

H.

hăb'it, *the general aspect of a plant.*
hăb'i tat, *the natural locality or growth of a wild plant.*
hāirs, *outgrowths of epidermal cells.*
hāir'y, *covered with hair; made of or resembling hair; hirsute.*
hal'berd-shaped, *hastate.*
hālved, *one half apparently deficient.*
hā'mate, *or* hā'mŭse', *having the end hooked or curved.*
hăm'u lose, *having a small hook.*
hap lo stēm'o nous, *having only one series of stamens.*
hăs'tate, *triangular, with the base-lobes abruptly spreading, as in a halberd.*
hĕad, *capitulum; a form of inflorescence.*
heärt'-shaped, *a form which would be presented by the section of a sheep's heart, if cut in halves.*
heärt'-wood, *the wood near the central part of an exogenous tree or shrub.*
hĕl'i coid, *coiled like a helix or the shell of a snail.*
hĕl'met, *the hood-formed upper sepal or petal of some flowers, as of the Monkshood.*
hĕl'vo lous, *grayish-yellow.*
hĕm'i carp, *half-fruit; one carpel of an umbelliferous plant.*
hep tăg'y nous, *having seven pistils or styles.*
hep tăm'er ous, *having the parts in sevens.*
hep tăn'drous, *having seven stamens.*
hĕrb, *a plant whose stem is not persistently woody, and does not remain permanent, but dies at least down to the ground after flowering.*
her bā'ceous, *green and cellular in texture.*
her bā'ri um, *a collection of dried plants, for the use of students of botany.*
her măph'ro dite (flower), *having both stamens and pistils.*
hes per Ĭd'i um, *a berry with a thick rind, as the orange, etc.*
het er o cär'pous, *bearing fruit of two kinds or shapes.*
het er o çĕph'a lous, *having heads of two sorts on the same plant.*
het er ŏg'a mous, *having two sorts of flowers on the same head.*
het er ŏg'o ny, *having stamens or pistils of two sorts.*
het er o mŏr'phous, *of two or more shapes.*
het er ŏph'yl lous, *having two sorts of leaves on the same stem.*
het er ŏt'ro pous, *having the embryo oblique or transverse to the funiculus; amphitropous.*
hex ăg'o nal, *six-sided.*
hex ăm'er ous, *in six parts.*
hex ăn'drous, *having six stamens.*
hi ber nāc'u lum, *a winter bud.*
hi'e mal, *occurring in winter.*
hī'lal, *belonging to the hilum.*

hī'lum, *the eye or scar of the seed.*
hip po crĕp'i form, *horse-shoe shaped.*
hir sūte', *hairy, with rather long hairs.*
hir tĕl'lus, *minutely hirsute.*
hĭs'pid, *bristly; having stiff hairs.*
his tŏl'o gy, *the science of cells and tissues.*
hōar'y, *frost-colored; grayish-white.*
ho mŏd'ro mous, *running in one direction.*
ho mŏg'a mous, *having all the flowers alike.*
ho mo gē'ne ous, *of the same kind or nature.*
ho mŏl'o gous, *of the same type.*
ho mŏt'ro pous, *having the radicle of the seed directed toward the hilum.*
hood, *the helmet-shaped upper petal of some flowers.*
hooked, *hook-shaped; hamate.*
hor tĕn'sial, *fit for a garden.*
hor'tus sĭc'cus, *an herbarium; a collection of dried plant specimens.*
hū'mi fuse, *spread over the surface of the ground; procumbent.*
hy'a line, *transparent, or nearly so.*
hy'brid, *a cross-breed between two species.*
hy'dro phyte, *water plants.*
hy mē'ni um, *the spore-bearing surface of some fungi.*
hy păn'thi um, *a hollow flower-receptacle, as of a rose.*
hy per bō're an, *inhabiting northern regions.*
hyp o cra tĕr'i form, *salver-shaped.*
hyp o gē'an, *growing under ground.*
hy pŏg'y nous, *inserted below the pistil.*
hys ter ăn'thous, *having the blossoms develop earlier than the leaves.*

I.

i co săn'drous, *having twenty stamens inserted in the calyx.*
ĭm'bri cate, *imbricated, overlapped so as to break joints, like shingles on a roof.*
im mär'gin ate, *having no rim or border.*
im mērsed', *growing wholly under water.*
ĭm'par i pĭn'nate, *pinnate with a single leaflet of the apex.*
im pĕr'fect flow'er, *a flower wanting either stamens or pistils.*
in āne', *empty; void of an anther which produces no pollen.*
in ap pen dĭc'u late, *not appendaged.*
in ca nĕs'çent, *or in cā'nous, hoary; having a soft white pubescence.*
in cär'nate, *flesh-colored.*
in cīsed', *divided deeply, as if cut.*
in clūd'ed, *inclosed or confined within; as short stamens in a corolla.*
in com plēte', *flower, wanting calyx or corolla.*
in crŭs'tate, *this Lewd.*
ĭn'cu bous, *having the tip of one leaf lying flat over the base of the next above.*
in cŭm'bent, *having the radicle lying against the back of one of the cotyledons.*
in cŭrved', *gradually curving inward.*

in děf'i nite, *too numerous or variable for specific enumeration.*
in děf'i nite in flo rěs'çençe, or in de těr'mi nate inflorescence, *a process of inflorescence in which the flowers all arise from axillary buds, the terminal bud continuing to grow, and extending the stem indefinitely.*
in de his'çent, *not opening.*
in de těr'mi nate, *see* indefinite.
in dĭg'e nŏus, *native to a country.*
in du měn'tum, *any hairy covering or pubescence which forms a coating.*
in dū'pli cate, *having the edges bent abruptly toward the axis.*
in dū'si um, *the shield of the fruit dots (sori) in many ferns.*
in dū'vi ate, *clothed with old and withered parts.*
in e qui lăt'er al, *unequal-sided, as the leaf of a Begonia.*
in ěr'mis, *devoid of prickles or thorns.*
In fěr'tile, *not producing seed or pollen, as the case may be.*
in flāt'ed, *turgid and bladdery.*
in flěct'ed, *bent inward; inflexed.*
in flo rěs'çençe, *mode of flowering, or the arrangement of flowers on a plant.*
In fra-ăx'il la ry, *situated beneath the axil.*
in fun dĭb'u li form, *funnel-shaped.*
In'nate, *growing on the top of the part that sustains it.*
in no vā'tion, *a young shoot or new growth.*
in sěrt'ed, *situated upon, growing out of, or attached to some part.*
in sěr'tion, *the attachment of one part to another.*
In'te gral, *entire, not lobed.*
in těg'u ment, *a coat or covering.*
in těr'ca la ry, *inserted or introduced among others.*
in ter çěl'lu lar (*passages, spaces*), *lying between the cells or cellules.*
in ter fo li ā'ceous, *situated between opposite or whorled leaves.*
in'ter node, *the space between two nodes.*
in ter pět'i o lar, *between the petioles.*
in ter rupt'ed ly pĭn'nate, *pinnate without a terminal leaflet.*
In'tĭne, *innermost coating of a pollen grain.*
in tra fo li ā'ceous (*stipules,* etc.), *growing between the leaf or petiole and the stem.*
in trōrse'(*anthers*), *turned inward, or toward the axis.*
in truse', *projected or pushed inward.*
in tus sus çěp'tion, *the interposition of new vital or formative material among the particles already in existence, as in the growth of a cell wall by the introduction of new matter throughout the structure, and not by adding to the surface.*
in věrse', *or* in vert'ed, *having the apex in the opposite direction to that of the organ it is compared with.*

in vŏl'u çel, *a partial or small involucre.*
in vo lū'cel late, *furnished with an involucel.*
in vo lū'crate, *furnished with an involucre.*
In'vo lu cre, *a cluster of bracts around the base of a flower.*
In'vo lute, *rolled inward.*
ir rěg'u lar flowers, *flowers whose like parts differ either in size or shape.*
i sŏm'er ous, *composed each of an equal number of parts, as the members of the several circles of a flower.*
i so stěm'o nous, *having the stamens equal in number to the sepals or petals.*

J.

joint'ed, *having joints or separable pieces.*
jū'gum, *one of the ridges commonly found on the fruit of umbelliferous plants; a pair of opposite leaflets.*
ju'li form, *or* ju lū'çent, *having the form of a catkin or julus.*

K.

keel, *the two lowest petals of the corolla of a papilionaceous flower united and inclosing the stamens and pistil; a carina.*
keeled, *having a longitudinal prominence on the back; carinate.*
kěr'nel, *the whole body of the seed within the coats.*
kēy-fruit, *a dry, indehiscent, usually one-seeded, winged fruit; a samara.*
kĭd'ney-shaped, *having the shape of a kidney; reniform.*

L.

la běl'lum, *the lower petal of an orchidaceous flower.*
lā'bi ate, *lip-shaped.*
lā'bi ā'ti flō'rous, *having flowers with labiate corollas.*
lăç'er ate, *torn irregularly by deep incisions; jagged.*
la çĭn'i ate, *slashed into deep, narrow, irregular lobes.*
lac tēs'çent, *containing a thick milk-like fluid or juice.*
lac tĭf'er ous tĭs'sue, *a tissue whose cells and ducts bear milk-like fluid.*
lăc'u nose, *having lacunæ or holes; furrowed; pitted.*
la cūs'trĭne, *growing in lakes.*
lēv'i gate, *smooth, as if polished.*
la gē'ni form, *bottle-shaped.*
la gō'pous, *densely covered with long, soft hairs.*
lăm'el lar, *or* lăm'el late, *consisting of flat plates or lamellæ.*
lā'nate, *wooly; clothed with long, soft, entangled hairs.*

GLOSSARY. 327

lan'ce o late, *lance-shaped.*
la nū'gi nose, *covered with down, or* **fine soft** *hair.*
lap pā'ceous, *covered with forked points.*
lā'tent, *concealed or undeveloped.*
lāt er al, *belonging to the side.*
lāt er ī'tious, *brick-colored.*
lā'tex, *the turbid or milky juice of plants.*
lat i fō'li ous, *broad-leaved.*
leaf, *a colored expansion, growing from the stems* **or** *branches of a plant.*
leaf bud, *a bud that develops into* **a leaf or leafy branch.**
leaf'let, *one part of a compound leaf.*
leaf scär, *a cicatrix on a stem from* **which a leaf has fallen.**
leath'er y, *having the consistency of leather; coriaceous.*
lēg'ume, *a seed vessel which opens by both a ventral and dorsal opening, as the bean, pea, etc.*
le gū'mi nous, *belonging to the legumes.*
lĕn'ti çel, *a small, oval, rounded spot upon a stem or branch, from which the underlying tissues may protrude, or roots may issue, either in the air or when the stem or branch is covered with water.*
len tic'u lar, *resembling a lentil in size and form.*
len tīg'i nose. *bearing numerous dots resembling freckles.*
lĕp'rous, *covered with scurfy scales.*
lī'ber, *the inner bark lying next to the wood.*
lī'chen, *a flowerless plant growing upon rocks, trees, and various bodies.*
līd, *the cover of the spore case of mosses; the top of an ovary which opens transversely; an operculum.*
līg'ne ous sys'tem, *woody system.*
līg'u late, *strap-shaped.*
līg'ule, *a stipule of grasses.*
lil i ā'ceous, *like a lily.*
līmb, *border of a petal or sepal.*
līm'bate, *bordered, as when one color is edged with another.*
line, *the twelfth of an inch.*
līn'e ar, *long and narrow.*
līn'e ate, *marked longitudinally with depressed parallel lines.*
līn'gui form, *ligulate.*
līn'gu late, *tongue-shaped.*
līp, *one of the lobes of a labiate corolla.*
līt'to ral, *belonging to the shore.*
līv'id, *clouded with bluish brown or gray.*
lō'bate, *lobed.*
lo cĕl'late, *divided into secondary compartments or cells.*
lŏc'u lar, *relating to the cell or compartment of an ovary.*
loc u li ci'dal, *dehiscent through the middle of the back of each cell.*
lo cūs'ta, *the spikelet* **of a flower** *cluster of grasses.*

lŏd'i cule, *one of the scales answering* **in** *grass flowers to the perianth.*
lō'ment, *a jointed legume.*
lo men tā'ceous, *like a loment; having fruits like loments.*
lō'rate, *strap-shaped; ligulate.*
lū'nate, *crescent-shaped.*
lū'pu lin, **a** *fine, yellow, resinous powder, found upon strobiles or fruit of hops.*
lū'te ous, *yellowish; more or less buff.*
lu tĕs'cent, **of a** *yellowish color.*
lȳ'rate, *lyre-shaped, or spatulate and oblong, with* **small lobes** *toward the base.*

M.

māc'ro spore, *a large spore of certain flowerless plants.*
mac'u late, *marked with spots or blotches.*
māle, *staminate.*
mam mōse', *breast-shaped.*
mar cĕs'cent, *withering, but persistent.*
mär'gin al, *pertaining to the margin or border.*
mär'gin ate, *having a margin distinct in appearance or structure.*
mär'i time, *belonging to seacoasts.*
mär'mo rate, *variegated like marble.*
mēal'y, *farinaceous.*
mē'di al, or mē'di an, *running through the middle;* **belonging** *to the middle.*
me dŭl'la, *pith; soft cellular tissue occupying the center of a stem or branch.*
mĕd'ul la ry rays, *rays of cellular tissue seen in a transverse section of exogenous wood, which pass from the pith to the bark.*
mĕd'ul la ry shēath, *the tube formed by the spiral vessels around the central column of pith.*
mei o stĕm'o nous, *having fewer stamens than the parts of the corolla.*
mem bra nā'ceous, mem brā'nous, *thin, and rather soft or pliable, as* **the leaves of the** *Rose, Peach-tree, and Aspen* **Poplar.**
me nīs'coid, *crescent-shaped.*
mēr'i carp, *one carpel of a cremocarp of an umbellifer.*
mer is măt'ic, *dividing into cells or segments by the formation of internal partitions.*
mĕs'o carp, *the middle layer of a pericarp, consisting of three distinct layers.*
mes o phlœ'um, *the middle or green bark.*
me tăb'o lism, *transformation of one kind of substance into another in assimilation.*
mī'cro pyle, *an opening in the outer coat of a seed through which the fecundating pollen enters the ovule.*
mī'cro spore, *an exceedingly minute* **spore** *found in certain flowerless plants.*
mĭd'rib, or mĭd'vein, *the central vein of a leaf.*
milk'-vĕs'sels, *certain cells in the inner bark of plants containing milky* **juice.**

mĭn′i ate, *vermilion.*
mĭt′ri form, *having the form of a miter or peaked cap.*
mon a dĕl′phous, *having the stamens united in one body by the filaments.*
mo năn′drous, *having but one stamen.*
mo nĭl′i form, *jointed or constricted at regular intervals, to resemble a string of beads.*
mon o cär′pic, *fruiting but once.*
mon o chla mўd′e ous, *having a single floral envelope, either calyx or corolla.*
mon o cot y lē′don, *a plant having only one cotyledon or seed lobe.*
mo nœ′cious, *having stamens and pistils on the same plant.*
mo nŏg′y nous, *having only one style or stigma.*
mon o pĕt′al ous, *having but one petal.*
mo nŏph′yl lous, *having but one leaf.*
mon o sĕp′al ous, *having the calyx in one piece.*
mon o spĕr′mous, *having but one seed.*
mŏn′strous flow′ers, *flowers whose stamens have developed into petals.*
mor phŏl′o gy, *that branch of biology which deals with the structure of animals and plants, and treats of the forms of organs, describing their varieties, homologies, and metamorphoses.*
mŏs′chate, *exhaling the odor of musk.*
mū′cro, *a minute, sharp, abrupt point, as of a leaf.*
mū′cro nate, *ending abruptly in a sharp point.*
mu crŏn′u late, *tipped with a small point or points.*
mŭl′ti fid, *cut into many segments.*
mū′ri cate, *full of sharp points or prickles.*
mū′ri form, *resembling a wall of mason work.*
mus cŏl′o gy, *bryology; that part of botany which relates to mosses.*
mū′ti cous, *without a point; blunt.*
my cē′li um, *the white threads or filamentous growth from which a mushroom or fungus is developed.*

N.

nā′ked seeds, *seeds not in a seed vessel.*
nā′pi form, *turnip-shaped.*
nā′tant, *floating in water; submersed.*
năt′u ral ized, *growing spontaneously, but not native.*
na vĭc′u lar, *boat-shaped, as the glumes of many grasses.*
nĕck′lace-shaped, *looking like a string of beads.*
nĕc′tar, *honey.*
nec tar ĭf′er ous, *secreting honey; having a nectary.*
nĕc′tar y, *a vessel containing honey.*
nĕm′o ral, *or* nĕm′o rous, *pertaining to a wood or grove; woody; inhabiting groves.*
nĕrves, *veins.*

ner vōse,′ *conspicuously nerved.*
nĕt′ted, *or* nĕt-veined′, *having the veins interlaced so as to present the appearance of a net.*
neū′tral flow′er, *a flower without stamens or pistils.*
nĭt′id, *bright; lustrous; shining.*
ni′vai, *living in or near snow.*
nĭv′e ous, *snowy; snow-white.*
nŏd′ding, *nutant; having the summit bent over, as in the Snowdrop.*
nōde, *a joint of a stem.*
no dōse′, *knotty; having numerous or conspicuous nodes.*
nŏd′u lose, *having small nodes or knots; diminutively nodose.*
nō′men cla ture, *the technical names used in any particular branch of science or art.*
nōr′mal, *regular; according to rule.*
nō′tate, *marked with spots or lines, which are often colored.*
nu ca men tā′ceous, *resembling a small nut; bearing one-seeded, nut-like fruits.*
nu cĕl′lus, *nucleus; kernel.*
nū′ci form, *shaped like a nut.*
nu clē′o lus, *a dense rounded body within a nucleus.*
nū′cle us, *a kernel; an incipient ovule of soft, cellular tissue.*
nŭt, *the fruit of certain trees and shrubs, consisting of a hard and indehiscent shell inclosing the kernel.*
nū′taut, *nodding; having the top bent downward.*
nŭt′let, *a small nut; the stone of a drupe.*

O.

ob com pressed′, *flattened back and front.*
ob cŏr′date, *heart-shaped, with the attachment at the pointed end.*
oblăn′ce o late, *lanceolate, narrowing toward the point of attachment.*
ob lĭque′, *having unequal sides, as the leaves of an Elm.*
ŏb′long, *longer than broad, the sides being nearly parallel.*
ŏb′o vate, *egg-shaped, having the broad part at the apex.*
ob tūse′, *blunt at the apex.*
ŏb′verse, *having the base, or end next the attachment, narrower than the top.*
ŏb′vo lute, *overlapping.*
o cĕl′late, *marked with eye-like spots of color.*
ō′chre a, *a kind of sheath formed by two stipules uniting around a stem.*
och ro leū′cous, *yellowish-white.*
oc tăn′drous, *having eight distinct stamens.*
oc tŏg′y nous, *having eight pistils.*
ŏc′u la ted, *having spots or holes resembling eyes.*
of fĭc′i nal, *used in medicine,— therefore kept in the shops.*

GLOSSARY. 329

off'set, *a short, prostrate shoot, which takes root and produces a tuft of leaves.*
ol er ā'ceous, *esculent, as a pot herb.*
ol i gān'drous, *having few stamens.*
ol i vā'ceous, *olive-green; resembling the olive.*
ō'ō phore, *an alternately produced form of certain cryptogamous plants, which bear opposite fructifying organs.*
ō'ō pho rid'i um, *the spore-case* **containing** *the larger female spores.*
o pāque', *dull, not shining.*
o pĕr'cu lar, *having a lid.*
ŏp'po site, *set over against each other,* **but** *separated by the whole diameter of* **the** *stem; placed directly in front of* **another** *part or organ.*
ŏp pos'i ti fō'li ous, *placed opposite a leaf.*
or bĭc'u lar, *or* **or bĭc'u late**, *having a circular or nearly circular outline.*
or chi dā'ceous, *like an Orchid in form.*
ŏr'der, *a group below Class.*
ŏr'gan, *any member of a plant, as a leaf, a stamen, etc.*
or ga nŏg'ra phy, *a description of the organs of a plant.*
or'thŏs'ti chous, *straight-ranked.*
or thŏt'ro pous, *having the axis of an ovule or seed straight from the hilum or chalaza to the orifice.*
ŏs'se ous, *bony; hard, as the peach-stone.*
out'growth, *growth from the surface of a leaf, petal, etc.*
ō'val, *shaped like the longitudinal section of an egg.*
ō'va ry, *that part of the pistil containing the ova.*
ō'vate, *oval.*
ō'void, *resembling an egg in shape.*
ō'vule, *the young seed.*

P.

păl'ate, *a projection of the lower lip of a* **labiate** *corolla into the* **throat,** *as in Snap-dragon, etc.*
pā'le ā, *chaff, or chaff-like bract.*
pa le ā'ceous, *chaffy; having palea.*
pa lē'o la, *a diminutive palea.*
pa lē'o late, *having paleolæ, or narrow paleæ.*
pă'let, *same as palea.*
păl'mate, *lobed so that the sinuses point to the apex.*
pal măt'i fid, *palmate, with the divisions separated but little more than half way to the common center.*
pal măt'i lobed, *palmate, with the divisions separated not half way to the common center.*
pal măt'i sect, *divided down to the midrib.*
păl'mi veined, *having veins or nerves extending toward the apex.*
păl'u dose, *living in marshes.*
pan dū'ri form, *fiddle-shaped.*

păn'i cle, *a branching raceme.*
păn'i cled, *or* **pan ĭc'u late**, *having panicles.*
păn'nose, *covered with a felt of woody hair.*
pā'per y, *of about the consistence of letter paper.*
pa pĭl'io nā'ceous, *resembling the butterfly.*
pa pĭl'la (pl. **pa pĭl'læ**), *little nipple-shaped protuberances.*
păp'il late, *or* **păp'il lose**, *covered with papillæ; resembling papillæ.*
păp'pus, *the scales, awns, or bristles which represent the calyx in Compositæ.*
pap y rā'ceous, *of the consistence of paper; papery.*
păr'al lel-veined, *having the veins or nerves extending from the base of the leaf to the apex, parallel to the midvein.*
pa răph'y sis, *a minute-jointed filament among* **the** *archegonia and antheridia of mosses.*
păr'a site, **a** *plant obtaining nourishment immediately from another plant, to which it attaches itself.*
pa rĕn'chy ma, *soft* **cellular substance of a tissue**, *like the pulp* **of leaves, having no wood** *fibre.*
pa rĕn'chy mal, *consisting of parenchyma.*
par en chȳm'a tous, *pertaining to the parenchyma of a tissue or organ.*
pa ri'e tal, *attached to the main wall of the ovary.*
păr i pĭn'nate, *having an equal number of leaflets on each side, with no odd leaflet.*
pärt'ed, *deeply divided into parts.*
pär'the no gĕn'e sis, *the production of seed* **without** *fertilization.*
pär'tial ĭn'vo lu cre, *a secondary or small* **involucre;** *involucel.*
pär'tial pe dŭn'cle, *a branch of a peduncle.*
pär'tial pĕt'i ole, *a division of a main leaf-stalk, or the stalk of a leaflet.*
par'tial ŭm'bel, *an umbellet.*
par tĭ'tion, *a wall in a capsule, anther, etc.*
pa tĕl'li form, *disk-shaped, like the patella, or knee-pan.*
păt'ent, *wide open; spreading.*
pāt'u lous, *half open; expanded.*
pĕar'-shaped, *obovoid, and larger above.*
pĕc'ti nate, *having teeth like a comb; finely pinnatifid.*
pĕd'ate, *shaped like a bird's foot.*
pĕd'i cel, *a stalk which supports one flower or fruit, whether solitary or one of many ultimate divisions of a common peduncle.*
pĕd'i cĕlled, **ped i cĕl'late**, *having a pedicel; supported on a pedicel.*
pe dŭn'cle, *a flower-stalk supporting a single flower or flower-cluster.*
pe lō'ri a, *an abnormal return to regularity and symmetry in an irregular flower, commonest in Snapdragon.*
pe lō'ric, *abnormally regular or symmetrical.*
pĕl'tate, *shield-shaped.*

pend'ent, or pen'du lous, *supported from above; suspended; hanging; drooping.*

pen i cĕl'late, *furnished with a pencil of fine hairs; ending in a tuft of hairs.*

pĕn'nate, pinnate; *having several leaflets arranged on each side of a common petiole.*

pĕn'ni nerved, *feather-veined.*

pen tăm'er ous, *five-parted; having the parts in fives.*

pen tăn'drous, *having five stamens.*

per ĕn'ni al, *living several years.*

pér'fect flow'er, *a flower having both stamens and pistils.*

per fō'li ate, *having the basal part produced around the stem.*

pĕr'fo rate, *pierced with holes or transparent dots resembling holes, as an Orange leaf.*

pĕr'i anth, *calyx or corolla, or both; the leafy parts of a flower surrounding the stamens and pistils.*

pĕr'i carp, *the ripened ovary; the covering of the seed.*

per i căr'pic, *belonging to the pericarp.*

pĕr'i gone, or per i gō'ui um, *an organ inclosing the essential organs of a flower; a perianth.*

per i gȳn'i um, *the bristles, scales, or more or less inflated sack which surrounds the pistil, as in Carex.*

pe rig'y nous, *surrounding the pistil; having a tubular ring or sheath surrounding the pistil, on which the various parts of the flower are inserted.*

per i phĕr'ic, *around the outside or periphery of any organ.*

pĕr'i sperm, *the albumen of a seed, especially that part formed outside the embryo sac.*

pĕr'i stome, *the fringe of teeth to the spore case of mosses.*

per sĭst'ent, *remaining long in place.*

pér'son ate, *masked by a closing of the throat of the corolla, as in the Snapdragon.*

per tūse', *punched; pierced with holes; slit.*

pĕr'u la, *a scale of a leaf bud.*

pĕr'u late, *furnished with scales.*

pĕt'al, *one of the leafy expansions of the corolla.*

pe tăl'o dy, *metamorphosis of stamens or pistils into petals, as in double flowers.*

pĕt'al oid, *pertaining to a petal; resembling a petal.*

pet'i o late, *having a stalk or petiole.*

pĕt'i ole, *a leaf-stalk; foot-stalk of a leaf connecting the leaf with the stem.*

pet i ŏl'u late, *supported by its own petiolule.*

pĕt'i o lule, *a small petiole, or the petiole of a leaflet.*

phæ no gā'mi a, or phan e ro gā'mi a, *name of that division of the vegetable kingdom which bears visible flowers.*

phā'lanx (*pl.* pha lăn'ges), *a group or bundle of stamens.*

phy cŏl'o gy, *the science of Algæ, or seaweeds.*

phyl lo clā'di um, *a flattened stem or branch which more or less resembles a leaf, and performs the functions of a leaf.*

phyl lō'di um, *a petiole dilated into the form of a blade.*

phȳl'lome, *a foliar part of a plant; an organ homologous with a leaf, or produced by the metamorphosis of a leaf.*

phȳl'lo tax y, *the order or arrangement of leaves on a stem.*

phys i o lŏg'i cal bŏt'an y, *that division of the science of botany which treats of the functions of plants.*

phy tŏg'ra phy, *the science of describing plants in a systematic manner.*

phy tŏl'o gy, *an account of the composition of plant organs and the substances that compose them.*

pi'le us, *the expanded upper portion of many of the fungi.*

pi lĭf'er ous, *bearing a slender bristle or hair; beset with hairs.*

pi lōse', *covered with long slender hairs.*

pĭn'na, *a primary division, with its leaflets, of a bipinnate or tripinnate leaf.*

pĭn'nate, *composed of several leaflets, or separate portions, arranged on each side of a common petiole.*

pĭn'nate ly lōbed, *lobed in a pinnate manner.*

pin năt'i fid, *divided in a pinnate manner, the divisions not reaching to the midrib.*

pinnăt'isect, *pinnately divided to the midrib.*

pĭn'nule, *one of the small divisions of a decompound frond or leaf.*

pi'si form, *resembling a pea in size or shape.*

pĭs'til, *organ of a flower, made up of ovary, style, and stigma, or ovary and stigma.*

pĭs'til late, *having a pistil or pistils, — usually said of flowers having a pistil but no stamens.*

pis til lĭd'i um, *archegonium; the organ in mosses which is analogous to a pistil in flowering plants.*

pĭtch'er, *a tubular or cup-like appendage or expansion of the leaves of certain plants.*

pĭth, *the soft tissue in the center of the stems of dicotyledonous plants.*

pĭt'ted, *having depressions or excavations.*

pĭt'ted cĕlls, *cells with spots or depressions on their walls.*

pla çĕn'ta, *the part of a pistil or fruit to which the ovules or seeds are attached.*

pla çĕn'ti form, *having the shape of a circular thickened disk, somewhat thinner about the middle.*

plait'ed, *folded; doubled over.*

plănt, *an organized body possessing vitality but not sensation.*

plănt grōwth, *the manner in which a plant is built up.*

plat y phȳl'lous, *broad-leaved.*

plei ŏph'yl lous, *having several leaves.*

pli'cate, *plaited like a fan; folded.*

GLOSSARY.

plūm'be ous, *resembling lead* in color.
plu mōse', *feathery.*
plū'mule, *the first bud or gemule of* a young plant; *the bud,* or *growing point,* of the embryo above the *cotyledons.*
plu ri fō'li o late, having *several or many leaflets.*
pŏd, *a capsule, especially a legume.*
pod o cēph'a lous, *having a head of flowers on a long peduncle.*
pŏd'o sperm, *the stalk of a seed or ovule.*
point'less, *destitute of any pointed tip,* such as a *mucro, awn, acumination, etc.*
pŏl'len, *the fructifying cells contained in the anthers.*
pŏl'len māss, *the united mass of pollen, as in the Milk-weed and Orchis.*
pŏl'len tūbe, *the slender tube sent down through the style of the pistil, through which the protoplasm of the pollen cell is conveyed to the ovum.*
pol li nā'tion, *the act of furnishing pollen to the stigma.*
pol lĭn'i um (*pl.* pol lĭn'i a), *a mass of pollen.* See pollen mass.
pol y dĕl'phous, *having the stamens* in several *groups.*
pol y ăn'drous, *having many stamens, — more than twelve.*
pol y cär'pic, *term used by* De Candolle *in the sense of perennial.*
pol y cot y lē'dou ous, having many (*more than two*) *cotyledons, as* Pines.
po lȳg'a mous, *having* both *hermaphrodite and unisexual flowers.*
po lȳg'y nous, *with many pistils or* styles.
po lȳm'er ous, *having many* parts or members in each set.
pol y mōr'phous, *of several or varying forms.*
pol y pĕt'al æ, *a group of dicotyledonous plants having separate petals forming a* circle inside *the calyx.*
pol y pĕt'al ous, *having several or* many separate petals.
pol y sĕp'al ous, *having the sepals* separate from each *other.*
pol y spēr'mous, *many-seeded.*
pōme, *a fruit like an apple.*
po mĭf'er ous, *pome-bearing.*
pŏr'rect, *outstretched.*
pos tē'ri or, *next the axis.*
pouch, *the silicle or short pod, as of Shepherd's-purse.*
pre cō'cious, *flowering before the leaves.*
pre fō'li ā'tion, *vernation.*
pre mōrse', *ending abruptly.*
prick'les, *slender thorn-like processes.*
prick'ly, *bearing prickles or sharp projections.*
pri'mine, *the outermost of the two integuments of an ovule.*
pri mōr'di al, *earliest formed.* Primordial *leaves are the first after the cotyledons.*
pris mat'ic, *prism-shaped.*

prŏc'ess, *any projection from the surface or edge of a body.*
pro clĭm'bent, *trailing; prostrate.*
pro cŭr'rent, *running through, but not projecting.*
pro dūced', *extended more than usual.*
pro lĭf'er ous, *bearing offspring, — applied to a flower* ... *which another is produced, or* to a branch *or frond from which another arises.*
pro pā'gu lum, *a runner terminated by a germinating bud.*
pros ĕn'cȳ ma. *plant tissue made up of lengthened cells.*
prŏs'trate, *lying flat on the ground.*
pro tăn'drous, pro ter ăn'drous, *having the stamens come to maturity before the pistil.*
pr ter... 'thous, *having flowers which appear before the leaves.*
pro ter ŏg'y nous, pro tŏg'y nous, *having the pistils come to maturity before the stamens.*
pro t'ăl'li v... , pro t'ăl'lus, *the minute primary growth from the spore of ferns, which bears the true sexual organs.*
pro tŏph'y ta, *one of the primary divisions of vegetable life, containing the smallest and simplest plants.*
prō'to plasm, *the primary* organic substance of plants.
pru'i nose, *covered with dust or bloom, so as to give the appearance of frost.*
pter i dŏph'y ta, *a class of flowerless plants, embracing ferns, horse-tails, club mosses, etc.*
pu bĕr'u lent, *very minutely downy.*
pu bĕs'cent, *covered with fine short hairs.*
pul ver ā'ceous, *or* pŭl'vēr u lent, *having a finely powdered surface.*
pŭl'vi nate, *having the form of a cushion.*
pūnc'tate, *dotted with small spots of color,* or *with minute depressions or pits.*
punc tic'u late, *minutely punctate.*
pŭn'gent, *prickly-pointed;* hard and sharp.
pu nĭ'ceous, *of a bright red color.*
pur pū're al, *of a purple color, or bluish red.*
py rām'i dal, *in the form of a cone or pyramid.*
py rē'na, *or* py'rene, *a nutlet resembling a seed; the kernel of a drupe.*
pȳr'i form, *in the form of a pear.*
pyx'i date, *furnished with a lid.*
pyx'is, *a box which divides circularly into an upper and lower half, the former being a kind of lid.*

Q.

quad rāy'gu lar, *four-angled.*
quad ri fō'li ate, *four-leaved.*
quad rĭj'u gate, *with four pairs of leaflets.*
quad ri lăt'er al, *having four sides.*
qui'nate, *growing in sets of five.*

quin cŭn'çial, *having the leaves of a pentamerous calyx or corolla so imbricated that two are exterior, two interior, and the fifth has one edge exterior and one interior.*
quīn'tu ple, *five-fold.*

R.

race, *a variety of such fixed character that it may be propagated by seed.*
ra çēme', *a flower cluster with an elongated axis and many one-flowered lateral pedicels.*
rāç'e mose, *growing in the form of a raceme.*
rā'chis, *or* rhā'chis, *the principal axis in a spike, raceme, panicle, or corymb.*
rā'di al, *consisting of, or like, radii or rays.*
rā'di ant, *having a ray-like appearance.*
rā'di ate, *diverging from a common center.*
răd'i cal, *belonging to or proceeding from the root.*
răd'i cant, *taking root on or above the ground; rooting from the stem.*
răd'i cle, *the rudimentary stem of a plant which supports the cotyledons in the seed, and from which the root is developed downward; a rootlet.*
rā'mal, *pertaining to a branch.*
ram en tā'ceous, *beset with thin brownish scales (ramenta), as the scales of many ferns.*
rām'i fi cā'tion, *process of branching.*
rām'u lose, *having many small branches.*
rā'phe, *the continuation of the seed stalk along the side of an anatropous ovule or seed, forming a ridge or seam.*
rays, *radiating branches of an umbel.*
re çĕp'ta cle, *the apex of the flower stalk, from which the organs of the flower grow, or into which they are inserted.*
rĕc'ti nerved, *having the nerves or veins straight.*
re cûrved', *curved in an opposite direction; bent back.*
re dū'pli cate, *valvate, with the margins curved outwardly, — said of the æstivation of certain flowers.*
re flĕxed', *bent backward excessively.*
re frăct'ed, *bent backward angularly, as if half-broken.*
rĕg'ma, *a dry fruit consisting of three or more cells, each of which at length breaks open at the inner angle.*
rĕg'u lar, *having all the parts of the same kind alike in size and shape.*
rĕn'i form, *kidney-shaped.*
re pănd', *having a slightly undulating margin.*
rē'pent, *prostrate and rooting.*
rē'plum, *the framework of some pods (as of the Prickly Poppy and Cress), persistent after the valves fall away.*

rĕp'tant, *repent; creeping.*
res pi rā'tion, *breathing; the absorption by plants of oxygen, the oxidation of assimilated products, and the release of carbon dioxide and watery vapor.*
re sū'pi nate, *inverted; appearing to be upside down or reversed.*
re tĭc'u late, *netted.*
rĕt'i nerved, *having reticulate veins.*
rē'tro flexed, *bent or turned abruptly backward.*
re trõrse', *bent backward or downward.*
re tūse', *having the end rounded and slightly indented.*
rĕv'o lute, *rolled backward or downward.*
rhā'chis, *see* rachis.
rhā'phe, *the continuation of the seed stalk along the side of an anatropous ovule or seed, forming a ridge or seam.*
rhăph'i des, *minute, transparent, often needle-shaped crystals, found in the tissues of plants.*
rhi zăn'thous, *producing flowers from a rootstock, or apparently from a root.*
rhi zō'ma, *or* rhi zōme', *a rootstock; a stem which has the appearance of a root.*
rhŏm'bic, *shaped like a rhomb.*
rhom boid'al, *shaped like a rhomboid.*
rĭbs, *the chief veins of a leaf; ridges.*
ri mōse', *full of fissures or chinks.*
rĭn'gent, *gaping, like an open mouth.*
rings of wood, *circular rings which appear in a cross section of an exogenous stem.*
ri pā'ri ous, *growing along river banks.*
root, *the descending axis of a plant; the part of a plant that grows downward into the ground.*
rōōt' cap, *a mass of dead cells which cover and protect the growing cells at the end of a root.*
rōōt'lets, *single roots or root branches.*
rōōt'stock, *a perennial underground stem, producing leafy stems or flower-stems from year to year.*
ro sā'ceous, *like a rose in shape or appearance.*
rōs'tel late, *having a rostellum, or small beak; terminating in a beak.*
ros tĕl'lum, *a small beak-like extension of some part.*
rōs'trate, *beaked; having a process resembling the beak of a bird.*
rō'su late, *arranged in little rose-like clusters, — said of leaves and bracts.*
rō'tate, *having the parts spreading out like a wheel; wheel-shaped.*
ro tā'tion, *circulation of fluids in the cell.*
rō'tund, *round or roundish in outline.*
ru'bi cund, *red; ruddy.*
ru'di ment, *an imperfect organ or part; a minute part.*
ru fĕs'cent, *reddish; tinged with red.*
ru gōse', *wrinkled; having the veinlets sunken and the spaces between them elevated.*

GLOSSARY. 333

ru'mi na ted, *having a hard albumen penetrated by irregular channels filled with softer matter.*

run'ci nate, *pinnately* **cut, with the lobes sloping downwards.**

run'ner, *a slender prostrate branch, rooting at the end.*

S.

sab'u lose, *growing in sandy places.*

sac, *any closed membrane, or* **a deep purse-shaped cavity.**

sac'cate, *sac-shaped.*

sag'it tate, *arrow-shaped.*

sal su'gi nous, *growing in brackish places, or salt marshes.*

sal'ver-shaped, *tubular, with a spreading border.*

sa ma'ra, *a winged fruit or seed vessel.*

sam'a roid, *resembling a samara, or winged seed vessel.*

sap, *the watery fluid taken up by the root, and moved through the vessel up to the leaves.*

sap'wood, *the last growth of wood in an exogen.*

sar'co carp, *the fleshy part of a drupaceous fruit.*

sar'ment, *a prostrate, filiform* **stem or runner,** *like the Strawberry.*

sar men ta'ceous, *bearing* **sarments or runners,** *either spreading or procumbent.*

sar men tose', *long and filiform, and almost naked,* **or** *having leaves* **only at the joints** *where there* **are roots;** *bearing sarments; sarmentaceous.*

saw'-toothed, *serrate.*

sca'brous, *rough; scaly.*

sca lar'i form *(cells), resembling a ladder;* **having** *transverse bars or* **markings,** *like* **the rounds of** *a ladder.*

scale, *a thin, scarious body.*

scal'loped, *having the edge or border cut or marked with segments of circles.*

scal'y, *furnished with scales, or* **scale-like in** *texture.*

scan'dent, *climbing.*

scape, *a* **flower stalk springing** *from the ground.*

sca'pi form, *resembling a scape.*

scar, *a mark left upon a stem or branch by the fall of a leaflet or frond, or upon a seed, by the separation of its support.*

sca'ri ose, *or* sca'ri ous, *thin, dry,* **membranous** *and not green.*

scat'tered, *irregular in position; having* **no** *regular order; sometimes used for alternate.*

sci'on, *a young shoot used for grafting.*

scle ren'chy ma, *hard, stony tissue.*

scle'rous, *hard; bony; indurated.*

scob'i form, *resembling sawdust.*

scro bic'u late, *having numerous small shallow depressions or hollows; pitted.*

scurf, *minute scales on the surface of many leaves, as in the Goosefoot.*

scur'fi ness, *quality of being scurfy.*

scu'tate, *or* scu'ti form, *buckler-shaped; shield-shaped; round, or nearly round.*

scu'tel late, *or* scu tel'li form, *saucer-shaped,* **or** *platter-shaped.*

sea'green, *light bluish-green; glaucescent.*

se'cund, *arranged on one side only; turned only one way.*

sec'un dine, *the second coat or integument of an ovule; tegmen.*

seed, *matured ovule.*

seg'ment, *a subdivision* **or lobe of any cleft** *body.*

seg're gate, *separated from others of the same kind.*

sem i lu'nar, *shaped like a half moon.*

sem'i nal, *pertaining to, containing, or consisting of, seed or semen.*

sem i nif'er ous, *seed-bearing; producing seeds.*

sem i-sag'it tate, *partly sagittate.*

sem per vi'rent, *always fresh; evergreen.*

sen'a ry, *containing six; in sixes.*

se'pal, *one of the foliaceous parts of the calyx.*

sep'al oid, *sepal-like.*

sep'a ra ted flowers, *those having stamens or pistils only.*

sep'tate, *divided by partitions.*

sep'ten ate, *having parts in sevens; heptamerous.*

sep'ti ci dal, *dividing the partitions; said of* **a** *method of dehiscence in which the pod splits through the partitions and is divided* **into** *its component carpels.*

sep tif'ra gal, *breaking from the partitions; said of a method of dehiscence in which* **the valves of** *a pod break away from the partitions, and these remain attached* **to** *the common axis.*

sep'tum, *a partition between two spaces.*

se'ri al, *or* se'ri ate, *in rows; as biserial, in two* **rows.**

se ri'ceous, *silky.*

se rot'i nous, *appearing or blossoming later in the season than is customary with allied species.*

ser'rate, *notched on the edge like a saw.*

ses'sile, **resting** *directly upon the main stem or branch, without a petiole or footstalk.*

se ta'ceous, *bristle-like; set with bristles.*

se'tæ, *bristles.*

se'tous, *or* se tig'er ous, *covered with bristles.*

set'u la, *a diminutive bristle.*

set'u lose, *provided with setulæ.*

sex an'gu lar, *six-angled.*

sheath, *the base of a leaf when covering a stem or branch.*

sheath'ing, *inclosing with a sheath.*

shield'-shaped, *scutate; peltate.*

shrub, *a woody perennial plant less than fifteen feet in height.*
sig'moid, *curved in two directions, like the letter s or the Greek sigma.*
sil'i cle, *the short pod of many cruciferous plants.*
sil'ique, *the long pod of many cruciferous plants.*
sil'i quose, *bearing siliques (as the crucifers).*
silk'y, *glossy, with a coat of fine and soft, close-pressed, straight hairs.*
sil'ver y, *shining white or bluish-gray, usually from a silky pubescence.*
sim'ple, *of one piece; not compound.*
sin'is trorse, *twining from right to left.*
sin'u ate, *having a wavy margin or edge.*
si'nus, *a recess or bay; the reëntering angle between two lobes or projections.*
sleep of plants, *a state of plants, usually at night, when their leaflets approach each other, and the flowers close and droop, or are covered by the folded leaves.*
sob o lif'er ous, *bearing shoots from near the ground (sŏb'o les).*
sol'i ta ry, *growing alone or singly.*
sor'did, *dull or dirty in hue.*
so re'di ate, *bearing patches of granular bodies on the surface.*
so ro'sis, *a fleshy multiple fruit, as the Mulberry.*
so'rus, *a fruit dot of ferns.*
spa di'ceous, *chestnut-colored; bearing flowers on a spadix.*
spa'dix, *a spike with a fleshy axis.*
span, *the distance between the tip of the thumb and the little finger when the hand is outstretched, — about six or seven inches.*
spa thā'ceous, *having or resembling a spathe.*
spathe, *a large bract, or a pair of bracts, inclosing a flower cluster.*
spat'u late, *shaped like a druggist's spatula.*
spe'cies, *the unit in natural history classification; a group of individuals believed to be descended from common ancestors, agreeing in essential characteristics, and capable of continued fertile reproduction.*
sper'ma to phore, *or spĕr'mo phore, or sper'mo spore, one of the names of the placenta.*
spi'cate, *or spi'ci form, resembling a spike; spike-shaped.*
spike, *an inflorescence in which the flowers are sessile on a lengthened axis.*
spike'let, *a little spike, as in grasses.*
spin'dle-shaped, *tapering from the middle both ways.*
spine, *a woody thorn.*
spi nes'cent, *armed with spines; becoming hard and thorny; tapering gradually to a rigid, leafless point.*
spi nif'er ous, *or spi'nose, thorny; full of spines.*
spi'ral arrangement (of leaves), *an arrangement wherein the leaves are alternately arranged around the stem or branch.*

spi'ral cells *or vessels, long, slender cells arranged in a coil.*
sponge'let, *or spon'gi ole, a supposed sponge-like expansion of the tip of a rootlet for absorbing water.*
spo rad'ic, *widely dispersed.*
spo ran'gi um, *a spore case in cryptogamous plants.*
spore, *a reproductive grain in flowerless plants, analogous to seeds in flowering plants.*
spore case, *a sporangium.*
spo'ro phore, *the generative organ in certain plants which reproduces asexually.*
sport, *a newly appeared variation.*
spor'ule, *a little spore; a spore.*
spu mes'cent, *appearing like froth.*
spur, *a stiff, sharp spine; a slender projecting appendage.*
squa mā'ceous, **squā'mate**, *or* **squa mōse'**, *covered with or consisting of scales; resembling a scale.*
squā'mi form, *having the shape of a scale.*
squăm'u late, *or* **squăm'u lose**, *having little scales.*
squar rōse', *divided into shreds or jags; having widely divaricating scales, as the involucral scales of the Composita.*
stalk, *the stem, petiole, peduncle, etc., of a plant.*
stā'mens, *the organs that produce pollen, consisting of filament and anther.*
stăm'i nate, *furnished with stamens; producing stamens. A staminate flower is one having stamens, but lacking pistils.*
stăm'i nō'di um, *a stamen without an anther; an organ resembling an abortive stamen.*
stănd'ard, *the upper petal, or banner, of a papilionaceous corolla.*
starch, *a widely diffused vegetable substance found especially in seeds, bulbs, and tubers, from which it is extracted as a white granular or powdery substance, without taste or odor.*
stā'tion, *the particular situation in which a plant occurs.*
stĕl'late, *or* **stĕl'lu lar**, *starry, or star-like; spreading out from a common center, like a star.*
stĕm'less, *destitute, or apparently destitute, of a stem.*
ste nŏph'yl lous, *having narrow leaves.*
stĕr'ile, *barren; not bearing seeds; unproductive.*
stig'ma, *the part of the pistil, usually the end, fitted to receive the pollen.*
stig măt'ic, *of or pertaining to a stigma.*
stĭngs, *stinging hairs; hairs sufficiently rigid to perforate animal tissue, and of which, having entered, the apex breaks off, discharging an irritating fluid.*
stipe, *the stalk of the ovary or ovaries; the stem of a Mushroom.*

GLOSSARY. 335

sti'pel, *an appendage to a leaflet* **corresponding to a** *stipule in a leaf.*
sti pel'late, *furnished with stipels, as in the bean tribe.*
stip'i tate, *supported on a stipe.*
stip'u late, *furnished with stipules.*
stip'ule, *an appendage, or little leaf, on each side of the base of a petiole or leaf.*
stŏck, *a word used for race or source; also, for any root-like base from which the herb grows up.*
stō'lon, *a branch at the base of a plant which roots easily.*
stō'lo nĭf'er ous, *producing stolons.*
stō'ma, *a mouth; one of the openings in the epidermis of a leaf; a* **breathing pore.**
stra mĭn'e ous, *straw-like, or straw-colored.*
străp'-shaped, *flat, narrow, and straight.*
stri'ate, *or* stri'at ed, *marked with slender longitudinal bars or stripes.*
strĭct, *erect and very straight.*
stri gōse', *set with stiff, straight bristles.*
strŏb'ile, *a multiple fruit in the form of a cone, as of the Hop and Pine.*
strom bū'li form, *twisted, like a spiral shell.*
stro'phi o late, *furnished with a strophiole or caruncle about the hilum.*
strō'phi ole, *a crest-like excrescence about the hilum of certain seeds.*
strŭc'tur al bŏt'an y, *the science which treats of the organs or parts of plants, of their forms and uses.*
stru'ma, *a wen; a swelling or protuberance of any organ.*
stu pōse', *composed of or having tufted filaments like tow.*
stȳle, *that part of the pistil between the ovary and the stigma.*
sty lĭf'er ous, *bearing one or more styles.*
stȳ'loid, *having the form of or resembling a style.*
sty lo pō'di um, *an epigynous disk, or enlargement at the base of the style.*
sŭb' class, *a natural group, more important than an Order.*
su bē're ous, *or* sū'ber ose, *having a corky texture.*
sub ŏr'der, *a group of genera a little lower in rank than an Order, and of greater importance than a Tribe or Family.*
sŭb' tribe, *a division of a Tribe.*
sū'bu late, *awl-shaped; very narrow, and tapering gradually to a fine point from a broadish base.*
suc cīse', *appearing as* **if cut off** *at the extremity.*
sŭc'cu bous, *having the leaves so placed that the upper part of each* **one is covered by** *the base of the leaf above.*
sŭc'cu lent, *very juicy and* **cellular.**
sŭck'er, *a shoot coming from a part of the stem beneath the ground.*

suf fru tes'cent, *slightly woody at the* **base.**
suf fru'ti cose, *woody in the lower part of the stem; more woody than suffrutescent.*
sŭl'cate, *furrowed.*
su pē'ri or, *above the ovary.*
su pē'ri or că'lyx, *calyx adherent to ovary.*
su pē'ri or ō'va ry, *ovary free from calyx.*
su per nū'mer ar y *(buds), exceeding the number stated or prescribed.*
sū'per vo lūte', *rolled up from the sides.*
sū'pine, *lying flat, with face upward.*
sū'pra-ăx'il la ry, *situated above the axil.*
sū'pra-de cŏm'pound, *divided many times.*
sūr'cu lose, *producing suckers, or shoots resembling suckers.*
sus pĕnd'ed, *hanging* **downward.**
sū'tur al, *pertaining to a suture; taking place at a suture.*
sū'ture, *the line of junction of two contiguous parts grown together.*
sy cō'nus, *a collective, fleshy fruit, in which the ovaries are hidden within a hollow receptacle, as in the Fig.*
syl vĕs'tri an, *growing in woods; sylvan.*
sȳm'me try, *equality in the number of parts of the successive circles in a flower; likeness in form and size of floral organs of the same kind; regularity.*
sym pĕt'al ous, *having the petals united; gamopetalous.*
sym phȳl'lous, *with perianth leaves united.*
sȳm'pode, *or* sym pō'di um, *a stem resembling a simple axis, but composed of superposed branches, as the stem of the grapevine.*
syn ăn'ther ous, *having the stamens united by their anthers.*
syn cär'pi um, *an aggregate fruit, in which the ovaries cohere in a solid mass.*
syn cär'pous, *composed of several carpels, united into one ovary.*
syn ge nē'sious, *having the stamens attached to each other so as to form a ring.*
sȳn'o nym, *an equivalent name.*
syn sĕp'al ous, *having united sepals; gamosepalous.*
sys tem ăt'ic bŏt'an y, *that department of botany which pertains to the classification of plants.*

T.

tāil, **any long and** *slender prolongation of an organ.*
tā'per-point'ed, *acuminate.*
tăp'root, *a simple descending root.*
taw'ny, *a dull, yellowish brown.*
tax ŏn'o my, *that division of natural science which treats of the classification of animals and plants.*
tĕg'men, *the inner seed coat.*
tĕn'dril, *a shred-like process, which helps the plant to cling to other plants.*

336 GLOSSARY.

ter a tŏl'o gy, *that branch of biological science which treats of abnormal and unusual formations.*
te rēte', *cylindrical and slightly tapering; columnar.*
tĕr'mi nal, *situated at the end or apex.*
ter mi nŏl'o gy, *the terms used in a business or science; nomenclature; technical terms.*
tĕr'nate, *in threes.*
tĕs'sel la ted, *checkered; marked like a checker-board.*
tĕs'ta, *the outer seed coat.*
tes tā'ceous, *having a dull red-brick or brownish-yellow color.*
tĕt'ra dȳn'a mous, *having four long and two short stamens.*
te trăg'o nal, *having four prominent longitudinal angles.*
te trăg'y nous, *having four pistils.*
tĕt'ra spore, *a non-sexual quadruple spore.*
thăl'a mi flō'rous, *bearing the stamens and petals directly on the torus or thalamus.*
thăl'a mus, *the receptacle of a flower; a torus; a thallus.*
thăl'lo gen, or **thăl'lo phyte,** *one of a large class or division of the vegetable kingdom, including all flowerless plants, composed of cellular tissue, and showing no distinction of root, stem, and leaf.*
thăl'lus, *a mass of cellular tissue, usually in the form of a flat stratum or expansion, instead of stem and leaves.*
thē'ca, *sheath; case; spore case.*
thē'ca phore, *a surface or receptacle bearing a theca or thecæ; the stipe upon which a simple pistil is sometimes borne, being the petiole of the carpellary leaf.*
thōrn, *a hard and sharp-pointed projection from a woody stem; a spine.*
thrōat, *orifice of a monopetalous corolla.*
thyrse, or **thyr'sus,** *a dense egg-shaped panicle, as in the Lilac.*
tĭs'sues, *the materials of which plants are composed.*
tō'men tose, *covered with short, matted, woolly hairs.*
tongue'-shaped, *long and flat, but thickish and blunt.*
tōothed, *furnished with teeth or sharp projections of any sort on the margin; as a saw-toothed margin.*
tŏp'-shaped, *inversely conical.*
tŏr'u lose, *cylindrical, with alternate swellings and contractions.*
tō'rus, *the axis on which all the parts of a flower, except the calyx, are seated.*
trā'che a, *a spiral duct.*
trā chy spĕr'mous, *rough-seeded.*
trans vērse', *across; being right and left, instead of up and down.*
tree, *a woody plant, branching so as to form a symmetrical head, growing to the height of twenty feet, or higher.*

tri a dĕl'phous, *having stamens joined by filaments into three bundles.*
tri ăn'drous, *having three distinct and equal stamens in the same flower.*
tribe, *a group higher than a Genus.*
tri chŏm'a tous, *having the nature of hair or pubescence.*
trĭch'ome, *a hair on the surface of a leaf or stem, or any modification of a hair.*
tri chŏt'o mous, *three-forked; trifurcate.*
tri cŏc'cous, *having three roundish one-seeded carpels.*
tri'col ored, *having three colors.*
tri cŭs'pid ate, *ending in three points; three-pointed; tridentate.*
tri ĕn'ni al, *lasting three years.*
tri fā'ri ous, *facing three ways; in three vertical ranks.*
tri'fid, *three-cleft; split to the middle into three parts.*
tri fō'li ate, *with three leaves or leaflets.*
tri fūr'cate. *three-forked; trichotomous; triangular.*
trĭg'y nous, *having three pistils or styles.*
tri lō'bate, *having three lobes.*
tri lŏc'u lar, *having three cells or cavities.*
tri'mer ous, *having the parts in threes.*
tri nĕrv'ate, *three-nerved, or with three slender ribs.*
tri œ'cious, *having three sorts of flowers on the same or different plants, as in the Red Maple.*
tri pärt'i ble, *divisible into three parts.*
trīp'ar tite, *divided into three parts; more deeply split than tripid.*
tri pĕt'al ous, *having three petals.*
trĭph'yl lous, *having three leaves.*
tri pĭn'nate, *thrice pinnate.*
trī'ple veined, *having three veins or nerves.*
tri quē'trous, *three-sided; three-angled.*
tri sē'ri al, or **tri sē'ri ate,** *arranged in three vertical or spiral rows.*
trĭs'tich ous, *arranged in three vertical rows.*
tri stig măt'ic, or **tri stĭg'ma tose,** *having three stigmas.*
tri sŭl'cate, *having three furrows or forks.*
tri tĕr'nate, *thrice ternate.*
trĭv'i al name, *the specific name.*
trŏch'le ar, *pulley-shaped.*
trŭm'pet-shaped, *tubular, and enlarged at or toward the summit.*
trŭn'cate, *cut off at the tip.*
trŭnk, *the main stem.*
trÿ'ma, *a drupe, or drupaceous nut, with a fleshy exocarp.*
tūbe, *a hollow, elongated body, usually cylindrical; applied especially to a gamopetalous corolla or gamosepalous calyx.*
tūbe'-form, *tubular; trumpet-shaped.*
tū'ber, *a fleshy underground stem, or branch, with buds.*
tu bĕr'cu lar, *having the form of a tuber; bearing tubercles.*

GLOSSARY. 337

tu bĕr'cu late, *covered with warts or tubercles.*
tu ber lf'er ous, *bearing or producing tubers.*
tū'ber ose, *consisting of or bearing tubers; resembling a tuber.*
tū'bu lar co rŏl'la, *a corolla having the form of a tube.*
tū'bu li flō'rous, *having the flowers of a head all with tubular corollas.*
tū'mid, *swollen or inflated.*
tū'ni cate, *covered or coated with layers, as a bulb.*
tûr'bi nate, *shaped like a top or inverted cone.*
tū'ri on, *a shoot or sprout from the ground.*
twin'ing, *ascending by coiling round a support.*
tӯpe, *the ideal pattern.*
tӯp'i cal flow'er, *a flower which serves for a pattern.*

U.

u līg'i nose, *growing in muddy or swampy places.*
ŭm'bel, *an inflorescence in which the pedicels all spring from the same point, like the ribs of an umbrella.*
ŭm'bel late, *bearing umbels.*
ŭm'bel let, *a small or partial umbel.*
um bĭl'i cate, *having a sharp depression at one end.*
ŭm'bo nate, *bossed; having a conical or rounded projection, like a boss (umbo).*
um brăc'u li form, *umbrella-shaped.*
un ärmed', *having no stings or thorns.*
ŭn'cial, *an inch in length.*
ŭn'ci nate, *hooked; bent at the tip in the form of a hook.*
ŭn'der shrub, *a low shrub.*
ŭn'du late, *wavy.*
un ē'qual ly pĭn'nate, *pinnate, with an odd terminal leaf.*
un guĭc'u late, *furnished with hooks or claws.*
un i cĕl'lu lar, *composed of a single cell.*
un i flō'rous, *having only one flower.*
un i fō'li ate, *having only one leaf.*
ū'ni form, *having always the same form.*
ū ni lăt'er al, *one-sided; on one side only.*
ū ni lŏc'u lar, *having only one cell or cavity.*
ū ni sē'ri al, *having only one row or series.*
ū ni sĕx'u al, *having stamens or pistils only.*
ū'ni valved, *having but one valve.*
un sym mĕt'ri cal flow'ers, *flowers in which similar parts are of different size or shape, or the parts of successive circles differ in number.*
ûr'ce o late, *shaped like a pitcher or urn.*
ū'tri cle, *a little bladdery seed-vessel; a little sac or vesicle, as the air cell of a sea-weed.*
ū trĭc'u lar, *resembling a small bladder or bag.*

PR. FL. — 23

V.

văg'i nate, *invested with a sheath.* Vaginate leaf, one invested by the tubular base of a leaf.
vălv'ate, *opening as by doors.*
vălve, *one of the pieces into which a capsule naturally separates when it bursts; a small portion of certain anthers, opening like a trap-door.*
vălv'u lar, *having valves; serving as a valve.*
va ri'e ty, *a particular form of species.*
văs'cu lar cryp'to gams, *cryptogams that have vascular tissue.*
văs'cu lar tĭs'sue, *tissue furnished with vessels or ducts.*
vault'ed, *arched.*
veins, *the system of branching vascular woody tissue seen in leaves.*
vein'lets, *or* vein'u lets, *little veins, or smaller branches in the network of a leaf.*
vē'late, *having a veil; veiled.*
ve lū'ti nous, *velvety.*
vĕn'e nate, *poisoned.*
ve nōse', *having numerous or conspicuous veins; veiny.*
vĕn'tral, *pertaining to that side of an organ of a flower which looks towards the axis or center of the flower; opposite of dorsal.*
vĕn'tri cose, *swelling out on one side, or unequally.*
vĕu'u lose, *having veinlets.*
ver mĭc'u lar, *worm-like; shaped like a worm.*
vĕr'nal, *appearing in the springtime.*
ver nā'tion, *the arrangement of leaves within the leaf bud; prefoliation.*
vĕr'ni cose, *having a brilliantly polished surface.*
vĕr'ru cose, *covered with warts; tuberculate; warty.*
vĕr'sa tile (*anther*), *swinging or turning on its support.*
ver'tex, *summit; apex.*
vĕr'ti cal, *up and down; parallel with the axis.*
ver ti cĭl lăs'ter, *a whorl of flowers, apparently of one cluster, but composed of two opposite axillary cymes.*
ver tĭc'il late, *whorled.*
ve sĭc'u lar, *bladdery.*
vĕs'per tine, *blossoming in the evening.*
vĕs'sels, *ducts.*
vex ĭl'lum, *the upper petal of a papilionaceous flower; the standard.*
vil lōse', *covered with long, fine hairs.*
vi mĭn'e ous, *producing long, slender twigs or shoots.*
vine, *a woody climbing or twining stem.*
vi rĕs'cent, *or* vi ri dĕs'cent, *greenish; turning green.*
vir'gate, *straight and slender; having the form of a straight rod.*
vĭs'cid, *or* vĭs'cous, *sticky; glutinous.*

vi tĕl'line, *pertaining to the yolk of the egg; yellow.*
vĭt'ta (*pl.* vĭt'tæ), *one of the oil tubes in the fruit of umbelliferous plants.*
vi vĭp'ar ous, *sprouting or germinating while attached to the parent plant.*
vŏl'u ble, *having the power or habit of turning or twining.*
vo lūte', *rolled up in any direction.*
vŏl'va, *a sac-like envelope of certain fungi, which bursts open as the plant develops.*

W.

wāv'y, *undulating on the border or surface.*
wăx'y, *resembling wax in texture or appearance.*
wĕdge'-shaped, *broad and truncate at the summit, and tapering down to the base.*
wheel-shaped, *expanding into a flat, circular border at top, with scarcely any tube, as a wheel-shaped corolla.*
whŏrl, *a circle of two or more leaves, flowers, or other organs, about the same part or joint of a stem.*
whŏrled, *arranged in whorls.*
wĭng, *any membraneous expansion; either of the two side petals of a papilionaceous flower.*
wĭnged, *furnished with a leaf-like appendage, as the fruit of the Ash and Elm.*
wĭtch grass, *a troublesome weed, with creeping underground stems.*
wood, *the hard or solid part of a stem or branch.*
wood cĕll, *a slender cylindrical or prismatic cell, usually tapering to a point at both ends, and the principal constituent of woody fiber.*

wood'y plant, *a shrub, tree, or plant in which the stems and branches are woody.*
wool'ly, *clothed with long and entangled soft hairs.*

X.

xăn'thic, *yellowish.*
xe nŏg'a my, *cross-fertilization; fertilization of a flower by pollen from a flower of another plant of the same species.*
xē'ro phil, *a plant that requires great heat and little moisture, and is specially adapted to arid regions.*
xȳ'lem, *that portion of a fibro-vascular bundle developed into wood cells.*
xȳ'lo carp, *a hard and woody fruit.*

Y.

yēast plant, *a unicellular plant which is the active agent of fermentation; it has the power of disintegrating starch and setting carbon dioxide free.*

Z.

zō'o phyte, *any one of the numerous species of invertebrate animals which more or less resemble plants in appearance or mode of growth.*
zō'o spore, *a spore provided with one or more slender cilia, by the vibration of which it swims in the water.*
zȳg'o spore, *a spore formed by the union of two or more cells.*

INDEX.

ABIES 29, 308, 309
  balsamea 309
  var. longifolia 309
  var. variegata 309
Acacia 18, 111, 112
  Catechu 112
  dealbata 112
  pycnantha 112
  Senegal 111, 112
  Seyal 112
  Suma 112
Acer 16, 91-93
  nigrum 91
  saccharinum 91
  var. nigrum 91
Achras sapota 176
Ægilops 253, 254
Agave 277
Alexandrian Senna 108
Allspice 20, 135, 136
Almond 19, 122
Aloe, aloes 28, 277, 278
  ferox 278
  socotrina 277
  spicata 277, 278
  vulgaris 277
American Arbor Vitæ 312
American Balsam Fir 309
American Beech 255
American Elm 231, 232
American Gooseberry 130
American Hazelnut 254
American Ipecacuanha 163
American Larch 309, 310
American Senna 108
American Strawberry 116
American Turkey Oak 249
Amorpha fruticosa 96
Ampelideæ 6, 16, 88-91
Amygdalus communis 122
Anacardiaceæ 7, 16, 93-96
Anacardium occidentale 16, 95
Ananassa sativa 28, 269, 270
Andropogon saccharatum 294, 295
Anemone 12, 30-33
  acutiloba 31
  Caroliniana 30
  cylindrica 30
  decapetala 30
  dichotoma 30, 31
  hepatica 31
  multifida 31, 32
  nemorosa 31, 32
  parviflora 32
  patens 32

Anemone patens, var. Nuttaliana 32
  Pennsylvanica 30, 31
  Virginiana 32, 33
Anemonella thalictroides 12, 33
Angiospermæ, apetalous dicotyledonous 6-8, 9, 10
  dicotyledonous 5, 12-27
  glumiferous monocotyledonous 11
  monocotyledonous 5, 28, 29
  petaloideous monocotyledonous 10
  spadiciflorous monocotyledonous 10
  sympetalous dicotyledonous 8, 9
Anise 21, 147, 148
Annatto 14, 63, 64
Anthemis 22, 165, 167
  arvensis 166
  nobilis 166
  var. flore pleno 166
Apetalæ 5
Apetalous dicotyledonous angiospermæ 9, 10
Apium graveolens 21, 146, 147
Apocynaceæ 96
Apopetalæ 5
Apopetalæ, inferior 12-21
Apopetalous dicotyledonous angiospermæ 6-8
Apple 19, 123-127
Apple, Carthage 139
  Crab 125
Apple-bearing Sage 204
Apple Pine 302
Apricot 18, 120
Arachis hypogæa 17, 99, 100
Arbor Vitæ 29, 312
Areca Catechu 29, 279, 280
Arrow-leaved Violet 60, 61
Arrowroot 28, 261-263
Asafœtida 21, 149-151
Asclepiadaceæ 96
Asparagus officinalis 28, 274
Astragalus 17, 97-99
  gummifer 98, 99
  tragacantha 98
Atropa 24, 191, 192
  belladonna 191
  Mandragora 191
Austrian Pine 301
Autumnal Marrow 144
Avena 29, 298, 299
  sativa 298, 299

BALM of Gilead 309
Balsam Fir 309
Banana 28, 265, 266
Banyan Tree 235
Baptisia tinctoria 96
Barbadoes aloes 278
Barberry 13, 42
Barley 29, 296-298
Bassia latifolia 176
  Parkii 176
Battledoor Barley 297
Beaked Hazelnut 254
Bean 17, 103-105
Beech 27, 255, 256
Beech of Europe 255
Beechwood Muskmelon 141
Beer 234, 267
Beet 26, 208, 209, 294
Bell Flower 170, 171
Bell Pepper 193
Bell-shaped Cranberry 173
Bengal Grass 300
Berberidaceæ 6, 13, 42, 43
Berberis vulgaris 13, 42
Bertholletia excelsa 20, 137
Beta 26, 208, 209
  maritima 209
  vulgaris 208, 209
  var. cicla 208
  var. mangel-würzel 208
  var. rapa 208
Betel Nut 29, 279, 280
Betel Pepper 214
Beverage-producing Plants—
  Apple (cider) 125
  Banana (beer and wine) 267
  Barley (whiskey) 298
  Cocoa 72
  Cocoanut Palm (milk, vinegar, and wine) 283
  Coffee 162
  Currant (wine) 129
  Grape (wine) 90
  Hops (beer) 234
  Indian Corn (whiskey) 286-288
  Maize (whiskey) 288
  New Jersey Tea 88
  Oats (whiskey) 299
  Peach (brandy) 122
  Pear (perry) 127
  Raspberry (brandy) 114
  Rye (whiskey) 296
  Tea 66
Bind-weed Jalap 184, 185
Birch-leaved Beech 255, 256

INDEX.

Bird-foot Violet 59
Bird Pepper 193
Bird's-eye Maple 91
Biting Crowfoot 54
Bitter Almond 122
Bitter Cassava 227, 228
Bitter Orange 81
Bixa orellana 14, 63, 64
Bixineæ 6, 14, 63, 64
Blackberry 18
Black Cap 113
Black Currant 128, 129
Black Huckleberry 172
Black Jack Oak 249
Black Larch 309, 310
Black Maple 91
Black Mustard 48, 49
Black Oak 248
Black Pepper 214, 215
Black Raspberry 113
Black-rock Muskmelon 141
Black Spanish Watermelon 142
Black Spruce 307
Black Tea 66
Black Walnut 239, 240
Blackwood 108, 109
Black-wooded Walnut Tree 239, 240
Blé Saracin 211
Blood Orange 79
Blueberry 172
Blue Cardinal Flower 170, 171
Blue Sweet Violet 57
Blumea balsamifera 218
Bog Fir 304
Bore Cole 47
Borneo Camphor 218
Borraginaceæ 8, 23, 182, 183
Box 27, 225
Bradford Watermelon 142
Brandy 114, 122
Brassica 13, 46-51
  alba 48, 49
  campestris 49-51
    var. rutabaga 50
  juncea 48, 49
  napus 51
  nigra 48, 49
  oleracea 47, 48
  rapa 50
    var. depressa 50
Brazil Nut 20, 137
Brazil Wood 17, 109
Bread-producing Plants —
  Arrowroot 262
  Barley 298
  Bitter Cassava 228
  Buckwheat 211
  Cocoanut Palm 283
  Date Palm 281
  Indian Corn 288
  Oats 299
  Rice 292
  Rye 296
  Sago Palm 285
  Tapioca 228
  Wheat 288
Bristly Crowfoot 37
British Oak 246, 247
Bromeliaceæ 10, 28, 268-270
Broom Corn 29, 294, 295
Broom Pine 305, 306
Buckthorn 16, 87, 88
Buckwheat 26, 211

Bugle-shaped Cranberry 173
Bulbous Crowfoot 35
Bull Pepper 193
Burr Oak 248, 249
Bush Bean 103, 104
Buttercup 12, 33-38
Butternut 27, 238, 239
Butter Tree 176
Buxus 27, 225
  sempervirens 225
    var. angustifolia 225
    var. suffruticosa 225

CABBAGE 47, 48
Cæsalpinia 17, 109
  Brazillensis 109
  Crista 109
Calisaya 157
Caltha 12, 39, 40
  leptosepala 40
  palustris 40
Camomile 22, 166, 167
Campanulaceæ 8, 22, 170, 171
Camphor Tree 27, 217, 218
Canada Pumpkin 145
Canada Violet 56, 57
Canna coccinia 263
  echinus 263
  edulis 263
  flaccida 263
  glauca 263
Cannabis sativa 27, 236-238
Cannon-ball Tree 137
Cantaloupe 140-142
Caoutchouc 27, 223, 224
Cape Aloes 278
Caper 13, 54, 55
Capparidaceæ 6, 13, 54-56
Capparis 13, 54-56
  ferruginea 55
  soldada 55
  spinosa 54, 55
Capsella Bursa-pastoris 13, 46, 47
Capsicum 24, 192-194
  annuum 192, 193
  cordifolia 193
  fastigiatum 193
  frutescens 193
  grossum 193
Caraway 156
Cardamom 28
Cardinal Flower 170
Carob Tree 17, 108
Carolina Anemone 30
Carolina Indigofera 96
Carolina Watermelon 142
Carrot 21, 154, 155
Carthage Apple 139
Carthamus tinctorius 22, 169, 170
Carum 21, 155, 156
  Carui 156
  petroselinum 155, 156
Carya alba 241
  sulcata 241
Cashew Nut 16, 95
Cassava, Bitter 227, 228
Cassia 17, 107, 108
  acutifolia 107, 108
  angustifolia 107, 108
  fistula 107
  Marilandica 107, 108
  obovata 107, 108

Castanea 27, 243, 244
  pumila 243, 244
  vesca 243, 244
    var. Americana 243
Castilloa elastica 223
Castor-oil Plant 27, 228, 229
Catmint 206, 207
Catnip 25, 206, 207
Cauld Kail 48
Cauliflower 47
Cayenne Pepper 24, 193
Ceanothus Americanus 16, 88
Cedar 29, 310-312
Celery 21, 146, 147
Celery-leaved Crowfoot 37, 38
Celery Parsley 155
Century-plant 277
Cephaëlis Ipecacuanha 22, 162, 163, 223
Ceratonia Siliqua 17, 108
Chamæcyparis 29, 314, 315
  pendula 314
  sempervirens 314
  thyoides 314
Charlotte Rothschild Pine-apple 269
Cheese Pumpkin 145, 146
Cheese-shaped Cranberry 173
Chenopodiaceæ 9, 26, 208-210
Cherry 18, 118
Cherry Currant 128
Cherry-shaped Cranberry 173
Chestnut 27, 243, 244
Chestnut Oak 250, 251
China Orange 79
Chinese Yam 28, 271, 272
Chrysanthemum 22, 167, 168
  carneum 167
  cinerariifolium 167
    var. rotundifolium 167
  corymbosum 167
  roseum 167
  Wilmoti 167
Cicuta 313
Cider 125
Cinchona 21, 157-159
  calisaya 157
  cordifolia 157
  lancifolia 157
  micrantha 157
  officinalis 157
  Pitayensis 157
  succirubra 157
Cinchonaceæ 96
Cinnamomum 27, 217-220
  camphora 217, 218
  cassia 219, 220
  iners 220
  obtusifolium 220
  pauciflorum 220
  Zeylanicum 219, 220
Cinnamon 27, 219
Citron 83
Citron Muskmelon 141
Citron Watermelon 142
Citrullus vulgaris 20, 142, 143
Citrus 15, 78-83
  aurantium 78-81
    var. sanguinea 79
  decumana 81, 82
  Limetta 82, 83
  Limonum 82
  Medica 83
  myrtifolius 81
  vulgaris 81

INDEX. 341

Clammy Rice	294	Common Marjoram	192	Cupressus sempervirens	314
Clearing Nut	181	Common Myrtle	173	thyoides	314
Clematis	12, 40-42	Common Potato	194	Cupulifera	9, 27, 242-256
crispa	40	**Common** Red Currant	128	Curaçoa Orange	81
cylindrica	40	Common Rice	294	Curcuma longa	28, 260, 261
ochroleuca	40	Common Sage	204	Curled Willow	257
Pitcheri	41	Common Sumach	93	Currant	19, 128-131
verticillaris	41	Common Virgin's Bower	41	Currants, Zante	89
viorna	41	Compositæ 8, 22, 165-170, 218		Cursed Crowfoot	37, 38
Virginiana	41	Cone Family	300-315	Custard Squash	144
Cloves	19, 134, 135	Coniferæ 5, 11, 29, 300-315		Cynanchum oleæfolium	168
Cluster Cucumber	146	Coniferous gymnosperms	11	Cypress	29, 314, 315
Coca	15, 76	Constantinople Hazelnut	254		
Coccus lacca	236	Convolvulaceæ 8, 24, 183-185			
Cochlearia	13, 52	Copper Beech	255		
armoracia	52	**Corchorus**	14, 73, 74	**D**ALBERGIA	17, 108, 109
officinalis	52	capsularis	73		
Cocoa	14, 70-72	olitorius	73	latifolia	109
Cocoanut	29	siliquosa	73	nigra	108, 109
Cocoanut Palm	282, 283	Cordage-producing Plants —		Damson Plum	117
Cocos nucifera	29, 282, 283	Aloe	278	Date Palm	29, 280, 281
Coffea	21, 160-162	Annatto	64	Daucus	21, 154, 155
acuminata	160	Banana	267	carota	154
Arabica	160	Cocoanut Palm	283	var. sativa	154
Australis	160	Flax	76	Deadly Nightshade	24, 191,
biflora	160	Manilla	268		192
Chamissonis	160	New Zealand Flax	277	Delicate Violet	56
ciliata	160	Sago Palm	284	Dew of the Sea	205, 206
densiflora	160	Coriander	21, 152, 153	Dichopsis gutta	23, 174, 175
Indica	160	Coriandrum sativum 21, 152,		Dicotyledonous angio-	
jasminoides	160		153	spermus	5, 12-27
kaduana	160	Cork	252, 253	Dioscorea	28, 271, 272
laurina	160	Cork Tree	252	alata	272
Liberica	160	Corky Elm	252	batatas	271, 272
longifolia	160	Corn, Broom	29, 294, 295	Japonica	272
magnolifolia	160	Indian	29, 286-298	sativa	271, 272
meridionalis	160	Corn Poppy	44	Dioscoreaceæ 10, 28, 271, 272	
Mexicana	160	Corylus	27, 254, 255	Diospyros	23, 176, 177
minor	160	Americana	254	ebenum	176, 177
Mozambicana	160	avellana	254	melanoxylon	176
nitida	160	colurna	254	Virginiana	176
nodosa	160	rostrata	254	Dog Violet	57
obovata	160	Cotton	14, 68-70	Downy-leaved Oak	248
occidentalis	160	Cotton Plant	67-70	Downy Yellow Violet	60
paguiodes	160	Cow Cabbage	47	Drinks. See Beverages.	
parvifolia	160	Cowslip	12, 39, 40	Drumhead Cabbage	47
pedunculata	160	Cow Tree	176	Dryobalanops Camphora	218
racemosa	160	Crab Apple	125	Dwarf Almond	122
rosea	160	Cranberry	23, 172-174	Dwarf Curled **Parsley**	155
semiexserta	160	Creeping Crowfoot	37	Dwarf Tobacco	187
sessilis	160	Creeping Spearwort	36	Dwarf Whortleberry 171, 172	
spicata	160	Crocus sativus	28, 270, 271	Dyer's Oak	248
stipulacea	160	Croton tiglium	27, 226	Dye-producing Plants —	
subsessilis	160	Croton-oil Plant	27, 226	Annatto	64
tetrandra	160	Crowfoot	12, 33-38	Black Oak	248
Travancorensis	160	Crowned Squash	144	Black Walnut	240
truncata	160	Cruciferæ	6, 13, 46-54	Brazil Wood	100
umbellata	160	Cucumber	20, 139, 140	Butternut	239
verticillata	160	Cucumis	20, 139-142	Cranberry	174
Wightiana	160	melo	140-142	Dyers' Oak	248
Zanguebarica	160	sativus	139, 140	Gall Oak	253
Coffee	21, 160-162	Cucurbita	20, 143-146	Indigo	97
Cold Chou	48	maxima	144, 146	Logwood	107
Cold Slaw	48	var. corona	144	Madder	164
Colophony	306	melopepo	145	Myrobalans	132
Coloring Indigofera	96	ovifera	144	Pomegranate	139
Combretaceæ 7, 19, 131, 132		var. medullosa	144	Safflower	170
Comfrey	23, 182, 183	pepo	145, 146	Saffron	271
Common Blue Violet	58	verrucosa	143, 144	Sumach	93
Common Cabbage	47	Cucurbitaceæ 6, 20, 139-146		Teak	198
Common Camomile	168	Cumin	21, 153	Turmeric	261
Common Field Pumpkin	145	Cuminum	21, 153	Valonia Oak	254
Common Fig	234, 235	cyminum	153	Violet	62
Common Flax	74	sativum	153	Woad	53
Common Hop	233, 234	Cuprea	255	Yellow-barked Oak	248
Common Juniper	311	Cupressus pendula	314	Yellow Berries	88

INDEX.

EARLY Crowfoot 35, 36
Early French Cucumber 140
Early Rice 291
Early Russian Cucumber 140
Ebenaceæ 9, 23, 176, 177
Ebony 23, 176
Elecampane 22, 165, 166
Elettaria 28, 263, 264
  cardamomum 263, 264
Elm 27, 231-233
Emetic Weed 170
English Cherry 118
English Elm 232
English Strawberry 116
English Violet 57
English Walnut 240, 241
Enville Pineapple 269
Erythroxylon 15, 76, 77
  coca 76
Eugenia 19, 20, 134-136
  caryophyllata 134, 135
  pimenta 135, 136
Euphorbia Ipecacuanha 27, 163, 222, 223
Euphorbiaceæ 10, 27, 222-229
European Larch 310
European Wine Grape 89

FABA sativa 104
Fagopyrum 26, 211, 212
  emarginatum 211
  esculentum 211
  Tartaricum 211
Fagus 27, 255, 256
  betuloides 255, 256
  colorata 255
  ferruginea 255
  obliqua 255
  sylvatica 255
Fascicle-rooted Crowfoot 35, 36
Fat Wood 305, 306
Feather Palms 279, 280
Fennel 21, 148, 149
Ferula nartbex 21, 149-151
Ficus 27, 234-236
  Bengalensis 235
  Carica 234, 235
  elastica 235, 235
Field Camomile 166
Field Pea 102
Field Turnip 49-51
Fig Tree 27, 234-236
Filbert 254
Fine Rosewood 108, 109
Fir 29, 308, 309
Flanders Spinach 210
Flat Dutch Cabbage 47
Flat-leaved Vanilla 259
Flat Squash 143
Flat Turnip 50
Flax 15, 74-76
Flax, New Zealand 276, 277
Flour 211, 288
Flowering Plants 5-11, 12-29
Fœniculum 21, 148, 149
  dulce 149
  vulgare 148, 149
Food Vegetables —
  Asparagus 276
  Bean 105
  Beet 209

Food Vegetables —
  Cabbage 48
  Caraway 156
  Carrot 155
  Fennel 149
  Jute 74
  Lentil 101
  Parsnip 152
  Pea 103
  Potato 195
  Rhubarb 213
  Rice 292
  Spinach 210
  Sweet Potato 184
  Turnip 51
  Yam 272
Four-valved Tobacco 187
Fox Grape 89
Fragaria 18, 116, 117
  vesca 116
  Virginiana 116
Fruits —
  Achras sapota 176
  Apple 125
  Apricot 120
  Banana 267
  Barberry 42
  Cashew Nut 95
  Cherry 119
  Cow Tree 176
  Cranberry 174
  Currant 129
  Date Palm 281
  Fig 236
  Gooseberry 131
  Grape 90
  High Blackberry 115
  Huckleberry 172
  Lemon 82
  Muskmelon 141
  Orange 80
  Peach 122
  Pear 126
  Persimmon 176
  Pineapple 270
  Plum 118
  Pomegranate 139
  Pumpkin 145
  Quince 127
  Raspberry 114
  Red Mulberry 231
  Shaddock 82
  Strawberry 117
  Watermelon 143

GALL Oak 253

Garden Buttercup 34
Garden Carrot 154
Garden Pea 101, 102
Garden Poppy 44
Garden Raspberry 113
Garden Sage 204
Gaylussacia 23, 171, 172
  dumosa 171, 172
  frondosa 172
  resinosa 172
Georgia Pine 305, 306
Ginger 28, 254, 285
Glastum 53
Glumiferæ 5
Glumiferous monocotyledonous angiosperms 11
Glycyrrhiza glabra 17, 105

Gooseberry 19, 128-131
Gossypium 14, 67-70
Gossypium arboreum 67, 68
  Barbadense 67
  herbaceum 67
Gourd 143-146
Gourd Squash 144
Gramineæ 11, 29, 285-300
Grape Vine 16, 88-91
Great-leaved Tobacco 187
Great lobelia 170, 171
Great-spurred Violet 61
Green Pine 305, 306
Green Tea 66
Ground Nut 99, 100
Ground Pea 99, 100
Guaiacum 15, 77, 78
  officinale 77, 78
  sanctum 78
Guaiacum Wood 77, 78
Guernsey Parsnip 151
Gum, Senegal 110
  Turkey 110
Gum Arabic 18, 111, 112
Gum Elastic 223
Gums, Resins, etc. —
  Asafœtida 150
  Balsam Fir 309
  Cashew Nut 95
  Comfrey 183
  Gum Arabic 112
  Gum Tragacanth 99
  Gutta Percha 175
  Hemp 237
  Larch 310
  Norway Spruce 308
  Oats 299
  Pitch Pine 304
  Prosopis juliflora 112
  Southern Pine 306
  Swiss Pine 305
  Tamarind 110
Gum Tragacanth 17, 98, 99
Gun-cotton 69
Gutta Percha 23, 174, 175
Gymnosperms 5, 11, 29

HACKMATAC 309, 310

Hæmatoxylon Campechianum 17, 105-107
Halberd-leaved Violet 57
Hamburg Large-rooted Parsley 155
Hand-shaped-leaf Crowfoot 38
Hand-shaped Violet 58
Haricot 103
Hazelnut 27, 254, 255
Heart-sease 61
Hemlock 29, 312-314
Hemlock Spruce 313, 314
Hemp 27, 236-238
Henbane 191
Hepatica acutiloba 31
  triloba 31
Herb Cotton 67
Hevea Braziliensis 27, 223, 224, 231
Hickory Nut 27, 241, 242
Hicoria 27, 241, 242
  olivæformis 242
  ovata 241, 242
  sulcata 241, 242

INDEX. 343

Hide-tanning Rhus 94
High Blackberry 114, 115
High Blueberry 172
Hoarhound 25, 207
Holly 85-87
Hooked Crowfoot 37
Hop, Common 27, 233, 234
Hordeum 29, 296-298
  distichum 296, 297
  hexastichon 297
  vulgare 296, 297
  Zeocriton 297
Horse Radish 13, 52
Hubbard Squash 144
Huckleberry 23, 171, 172
Humulus lupulus 27, 233, 234
Hymenæa courbaril 110
  verrucosa 110

ILEX Paraguayensis 15, 85-87
Ilicineæ 7, 15, 85-87
Illicium anisatum 148
India Ink 132
India Rubber 223, 224, 235
Indian Corn 29, 286-288
Indian Fig 235
Indian Rosewood 109
Indian Tobacco 22, 170, 171
Indigo 17, 96, 97
Indigofera 17, 96, 97
  argentea 96
  Caroliniana 96, 97
  tinctoria 96
Indigo Plant 96
Inferior apopetalæ 12-21
Ink-producing Plants —
  Banana 267
  Brazil Wood 109
  Cashew Nut 95
  Gall Oak 253
  Myrobalans 132
  Sumach 95
Inula Helenium 22, 165, 166
Ipecac, Ipecacuanha 22, 163, 223
Ipomœa 24, 183-185
  batatas 183, 184
  purga 184, 185
Iridaceæ 10, 28, 270, 271
Irish Potato 194
Iron Oak 249
Isatis tinctoria 13, 53
Ivy, Poison 93

JAPAN Quince 127
Jatropha manihot 227, 228
Jesuits' Bark 157
Juglandaceæ 9, 27, 238-242
Juglans 27, 238-241
  cinerea 238, 239
  nigra 239, 240
  regia 240, 241
  var. maxima 240
  var. serotina 240
  var. tenera 240
Juniper 310-312
Juniperus 29, 310-312
  Bermudiana 311
  communis 311
  var. alpina 311
  var. prostrata 311
  var. pyramidalis 311

Juniperus nana 311
  Virginiana 311, 312
  var. forma cylindrica 311
Jute 73

KIDNEY Bean 103

Kittatinny Blackberry 115

LABIATÆ 8, 24, 25, 198-207
Lance-leaved Violet 57
Larch 29, 309, 310
Large-fruited Cranberry 172, 173
Large Prickly Spinach 210
Large-ribbed Muskmelon 141
Large-rooted Beet 208
Larix 29, 309, 310
  Europea 310
  laricina 309, 310
Lauraceæ 9, 26, 27, 217-220
Lavandula 25, 198, 199
  spica 198
  var. alba 198
  var. latifolia 198
  stœchas 198
  vera 198
Lavender 25, 198
Lawton Blackberry 115
Leather 246, 248, 251, 310, 314
Leather Back 145
Leather Flower 41
Leguminosæ 7, 16-18, 96-112
Lemon 15, 82
Lens 17, 100, 101
  esculenta 100
  var. lutea 100
  var. Provence 100
Lentil 17, 100, 101
Lettuce-leaved Spinach 210
Lignum-vitæ 15, 77, 78
Liliaceæ 10, 28, 272-278
Lima Bean 103, 104
Lime 82, 83
Lime Juice 83
Linaceæ 6, 15, 74-77
Linseed Oil 76
Linum 15, 74-76
  angustifolium 75
  usitatissimum 74
  Virginianum 75
Liquorice 17, 105
Live Oak 252
Lobelia 22, 170, 171
  cardinalis 170, 171
  inflata 170, 171
  syphilitica 170, 171
Loganiaceæ 9, 23, 180-182
Logwood 17, 105-107
Long Beet 208, 209
Long-fruited Anemone 30
Long-leaved Pine 305, 306
Long-leaved Poppy 44
Long-necked Squash 143, 144
Long-spurred Violet 60
Lycopersicum esculentum 24, 185, 186
Lythraceæ 7, 20, 138, 139

MACE 216

Madder 22, 163, 164
Magnoliaceæ 148

Mahogany 15, 83, 84
Maize 286-288
Malvaceæ 6, 14, 67-70
Mandarin Orange 81
Mandragora officinalis officinarum 43
   191
Mandrake 13, 42, 43
Manihot 27, 227, 228
  api 227, 228
  utilissima 227, 228
Manilla 28, 267, 268
Many-cleft Anemone 31, 32
Maple 16, 91-93
Maple, Sugar 254
Maple Sugar 92
Maranta arundinacea 28, 261-263
Marjoram 25, 202
Marrubium vulgare 25, 207
Marsdenia tinctoria 96
Marsh Marigold 12, 39, 40
Marsh Violet 59
May Apple 13, 42, 43
Meadow Rue 12, 39
Meadow Violet 59
Meconium 45
Medicinal Plants —
  Allspice 136
  Aloe 278
  Anise 148
  Annatto 64
  Asafœtida 151
  Asparagus 276
  Balsam Fir 309
  Barberry 42
  Betel Nut 280
  Betel Pepper 214
  Bind-weed Jalap 185
  Black Pepper 215
  Blue Cardinal Flower 171
  Box 225
  Butternut 239
  Camomile 166, 167
  Camphor Tree 218
  Caper 56
  Caraway 156
  Cardamom 263
  Cashew Nut 95
  Cassia 221
  Castor-oil Plant 229
  Catnip 207
  Celery 147
  Cinchona Bark 159
  Cinnamon 219
  Cloves 135
  Coca 77
  Comfrey 183
  Common Hop 234
  Common Juniper 312
  Common Sumach 36
  Coriander 152, 153
  Cotton 70
  Croton-oil Plant 229
  Cumin 153
  Elecampane 166
  Emetic Weed 171
  Fennel 149
  Fig 236
  Flax 76
  Ginger 265
  Great Lobelia 171
  Gum Arabic 112
  Gum Tragacanth 99
  Henbane 192
  High Blackberry 115
  Hoarhound 207

344   INDEX.

Medicinal Plants
 Horse Radish 52
 Indian Tobacco 171
 Ipecac 163
 Jute 74
 Larch 310
 Lavender 199
 Lignum-vitæ 78
 Liquorice 105
 Logwood 106
 Mace 216, 217
 Madder 164
 Mahogany 85
 Mandrake 43
 Marjoram 202
 Mustard 49
 Myrobalans 132
 Myrtle 134
 Nutmeg 216, 217
 Parsley 156
 Pennyroyal 201
 Pepper 193
 Peppermint 201
 Pomegranate 139
 Poppy 45, 46
 Quince 127
 Rhubarb 213
 Safflower 170
 Sage 204
 Sandal-wood Tree 222
 Sarsaparilla 273
 Sharp-lobed Hepatica 33
 Slippery Elm 233
 Southern Pines 306
 Spearmint 201
 Strawberry 117
 Strychnine 182
 Tamarind 110
 Tansy 169
 Tapioca 228
 Thyme 203
 Turmeric 261
 Vanilla 260
 Violet 62
Mediterranean Wheat 288
Meliaceæ 7, 15, 83–85
Melon 142, 143
Mentha 25, 199–201
 piperita 200
 var. subhirsuta 200
 pulegium 201
 viridis 199, 200
Metroxylon 29, 283–285
 lævis 284
 sagu 284, 285
 spinosa 284
Milk Vetch 97–99
Millet 29, 300
Mimusops elata 176
Mitchell's Matchless Parsley 155
Monkey Nut 99, 100
Monocotyledonous angiosperms 5, 28, 29
Morello Cherry 118
Morus 27, 230, 231
 alba 230, 231
 var. multicaulis 231
 rubra 230, 231
Moss-cup Oak 248, 249
Mossy-cup White Oak 248, 249
Mountain Rice 291
Mountain Sumach 93
Mountain Sweet Watermelon 142

Much-divided-leaved Crowfoot 36
Mulberry 27, 230, 231
Musa 28, 265–268
 paradisiaca 265
 sapientum 265, 266, 267
 textilis 267, 268
 violacea 267
Muskmelon 20, 140–142
Mustard 13, 46–51
Myatt's Triple-curled Parsley 155
Myristica 26, 215–217
 fatua 216
 fragrans 216
 spuria 216
Myristicaceæ 9, 26, 215–217
Myrobalaus 19, 131
Myrtaceæ 7, 19, 20, 132–137
Myrteæ 132–137
Myrtle 19, 133, 134
Myrtus 19, 133, 134
 communis 133
 var. Belgica 133
 var. Bœotica 133
 var. mucronata 133
 var. Romana 133
 var. Tarentina 133

NAPA-BRASSICA 50
Naples Parsley 155
Narrow-leaved Flax 75
Nasturtium 13, 53, 54
 officinale 53, 54
Nectarine 121
Nepeta cataria 25, 206, 207
Nerium tinctorium 96
Nettle Family 230–238
New England Pine 302
New Jersey Tea 16, 88
New Rochelle Blackberry 115
New Zealand Flax 28, 276, 277
Nicotiana 24, 187–191
 macrophylla 187
 nana 187
 Persica 187
 quadrivalvis 187
 repanda 187
 rustica 187
 tabacum 187
Norway Spruce 308
Nut-gall 253
Nutmeg 26, 216
Nutmeg Muskmelon 141
Nuts —
 Almond 123
 Beech-nut 256
 Black Walnut 240
 Brazil Nut 137
 Butternut 239
 Cashew Nut 95
 Chestnut 244
 English Walnut 240, 241
 Hazelnut 255
 Hickory Nut 241, 242
 Peanut 100
 Pecan Nut 242
 Pine Nut 301
 Shag-bark 241, 242
 Shell-bark 241, 242
 Thick Shell-bark 241, 242
 Walnut 240–242
 White Walnut 241, 242

Nux Avellana 255
 Pontica 255
 Vomica 23, 180, 181

OAK 27, 243–254
Oak of the Barrens 249
Oats 29, 298, 299
Oblique-leaved Beech 255
Oblong-leaved Crowfoot 36
Odell's Large Watermelon 142
Oil of Tar 306
Oil-producing Plants —
 Allspice 136
 Anise 148
 Asafœtida 150
 Beech 256
 Black Walnut 239
 Brazil Nut 137
 Butternut 238
 Butter Tree 176
 Camomile 167
 Camphor Tree 218
 Caraway 156
 Cardamom 263
 Castor-oil Plant 229
 Cinnamon 219
 Cloves 135
 Cocoanut Palm 283
 Coriander 152
 Cotton 69
 Croton-oil Plant 226
 Cumin 153
 English Walnut 240, 241
 Fennel 149
 Flax 76
 Hemp 237
 Lavender 199
 Lime 83
 Mace 217
 Marjoram 203
 Mustard 49
 Nutmeg 217
 Oats 299
 Olive 180
 Orange 81
 Peanut 100
 Pennyroyal 201
 Peppermint 201
 Poppy 46
 Rape 51
 Rosemary 206
 Saffron 271
 Sandal-wood Tree 222
 Sesame 197
 Shag-bark 241, 242
 Shell-bark 241, 242
 Southern Pine 306
 Tansy 169
 Teak 198
 Thick Shell-bark 242
 Thyme 203
Oils 81, 149, 153, 156, 169, 179, 180, 199, 201, 222, 306
Oil Seed 195–197
Olea 23, 178–180
 Europeæ 178
 var. buxifolia 178
 var. ferruginea 178
 var. latifolia 178
 var. longifolia 178
 var. obliqua 178
 var. sylvestris 178
Oleaceæ 9, 23, 178–180
Olive 23, 178–180

INDEX. 345

Opium	45, 46	Papaver	13, 43-46	Pinus cembra, var Siberica	305
Opium **Poppy**	44	dubium	44	mitis	302, 303
Orange	15	orientale	44	monophylla	301
Orange, Sweet	78-81	Rhœas	44	nigra	301
Orange Gourd-Squash	144	somniferum	44	palustris	305, 306
Orchidaceæ	10, 28, 258-260	Papaveraceæ	6, 13, 43-46	var. excelsa	305
Origanum	25, 201-203	Paper	74, 268	resinosa	304
marjorana	202	Para Sarsaparilla	273	rigida	304
vulgare	202	Paraguay Tea	15, 85-87	rubra	304
Ornamental **Plants** —		Parsley	21, 155, 156	strobus	302
Acacia	112	Parsnip	21, 151, 152	sylvestris	303, 304
American Arbor **Vitæ**	312	Pasque-flower	32	variabilis	302, 303
American Beech	256	Pastinaca sativa	151, 152	Piper	26, 214, 215
American Elm	233	Pea	17, 101-103	betel	214
Anemone	32, 33	Pea, Ground	99, 100	Piper nigrum	214, 215
Austrian Pine	301	Peach	18, 121, 122	var. longum	214
Barberry	42	Peach, Wolf	186	Piperaceæ	9, 26, 213-215
Beech of Europe	256	Peanut	17, 99, 100	Pisum	17, 101-103
Black Spruce	307	Pear	19, 123-127	arvense	102
Box	225	Pear-shaped Cranberry	173	sativum	101, 102
Castor-oil Plant	229	Pear Tree	125, 126	Pitch	306
Clematis	42	Pearl Sago	285	Pitch Pine	304, 305, 306
Clove Tree	135	Pecan Nut	27, 242	Plantain	205
Common Juniper	311	Pedalineæ	8, 24, 195-197	Plum	18, 117-123
English Elm	233	Pekoe	66	Plum, Damson	117, 118
Hemlock Spruce	313	Pencil Cedar	311	Podophyllum peltatum	13,
Japan Quince	127	Pennsylvanian Anemone	30,		42, 43
Larch	310		31	Poison Ivy	93
Logwood	107	Pennsylvanian Crowfoot	37	Poison Oak	93
Marjoram	203	Pennyroyal	25, 201	Poisonous Plants —	
Moss-cup Oak	249	Pepper, Bell	193	Aloe	278
Myrtle	134	Betel	214	Box	225
Norway Spruce	308	Black	214, 215	Camphor	218
Pin Oak	250	Cayenne	193	Chrysanthemum	167, 168
Pomegranate	139	Red	192, 193	Hemlock	313
Poppy	44	Peppermint	200	Poison Ivy	93
Ranunculus	38	Perfume-producing Plants —		Poison Oak	93
Red Cedar	312	Butter Tree	176	Poison Sumach	93
Red Oak	251	Hemp	237	Red Cedar	312
Rosemary	206	Lavender	199	Strychnine	182
Sage	204	Mace	217	Tapioca	228
Scarlet Oak	248	Rosemary	206	White Cedar	315
Scotch Pine	304	Sandal-wood Tree	222	Poison Sumach	93
Shaddock	82	Perry	127	Pole Bean	103
Spanish Oak	248	Persian Berries	87, 88	Polygala tinctoria	96
Swiss Pine	305	Persimmon	23, 176	Polygalaceæ	96
Thyme	203	Peruvian Bark	21, 157	Polygonaceæ	9, 26, 210-213
Violet	62	Petaloideæ	5	Polygonum aviculare	96
Weeping Willow	258	Petaloideous monocotyled-		barbatum	96
White Pine	302	onous angiosperms	10	Chinense	96
White Spruce	308	Peucedanum	21, 151, 152	fagopyrum	211
Willow-leaved **Oak**	250	pastinaca	151, 152	perfoliatum	96
Yellow Pine	303	Phænogams	5-11, 12, 29	tinctorium	96
Oryza	29, 290, 291	Phase.lus	17, 103-105	Pomegranate	20, 138
coarctata	291	lunatus	103, 104	Pomum Punicum	139
glutinosa	291	vulgaris	103	Poppy	13, 43-46
mutica	291	var. nanus	103, 104	Post Oak	249
præcox	291	Phœnix dactylifera	29, 280,	Potato	24, 194
sativa	291		281	Potato, Sweet	24, 183, 184
Osier	28	Phormium tenax	28, 276, 277	Prickly Cayenne Pineapple	
Osier Willow	258	Picea	29, 307, 308		269
Ox Heart	118	alba	307, 308	Prickly-seeded Crowfoot	36
Oxycoccus	23, 172-174	balsamifera	308	Primrose-leaved Violet	60
macrocarpus	172, 173	excelsa	308	Providence Pineapple	269
palustris	173	nigra	307	Prunus	18, 19, 117-123
		Pimpinella anisum	21, 147,	Armeniaca	120
			148	avium	118
PALE Violet	61	Pine	29, 300-306	cerasus	118
		Pineapple	28, 269, 270	communis	122
Palm, Cocoanut	282, 283	Pin Oak	249, 250	var. amara	122
Date	29, 280, 281	Pinus	29, 300-306	domestica	117
Sago	29, 284, 285	Austriaca	301	nana	122
Palmæ	10, 29, 278-285	balsamifera	309	Persica	121, 122
Palms, Feather	279, 280, 281,	Canadensis bifolia	304	var. lævis	121
	282, 283, 284, 285	cembra	305	Pumpkin	20, 143-146
Pansy	61	var. pygmæa	305	Pumpkin Pine	302

INDEX.

Punica 20, 138, 139
  granatum 138
  var. alba 138
  var. plena 138
Puny Crowfoot 37
Puritan Squash 144
Purple Cabbage 47
Purple Sweet Violet 57
Purplish Meadow Rue 39
Pyrus 19, 123–127
  communis 125–127
  Cydonia 127
  Japonica 127
  malus 123–125

QUERCUS 27, 245–254
  ægilops 253, 254
  var. latifolia 253
  var. pendula 253
  alba 245
  var. pinnatifida 245
  var. repanda 245
  bicolor 247
  coccinea 247, 248
  var. tinctoria 248
  falcata 248
  Ilex 86
  infectoria 253
  macrocarpa 248, 249
  nigra 249
  obtusiloba 249
  palustris 249, 250
  pedunculata 246, 247
  Phellos 250
  var. latifolius 250
  var. sylvatica 250
  Prinus 250
  var. acuminata 250, 251
  var. monticola 251
  pseudococcifera 245
  robur 246, 247
  var. fastigiata 246
  var. foliis variegatis 246
  var. heterophylla 246
  var. pendula 246
  var. pubescens 246
  var. purpurea 246
  var. sessiliflora 246
  rubra 250, 251
  var. runcinata 251
  suber 252
  var. angustifolium 252
  var. dentatum 252
  var. latifolia 252
  virens 252
Quince 19, 127
Quince, Japan 127

RANDIA aculeata 96

Ranunculaceæ 6, 12, 30–42
Ranunculus 12, 33–38
  abortivus 33, 34
  var. micranthus 34
  acris 34
  alismæfolius 34
  ambigens 34
  aquatilis 35
  var. tricophyllus 35
  bulbosus 35
  Cymbalaria 35
  fascicularis 35, 36
  flammula 36

Ranunculus flammula, var.
  intermedius 36
  var. reptans 36
  micranthus 34
  multifidus 36
  var. terrestris 36
  muricatus 36
  oblongifolius 36
  palmatus 38
  parviflorus 36
  Pennsylvanicus 37
  pusillus 37
  recurvatus 37
  repens 37
  rhomboideus 37
  sceleratus 37, 38
  septentrionalis 38
Rape 51
Raspberry 18, 113, 114
Red Banana 265, 266
Red Cabbage 47
Red Cedar 311
Red Cherry 118
Red Cinchona Bark 157
Red Dutch Currant 128
Red Elm 232
Red Grape Currant 128
Red Mulberry 230, 231
Red Oak 251
Red Pepper 24, 192, 193
Red Pine 304, 305, 306
Red Raspberry 88
Red Root 88
Red Top Turnip 50
Red Wheat 288
Revalenta 101
Rhamnaceæ 16, 87, 88
Rhamnus 16, 87, 88
  chlorophorus 87
  infectorius 87, 88
  utilis 87
Rheum 26, 212, 213
  compactum 212
  palmatum 212
  rhaponticum 212
  undulatum 212
Rhomboid-leaved Crowfoot 37
Rhubarb 26, 212, 213
Rhus 16, 93–95
  copallina 93
  coriaria 94
  Cotinus 93, 94
  glabra 93, 94, 95
  Toxicodendron 93, 95
  typhina 93
  venenata 93, 95
Ribes 19, 128–131
  cynosbate 129, 130
  floridum 128
  Grossularia 129
  hirtellum 130
  nigrum 129
  oxyacanthoides 130
  rotundifolium 130
  rubrum 128
Rice 29, 290, 291
Ricinus 27, 228, 229
  communis 228, 229
Ring-leaved Willow 257
Rio Negro Sarsaparilla 273
Ripley Pineapple 269
Rock Chestnut Oak 251
Rock Maple 91
Rock Oak 251
Roman Camomile 166
Rosaceæ 7, 18, 19, 113–127

Rosemary 25, 205, 206
Rosewood 17, 108, 109
Rosin 306
Rosmarinus 25, 205, 206
  officinalis 205
  var. variegata 205
Round-leaved Violet 60
Round-lobed Hepatica 31
Rubia 22, 163, 164
  Chiliensis 164
  cordifolia 164
  tinctorum 163, 164
Rubiaceæ 9, 21, 22, 157–164, 223
Rubus 18, 113–115
  fruticosus 115
  Idæus 113
  occidentalis 113
  strigosus 88, 113, 114
  villosus 114, 115
  var. frondosus 115
  var. humifusus 115
Rue Anemone 12, 33
Russia Turnip 50
Russian Rhubarb 213
Rutabaga 50
Rutaceæ 7, 15, 78–83
Rye 29, 295, 296

SACCHARUM 29, 292–294
  atrorubens 293
  contractum 293
  dubium 293
  fragile 293
  officinarum 292, 293
  polystachyum 293
  rubicundum 293
Sacred Barley 297
Safflower 22, 169, 170
Saffron 28, 169, 170, 270, 271
Sage 25, 204
Sago, Pearl 285
Sago Palm 29, 284, 285
Salads —
  Beet 209
  Cabbage 48
  Caper 56
  Celery 147
  Cowslip 40
  Cucumber 140
  Fennel 149
  Ranunculus ficaria 38
  Tomato 186
  Water-Cress 54
Salicaceæ 9, 28, 256–258
Salix 28, 256–258
  annularis 257
  Babylonica 256
  var. crispa 257
  var. Napoleona 257
  var. vulgaris 256, 257
  viminalis 258
Salvia 25, 204
  officinalis 204
  var. variegata 204
  pomifera 204
Sandal-wood Tree 27, 220–222
Santalaceæ 9, 27, 220–222
Santalum 27, 220–222
  album 220, 221
  Freycinetianum 221
  myrtifolium 221
  yasi 221
Sapindaceæ 7, 16, 91–93

INDEX. 347

Sapotaceæ	9, 23, 174-176	Spices, Condiments, etc. —		Symphytum	23, 182, 183	
Sapling Pine	302	Cardamom	203, 204	officinale	182, 183	
Sarasin Wheat	211	Cassia	220	var. Bohemicum	182	
Sarsaparilla	28, 273, 274	Cinnamon	219			
Savoy Cabbage	47	Cloves	135			
Saxifragaceæ	7, 19, 128-131	Coriander	152			
Scarlet Oak	247, 248	Cumin	153	TACCA integrifolia	262	
Scotch Fir	303, 304	Fennel	149			
Scotch Pine	303, 304	Ginger	205	Tall Meadow Rue	39	
Scurvy Grass	52	Horse Radish	52	Tamarack	303, 310	
Sea Island Cotton	67	Mace	216, 217	Tamarind	18, 109-111	
Seaside Crowfoot	35	Mustard	49	Tamarindus Indica	18, 109-111	
Secale cereale	29, 295, 296	Nutmeg	216, 217			
Secor's Mammoth Blackberry	115	Parsley	156	Tanacetum vulgare	22, 168, 169	
Senna	17, 107, 108	Pepper	194			
Sesame	24, 196	Sage	204	Tangleberry	172	
Sesamum	24, 195-197	Spearmint	201	Tannin-producing Plants —		
Indicum	196	Thyme	203	Allspice	136	
orientale	196	Vanilla	209	Black Jack Oak	249	
Setaria	29, 300	Spicy Vanilla	259	Black Oak	248	
Italica	300	Spinach, Spinage	26, 209, 210	Chestnut Oak	251	
var. Germanica	300	Spinacia oleracea	26, 209, 210	Dyers' Oak	248	
Seville Orange	79, 81	Spotted Violet	58	Gall Oak	253	
Shaddock	81, 82	Spruce	29, 307, 308	Gum Arabic	112	
Shag-bark	241	Spruce, Hemlock	313, 314	Hemlock Spruce	314	
Sharp-lobed Hepatica	31	Spruce Pine	302, 303	Larch	310	
Shell-bark	241	Spurge	27, 222, 223	Moss-cup Oak	249	
Shepherd's Purse	13, 46, 47	Spurge Family	222-229	Myrobalans	132	
Sickle-leaved Oak	248	Squash	20, 143-146	Myrtle	134	
Silk	231	Stag-horn Sumach	93	Pin Oak	250	
Silver-leaved Indigofera	96	Staining Buckthorn	87, 88	Pomegranate	139	
Silver-leaved Oak	247	Star Anise	148	Red Oak	252	
Single-leaved Nut-pine	301	Sterculiaceæ	7, 14, 70-72	Rock Oak	251	
Six-rowed Barley	297	Stone Pine	305	Scarlet Oak	248	
Slippery Elm	232	Strap-leaved Turnip	50	Silver-leaved Oak	247	
Small Cranberry	173	Strawberry	18, 116, 117	Spanish Oak	248	
Small-flowered Anemone	32	Streaked Violet	58	Sumach	94	
Small-flowered Crowfoot	33, 34, 36	String Bean	103	Valonia Oak	254	
		Striped Violet	61	White Oak	246	
Smaller Spearwort	36	Strychnos	23, 180-182	Willow-leaved Oak	250	
Smilax	28, 273, 274	colubrina	181	Yellow-barked Oak	248	
medica	273	Ignatii	181	Tansy	22, 168, 169	
officinalis	273, 274	nux vomica	180, 181	Tapioca	27, 227, 228	
papyraceæ	273, 274	potatorum	181	Tar	304, 306	
Sarsaparilla	273, 274	tieuté	181	Tea	14, 64-67	
Smoke Tree	93	toxifera	181	Tea, New Jersey	16, 88	
Smooth Cayenne Pineapple	269	Student Parsnip	151	Paraguay	15, 85-87	
		Sugar Cane	29, 292-294	Teak	24, 197, 198	
Smooth Sumach	93	Sugar Loaf Cabbage	47	Tectona grandis	24, 197, 198	
Snake-wood	181	Sugar Maple	91, 294	Tephrosia apollinea	96	
Socotrine Aloes	278	Sugar-producing Plants —		tinctoria	96	
Solanaceæ	8, 24, 185-195	Beet	293	toxicaria	96	
Solanum	24, 194, 195	Broom Corn	295	Terminalia	19, 131, 132	
Commersonii	194	Cocoanut Palm	283	angustifolia	132	
iminite	194	Oats	299	bellerica	132	
tuberosum	194	Sugar Cane	294	catappa	132	
verrucosum	194	Sugar Maple	92	chebula	131	
Sorghum saccharatum	29, 294, 295	Sugar Pumpkin	145, 146	citrina	132	
		Sugar Tree	91	Ternstrœmiaceæ	7, 14, 64-67	
Sour Cherry	118	Sumach	16, 93-95	Textile Fabrics —		
Southern Pine	305, 306	Summer Wheat	288	Cocoanut Palm	283	
Spadicifloræ	5	Swamp Chestnut Oak	270	Common Hop	234	
Spadiciflorous monocotyledonous angiosperms	10	Swamp Spanish Oak	249, 250	Cotton	69	
		Swamp White Oak	247	Flax	76	
Spanish Oak	248	Swedish Turnip	50	Hemp	237	
Sparrow Grass	274-276	Sweet Marjoram	202	Jute	74	
Spearmint	25, 199, 200	Sweet Orange	78-81	Manilla	268	
Spearwort	34, 36	Sweet Potato	24, 183, 184	New Zealand Flax	277	
Spices, Condiments, etc. —		Sweet Potato Squash	144	Thalictrum	12, 39	
Allspice	136	Sweet Thyme	25, 203	anemonoides	33	
Anise	148	Sweet Violet	57	rivatum	39	
Asafœtida	151	Swietenia mahogani	15, 83-85	Cornuti	39	
Black Pepper	215	Swiss Pine	305	dioïcum	39	
Caper	56	Sympetalæ	5, 21-27	polygamum	39	
Caraway	150	Sympetalous dicotyledonous angiosperms	8, 9	purpurascens	39	
				var ceriferum	39	

INDEX.

Thea 14, 64-67
  Bohea 64
  viridis 64
Thebaicum 45
Theobroma 14, 70-72
  angustifolia 71
  bicolor 71
  cocoa 70, 71
  Guianensis 71
  microcarpa 71
  ovatifolia 71
  sylvestris 71
Thick Shell-bark 241
Thimble Berry 113
Thimble Weed 32, 33
Thuja occidentalis 29, 312
Thunderbolt Flower 196
Thyme 203
Thymelaceæ 278
Thymus 25, 203
  serpyllum 203
  vulgaris 203
Tiliaceæ 7, 14, 72-74
Tinnevelly Senna 108
Tobacco 24, 187-191
Tobacco, Indian 22, 170, 171
Tomato 24, 186
Trachylobium Horneman-
  nianum 110
Tree Cabbage 47
Tree Cotton 67, 68
Triticum 29, 288-290
  æstivum 288
  compositum 289
  hybernum 288
  vulgare 288, 289
  var. æstivum 288
  var. album 288
  var. hybernum 288
  var. nudum 288
  var. rubrum 288
Tropæolum 54
Tsuga Canadensis 29, 312-314
Turkey Rhubarb 213
Turmeric 28, 260, 261
Turnip 13, 46-51
Turnip-rooted Beet 208
Turpentine 304, 306
Turpentine, Spirits of 306
  Venice 310

U

ULMUS 27, 231-233
  alata 232
  Americana 231, 232, 233
  campestris 232, 233
  fulva 232
  racemosa 232
Umbelliferæ 6, 20, 21, 146-156
Upland White Oak 249
Urticaceæ 10, 27, 233, 230-238

V

VACCINIACEÆ 8, 23,
  171-174
Vaccinium macrocarpum 172,
  173
  oxycoccus 173
Valonia Oak 253, 254
Valparaiso Squashes 144
Vanilla 28, 259, 260
  aromatica 259
  planifolia 259
Van-Mais 38
Vegetable Marrow 144

Venetian Sumach 93, 94
Venice Turpentine 310
Verbenaceæ 9, 24, 197, 198
Versailles Currant 128
Vicia faba 104
Vine, Grape 88-91
Vinegar 283
Viola 14, 56-62
  blanda 56, 60
  Canadensis 56, 57
  canina 57
  var. sylvestris 57
  cordata 59
  cucullata 58, 59
  var. cordata 59
  var. reniformis 59
  var. striata 58
  hastata 57
  lanceolata 57, 60
  Muhlenbergii 57
  odorata 57
  palmata 58
  palustris 59
  pedata 58, 59, 60
  var. bicolor 60
  primulæfolia 60
  pubescens 60
  var. scabriuscula 60
  rostrata 60
  rotundifolia 60
  sagittata 60, 61
  var. ovata 61
  Selkirkii 61
  sororis 50
  striata 61
  tenella 61, 62
  tricolor 61, 62
  var. arvensis 61, 62
  villosa 59
Violaceæ 6, 14, 56-62
Violet 14, 56-62
Virginian Anemone 32, 33
Virgin's Bower 12, 40-42
Vitis 16, 88-91
  labrusca 89
  vinifera 89

W

WALNUT 27, 239-241
Warty Squash 143, 144
Water Cress 13, 53, 54
Watermelon 20, 142, 143
Water Oak 249, 250
Water Plantain 34
Wattle Barks 112
Wavy Tobacco 187
Way-adorner 41
Wayside Crowfoot 34
Weeping Elm 231, 232
Weeping Willow 256
Weymouth Pine 302
Wheat 29, 288, 289
Wheat, Sarasin 211
Whiskey 288, 296, 298, 299
White Cabbage 47
White Cedar 314
White Dutch Currant 128
White Elm 231, 232
White Mulberry 230, 231
White Mustard 48, 49
White Oak 245
White Pine 302
White Potato 194
White Spine Cucumber 140
White Spruce 307, 308

White Stone Turnip 50
White Sweet Violet 57
White Walnut 241
White Water Crowfoot 35
White Wheat 288
Whorled-leaved Virgin's
  Bower 41
Wicked Crowfoot 37, 38
Wild Carrot 154
Wild Flax 75
Wild Gooseberry 129, 130
Wild Red Raspberry 113, 114
Willow 28, 256-258
Willow-leaved Oak 250
Windflower 12, 30-33
Wine 90, 129, 267, 283
Winged Elm 232
Winter Squash 144
Winter Wheat 288
Woad 13, 53
Wolf Peach 186
Wood Anemone 32
Wood Crowfoot 37
Woods —
  American Arbor Vitæ 312
  Apple 125
  Austrian Pine 301
  Beech 256
  Black Jack Oak 249
  Black Oak 248
  Black Spruce 307
  Black Walnut 239, 240
  Box 225
  Brazil Wood 109
  British Oak 247
  Butternut 239
  Cherry 118
  Chestnut 244
  Chestnut Oak 251
  Cloves 135
  Cocoanut Palm 283
  Constantinople Hazel-
    nut 255
  Cork Tree 252
  Dyers' Oak 248
  Ebony 176
  English Elm 233
  English Walnut 241
  Hemlock Spruce 313, 314
  Larch 310
  Lignum-vitæ 78
  Live Oak 252
  Logwood 107
  Mahogany 85
  Moss-cup Oak 249
  Myrobalans 132
  Norway Spruce 308
  Olive 180
  Orange 80
  Pecan Nut 242
  Pin Oak 250
  Pitch Pine 304
  Post Oak 249
  Red Cedar 310
  Red Mulberry 231
  Red Oak 251, 252
  Rock Maple 92, 93
  Rock Oak 251
  Rosewood 109
  Sandal-wood 222
  Scarlet Oak 248
  Scotch Pine 304
  Shag-bark 242
  Shell-bark 242
  Silver-leaved Oak 247
  Single-leaved Nut-pine 301

INDEX. 349

Woods —
  Southern Pine   305
  Spanish Oak   248
  Strychnine   182
  Sugar Maple   92, 93
  Swamp Chestnut Oak   250
  Swiss Pine   305
  Tamarind   110
  Teak   198
  Thick Shell-bark   242
  Valonia Oak   254
  White Cedar   315
  White Oak   246
  White Pine   302

Woods —
  White Spruce   308
  Yellow-barked Oak   248
  Yellow Pine   303
Wrightia tinctoria   96
Wrinkled Cabbage   47

YAM   28, 271, 272

Yellow Banana   265
Yellow-barked Oak   248
Yellow Berries   87, 88
Yellow Cinchona Bark   157

Yellow Jute   14, 73, 74
Yellow Lentil   100
Yellow Pine 302, 303, 305, 306
Yellow Water Crowfoot   36
Yerba Maté   85–87

ZANTE Currants   89

Zea mays   29, 286–288
Zingiber officinalis   28, 264, 265
Zingiberaceæ   10, 28, 260–268
Zygophyllaceæ   7, 15, 77, 78